Zebrafish: A Model Organism for Human Health and Disease

Guest Editor

Jyotshna Kanungo

MDPI

Basel • Beijing • Wuhan • Barcelona • Belgrade • Novi Sad • Cluj • Manchester

Guest Editor
Jyotshna Kanungo
National Center for
Toxicological Research
US Food and Drug
Administration (USFDA)
Jefferson
USA

Editorial Office
MDPI AG
Grosspeteranlage 5
4052 Basel, Switzerland

This is a reprint of the Special Issue, published open access by the journal *International Journal of Molecular Sciences* (ISSN 1422-0067), freely accessible at: https://www.mdpi.com/journal/ijms/special_issues/zebrafish_model.

For citation purposes, cite each article independently as indicated on the article page online and as indicated below:

Lastname, A.A.; Lastname, B.B. Article Title. *Journal Name* **Year**, *Volume Number*, Page Range.

ISBN 978-3-7258-4202-5 (Hbk)
ISBN 978-3-7258-4201-8 (PDF)
https://doi.org/10.3390/books978-3-7258-4201-8

Contents

International Journal of
Molecular Sciences

MDPI

Editorial

Special Issue "Zebrafish: A Model Organism for Human Health and Disease"

Jyotshna Kanungo

Division of Neurotoxicology, National Center for Toxicological Research, U.S. Food and Drug Administration, Jefferson, AR 72079, USA; jyotshnabala.kanungo@fda.hhs.gov

In 1996, a special zebrafish (*Danio rerio*) issue of the journal *Development*, with a 481-page volume containing 37 papers from four different laboratories, published genetic and phenotypic details of hundreds of different mutants, obtained through mutagenesis screening. The collective data showed developmental defects in almost every organ or tissue of the early life stages of zebrafish. Those findings not only revealed a wide array of information in the areas of basic science but also provided a plethora of information to investigate many human diseases. As an emergent vertebrate animal model then, zebrafish has since become the second most important animal model next to rodents, and research using the model has become a flourishing field in studying human diseases [1]. A PubMed search on "zebrafish and human disease modeling" (26 March 2025) yields a total of 1489 papers showing an exponential increase in published papers in the last two decades (Figure 1). Zebrafish share 70% of protein-coding genes with humans. One of the major advantages of using zebrafish in research is that a single female lays hundreds of transparent eggs that are externally fertilized, which helps researchers increase the sample size for their experiments. Because of the transparency of the embryos, a longitudinal assessment of developing organs and reporter transgene expressions can be easily performed by imaging with microscopes. Additionally, zebrafish can breed at around three months of age, thus becoming a cost-effective animal model for research. As an alternative animal model, zebrafish align well with the three Rs (reduction, refinement, and replacement) that are meant to address the humane use of animals. With comparable genetic and physiologic make-up between zebrafish and mammals, including humans, zebrafish possess many advantages for biomedical and toxicological research. Nearly 10 compounds from zebrafish screens conducted by several different laboratories are in or about to enter the clinic [2].

The Special Issue "Zebrafish: A Model Organism for Human Health and Disease" contains nine papers covering various aspects of pathological conditions modeled in zebrafish and the underlying mechanisms highlighting strategic interventional avenues. These studies used both wild-type and genetically modified zebrafish to address tissue/organ-specific conditions and covered a wide array of topics such as the nervous system, inflammation, cancer, and the cardiovascular system. The authors provide in-depth insight into these domains, which can lead to understanding equivalent human conditions along with predictive scenarios that may be used for future drug discovery.

In the first paper, Ciapaite et al. described the generation of the first zebrafish model of hypophosphatasia (HPP), which exhibits multiple features characteristic of the human disease, thus providing a model system for not only the pathophysiology study of HPP in depth but also the testing of compounds that have therapeutic value [3]. HPP is a rare inherited metabolic disorder that adversely affects the development of bones and teeth due to a pathogenic variation in the enzyme tissue-nonspecific alkaline phosphatase (TNSALP), encoded by the *alpl* gene [4]. The authors created the *alpl* knockout zebrafish using clustered regularly interspaced short palindromic repeat (CRISPR)/CRISPR-associated protein 9 (CRISPR/Cas9) gene editing. There was decreased bone mineralization in the $alpl^{-/-}$ embryos, a hallmark of HPP. These embryos had depleted pyridoxal and its degradation product 4-pyridoxic acid, suggesting disturbances in broad vitamin B6-related metabolism. This study provides novel mechanistic insight into the development of HPP.

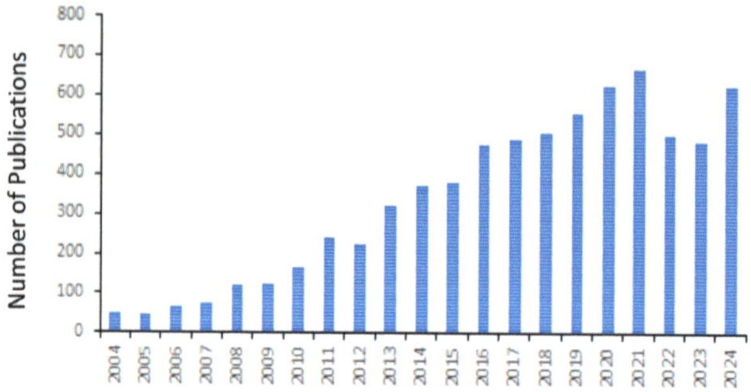

Figure 1. Number of publications on "Zebrafish and Human Disease Modeling" from 2004 to 2024.

In the second paper, Danila et al. described the differential effect of oxytocin on stimulus discrimination in albino and non-albino zebrafish [5]. The authors explored the potential of oxytocin in reducing psychosocial difficulties, such as anxiety, stress, stigmatization, and social isolation, including factors that affect the development and maintenance of interpersonal relationships, behaviors exhibited by individuals with albinism. The major goal of the study was to investigate how oxytocin can influence behavior, thus leading to an insight into interventional avenues and managing the negative effects associated with albinism [6]. The study used behavioral analyses of adult albino and non-albino zebrafish and concluded that oxytocin enhanced social interactions in albino zebrafish, suggesting a potential neurochemical mechanism. The authors propose that their study provides an encouraging starting point for future research that would have applicability to humans with similar behavioral disorders.

The third paper by Li et al. revealed the mechanism of how the deletion of *Slc1a4* suppresses axon regeneration in zebrafish [7]. Zebrafish are well known for their remarkable regenerative capacity [8]. In this study, the authors used $Slc1a4^{-/-}$ mutants created by the CRISPR-Cas9 genome editing and investigated Mauthner cell axon regeneration potential in those mutants. Since there is evidence that amino acids play a vital role in axon regeneration [reviewed in [9]], the study utilized the deletion of *Slc1a4*, an amino acid transporter that is involved in the uptake of amino acids. Although Slc1a4 deficiency did not affect Mauthner cell development or motor function, it suppressed Mauthner cell axon regeneration and negatively affected the associated functions. RNA-Seq analyses revealed that Slc1a4 may influence Mauthner cell axon regeneration via the p53 signaling pathway. The study also showed that Slc1a4 deficiency may inhibit Mauthner cell axon regeneration thorough suppression of Gap43 (Growth-Associated Protein 43), a marker and a critical factor of axonal growth and regeneration [reviewed in [10]].

Ricarte et al. [11] undertook an integrative approach that included behavioral analysis and neurotransmitter profiling to determine the effects of early-life exposure to valproic acid (VPA) both in the larval and adult zebrafish. Their study was intended to gain further insight into autism spectrum disorder (ASD) since prenatal exposure to VPA has been linked to an increased incidence of autism [reviewed in [12]]. The study showed that VPA-treated larvae exhibited hyperactivity and compromised visual and vibrational escape responses. The larvae had altered neurotransmitter levels showing increased glutamate but reduced acetylcholine and norepinephrine levels. VPA-treated embryos upon reaching adulthood exhibited larger shoal sizes and a decreased interest for their conspecifics, hallmarks of impaired social behavior in zebrafish. Their brains had significantly reduced dopamine and gamma-aminobutyric acid (GABA) levels. Data from this study's combined behavioral and biochemical approach validate zebrafish as an ideal model of ASD.

The zebrafish has emerged as a powerful vertebrate model for studying cancer [reviewed in [13]]. Mutations in the succinate dehydrogenase subunit B (SDHB) gene are associated with the highest malignancy rate [14]. In their paper, Miltenburg et al. present a zebrafish model to study SDHB-associated phaeochromocytomas and paragangliomas (PPGLs) [15]. Their study revealed that although adult SDHB mutant zebrafish did not show an obvious tumor phenotype and were anatomically and histologically not different from their wild-type counterparts, they had significantly increased succinate levels, a characteristic of SDHB-related PPGLs. The authors conclude that they characterized an adult in vivo zebrafish model that can be used to gain in-depth knowledge on the mechanism underlying the development of SDHB-associated PPGL pathology.

The angiotensin-converting enzyme (ACE) plays a crucial role in blood pressure regulation by converting angiotensin I to angiotensin II, a vasoconstrictor that narrows blood vessels, resulting in increased blood pressure [16]. A potential link between ACE deficiency and gastrointestinal tract inflammation was revealed when several studies showed compromised proliferation and apoptosis of intestinal epithelial cells in ACE-deficient mice [17–19]. In their paper, Wei et al. modeled intestinal inflammation by creating ACE-deficient zebrafish using the CRISPR/Cas9 approach [20]. The authors show that the deletion of *ACE* in zebrafish not only caused intestinal inflammation but also increased the expression of several inflammation marker genes. Histopathological data revealed significantly increased secretion of mucus in the intestines of the *ACE* mutants compared to their wild-type counterparts. The mutant zebrafish showed increased susceptibility to enteritis. The authors conclude that such findings validate the zebrafish model to study intestinal inflammation and gain insight into strategies for potential therapeutic interventions.

In their manuscript on zebrafish olfactory bulb injury and recovery, Rozofsky et al. [21] show that there is remarkable plasticity of adult dendritic arbor structures following injury, which reaffirmed the existing knowledge that zebrafish is an ideal model system to study nervous system injury and neuroregenerative processes. In an earlier study, their laboratory showed that deafferentation (loss or reduction in sensory input to the central nervous system) of the olfactory bulb altered mitral cell dendritic arbor morphology [22]. As a continuation of that investigation, the current study [21] reveals that mitral cell dendritic branch regeneration and growth was independent of overall growth-related changes. The adult zebrafish model provided a system to investigate both degeneration and subsequent regeneration of the mitral cell dendritic branch in the olfactory bulb. The authors suggest that due to a lack of sufficient information on the effects of deafferentation and the recovery of aspinous dendrites in adult brain structures, their work on the zebrafish model will help fill the knowledge gap and serve as a step toward an in-depth understanding of the processes involved in regeneration following specific injuries.

In a review paper, Yang et al. [23] provide a comprehensive report on the zebrafish congenital heart disease (CHD) models that are used in research to understand and reveal the mechanisms of CHDs. The findings from such studies offer potential strategies for the treatment and intervention of CHDs. The review also outlines the advantages and disadvantages of using the zebrafish model in CHD investigation. For example, while zebrafish and the human heart share a similar electrophysiology, the zebrafish heart does not have pulmonary circulation in the double-chambered heart due to the absence of lungs. Nonetheless, the zebrafish model has been successfully used in CHD investigations, leading to revelations of the mechanisms and filling in the knowledge gap.

Lastly, a review article by Gu and Kanungo [24] details how arsenic-induced supernumerary motor neurons in zebrafish, possibly through the sonic hedgehog (Shh) pathway [25], presents an opportunity to investigate ASD. The review summarized existing reports on potential cross talk among Shh signaling, cyclin-dependent kinase 5 (Cdk5) signaling, and mindbomb (Mib) in generating excess neurons. Generation of a specific type of neurons in excess can adversely affect the development of other types of neurons. Ultimately, this will negatively affect proper development of the nervous system, otherwise known as nervous system patterning, and cause excitation/inhibition imbalance, a hallmark of ASD [26]. While Shh is obligatory for motor neuron development [27], Cdk5 is predominantly expressed in neurons and plays a crucial role in neuron development and many neuronal functions [28]. Furthermore, *Mib1* has been identified as an autism risk gene [29]. The review article proposes a potential link among these molecules (Shh, Cdk5, and Mib) in the development of ASD.

In conclusion, this Special Issue presents a collection of seven original research papers and two review articles highlighting the successful use of the zebrafish model to investigate human health-related issues and diseases. The papers cover a range of topics, from cancer, neurobehavior, and regeneration, to chemical toxicity and various state-of-the-art techniques, including the CRISPR/Cas9 gene editing technology, providing greater insights on the respective research areas. Given the increasing popularity of zebrafish as an alternative animal model in biomedical and toxicological research, this Special Issue will help expand the understanding of the underlying mechanisms of specific human diseases and health-related issues.

Acknowledgments: This work was supported in part by the Food and Drug Administration's Perinatal Health Center of Excellence (PHCE) funding program administered by the National Center for Toxicological Research.

Conflicts of Interest: The author declares no conflict of interest. This manuscript reflects the views of the author and does not necessarily reflect those of the US Food and Drug Administration. Any mention of a commercial product is for clarification only; it is not an endorsement for its use.

References

1. Ablain, J.; Zon, L.I. Of fish and men: Using zebrafish to fight human diseases. *Trends Cell Biol.* **2013**, *23*, 584–586. [CrossRef] [PubMed]
2. Cully, M. Zebrafish earn their drug discovery stripes. *Nat. Rev. Drug Discov.* **2019**, *18*, 811–813. [CrossRef]
3. Ciapaite, J.; Albersen, M.; Savelberg, S.M.C.; Bosma, M.; Meijer, N.W.F.; Tessadori, F.; Bakkers, J.P.W.; van Haaften, G.; Jans, J.J.; Verhoeven-Duif, N.M. Broad Vitamin B(6)-Related Metabolic Disturbances in a Zebrafish Model of Hypophosphatasia (TNSALP-Deficiency). *Int. J. Mol. Sci.* **2025**, *26*, 3270. [CrossRef]
4. Mornet, E.; Taillandier, A.; Domingues, C.; Dufour, A.; Benaloun, E.; Lavaud, N.; Wallon, F.; Rousseau, N.; Charle, C.; Guberto, M.; et al. Hypophosphatasia: A genetic-based nosology and new insights in genotype-phenotype correlation. *Eur. J. Hum. Genet.* **2021**, *29*, 289–299. [CrossRef]
5. Danila, A.M.; Savuca, A.; Ciobica, A.S.; Gurzu, I.L.; Nicoara, M.N.; Gurzu, B. The Impact of Oxytocin on Stimulus Discrimination of Zebrafish Albino and Non-Albino Models. *Int. J. Mol. Sci.* **2025**, *26*, 2070. [CrossRef] [PubMed]
6. Ciobica, A.; Balmus, I.M.; Padurariu, M. Is Oxytocin Relevant for the Affective Disorders? *Acta Endocrinol.* **2016**, *12*, 65–71. [CrossRef] [PubMed]
7. Li, K.; Fan, D.; Zhou, J.; Zhao, Z.; Han, A.; Song, Z.; Tang, X.; Hu, B. Deletion of Slc1a4 Suppresses Single Mauthner Cell Axon Regeneration In Vivo through Growth-Associated Protein 43. *Int. J. Mol. Sci.* **2024**, *25*, 10950. [CrossRef]

8. Marques, I.J.; Lupi, E.; Mercader, N. Model systems for regeneration: Zebrafish. *Development* **2019**, *146*, dev167692. [CrossRef]
9. Wu, S.; Xu, J.; Dai, Y.; Yu, B.; Zhu, J.; Mao, S. Insight into protein synthesis in axon regeneration. *Exp. Neurol.* **2023**, *367*, 114454. [CrossRef]
10. Chung, D.; Shum, A.; Caraveo, G. GAP-43 and BASP1 in Axon Regeneration: Implications for the Treatment of Neurodegenerative Diseases. *Front. Cell Dev. Biol.* **2020**, *8*, 567537. [CrossRef]
11. Ricarte, M.; Tagkalidou, N.; Bellot, M.; Bedrossiantz, J.; Prats, E.; Gomez-Canela, C.; Garcia-Reyero, N.; Raldúa, D. Short- and Long-Term Neurobehavioral Effects of Developmental Exposure to Valproic Acid in Zebrafish. *Int. J. Mol. Sci.* **2024**, *25*, 7688. [CrossRef] [PubMed]
12. Valentino, K.; Teopiz, K.M.; Kwan, A.T.H.; Le, G.H.; Wong, S.; Rosenblat, J.D.; Mansur, R.B.; Lo, H.K.Y.; McIntyre, R.S. Anatomical, behavioral, and cognitive teratogenicity associated with valproic acid: A systematic review. *CNS Spectr.* **2024**, *29*, 604–610. [CrossRef]
13. Astell, K.R.; Sieger, D. Zebrafish In Vivo Models of Cancer and Metastasis. *Cold Spring Harb. Perspect. Med.* **2020**, *10*, a037077. [CrossRef]
14. Andrews, K.A.; Ascher, D.B.; Pires, D.E.V.; Barnes, D.R.; Vialard, L.; Casey, R.T.; Bradshaw, N.; Adlard, J.; Aylwin, S.; Brennan, P.; et al. Tumour risks and genotype-phenotype correlations associated with germline variants in succinate dehydrogenase subunit genes SDHB, SDHC and SDHD. *J. Med. Genet.* **2018**, *55*, 384–394. [CrossRef]
15. Miltenburg, J.B.; Gorissen, M.; van Outersterp, I.; Versteeg, I.; Nowak, A.; Rodenburg, R.J.; van Herwaarden, A.E.; Olthaar, A.J.; Kusters, B.; Conrad, C.; et al. Characterisation of an Adult Zebrafish Model for SDHB-Associated Phaeochromocytomas and Paragangliomas. *Int. J. Mol. Sci.* **2024**, *25*, 7262. [CrossRef]
16. Enseleit, F.; Hürlimann, D.; Lüscher, T.F. Vascular protective effects of angiotensin converting enzyme inhibitors and their relation to clinical events. *J. Cardiovasc. Pharmacol.* **2001**, *37* (Suppl. 1), S21–S30. [CrossRef] [PubMed]
17. Garg, M.; Burrell, L.M.; Velkoska, E.; Griggs, K.; Angus, P.W.; Gibson, P.R.; Lubel, J.S. Upregulation of circulating components of the alternative renin-angiotensin system in inflammatory bowel disease: A pilot study. *J. Renin Angiotensin Aldosterone Syst.* **2015**, *16*, 559–569. [CrossRef] [PubMed]
18. Haxhija, E.Q.; Yang, H.; Spencer, A.U.; Koga, H.; Sun, X.; Teitelbaum, D.H. Modulation of mouse intestinal epithelial cell turnover in the absence of angiotensin converting enzyme. *Am. J. Physiol. Gastrointest. Liver Physiol.* **2008**, *295*, G88–G98. [CrossRef]
19. Wildhaber, B.E.; Yang, H.; Haxhija, E.Q.; Spencer, A.U.; Teitelbaum, D.H. Intestinal intraepithelial lymphocyte derived angiotensin converting enzyme modulates epithelial cell apoptosis. *Apoptosis* **2005**, *10*, 1305–1315. [CrossRef]
20. Wei, M.; Yu, Q.; Li, E.; Zhao, Y.; Sun, C.; Li, H.; Liu, Z.; Ji, G. Ace Deficiency Induces Intestinal Inflammation in Zebrafish. *Int. J. Mol. Sci.* **2024**, *25*, 5598. [CrossRef]
21. Rozofsky, J.P.; Pozzuto, J.M.; Byrd-Jacobs, C.A. Mitral Cell Dendritic Morphology in the Adult Zebrafish Olfactory Bulb following Growth, Injury and Recovery. *Int. J. Mol. Sci.* **2024**, *25*, 5030. [CrossRef]
22. Pozzuto, J.M.; Fuller, C.L.; Byrd-Jacobs, C.A. Deafferentation-induced alterations in mitral cell dendritic morphology in the adult zebrafish olfactory bulb. *J. Bioenerg. Biomembr.* **2019**, *51*, 29–40. [CrossRef] [PubMed]
23. Yang, D.; Jian, Z.; Tang, C.; Chen, Z.; Zhou, Z.; Zheng, L.; Peng, X. Zebrafish Congenital Heart Disease Models: Opportunities and Challenges. *Int. J. Mol. Sci.* **2024**, *25*, 5943. [CrossRef]
24. Gu, Q.; Kanungo, J. Neurogenic Effects of Inorganic Arsenic and Cdk5 Knockdown in Zebrafish Embryos: A Perspective on Modeling Autism. *Int. J. Mol. Sci.* **2024**, *25*, 3459. [CrossRef]
25. Kanungo, J.; Twaddle, N.C.; Silva, C.; Robinson, B.; Wolle, M.; Conklin, S.; MacMahon, S.; Gu, Q.; Edhlund, I.; Benjamin, L.; et al. Inorganic arsenic alters the development of dopaminergic neurons but not serotonergic neurons and induces motor neuron development via Sonic hedgehog pathway in zebrafish. *Neurosci. Lett.* **2023**, *795*, 137042. [CrossRef] [PubMed]
26. Fang, W.Q.; Chen, W.W.; Jiang, L.; Liu, K.; Yung, W.H.; Fu, A.K.Y.; Ip, N.Y. Overproduction of upper-layer neurons in the neocortex leads to autism-like features in mice. *Cell Rep.* **2014**, *9*, 1635–1643. [CrossRef]
27. Oh, S.; Huang, X.; Liu, J.; Litingtung, Y.; Chiang, C. Shh and Gli3 activities are required for timely generation of motor neuron progenitors. *Dev. Biol.* **2009**, *331*, 261–269. [CrossRef]
28. Pao, P.C.; Tsai, L.H. Three decades of Cdk5. *J. Biomed. Sci.* **2021**, *28*, 79. [CrossRef]
29. De Rubeis, S.; He, X.; Goldberg, A.P.; Poultney, C.S.; Samocha, K.; Cicek, A.E.; Kou, Y.; Liu, L.; Fromer, M.; Walker, S.; et al. Synaptic, transcriptional and chromatin genes disrupted in autism. *Nature* **2014**, *515*, 209–215. [CrossRef]

International Journal of
Molecular Sciences

MDPI

Article

Broad Vitamin B_6-Related Metabolic Disturbances in a Zebrafish Model of Hypophosphatasia (TNSALP-Deficiency)

Jolita Ciapaite [1,*], Monique Albersen [1,†], Sanne M. C. Savelberg [1], Marjolein Bosma [1], Nils W. F. Meijer [1], Federico Tessadori [2], Jeroen P. W. Bakkers [2,3], Gijs van Haaften [1], Judith J. Jans [1] and Nanda M. Verhoeven-Duif [1]

1 Department of Genetics, University Medical Center Utrecht, 3584 EA Utrecht, The Netherlands; m.albersen@amsterdamumc.nl (M.A.); m.bosma@umcutrecht.nl (M.B.); n.w.f.meijer@umcutrecht.nl (N.W.F.M.); g.vanhaaften@umcutrecht.nl (G.v.H.); j.j.m.jans@umcutrecht.nl (J.J.J.); n.verhoeven@umcutrecht.nl (N.M.V.-D.)
2 Hubrecht Institute-KNAW and University Medical Center Utrecht, 3584 CT Utrecht, The Netherlands; f.tessadori@hubrecht.eu (F.T.); j.bakkers@hubrecht.eu (J.P.W.B.)
3 Department of Medical Physiology, University Medical Center Utrecht, 3584 CM Utrecht, The Netherlands
* Correspondence: j.ciapaite@umcutrecht.nl
† Current address: Department of Laboratory Medicine, Amsterdam UMC location AMC, 1105 AZ Amsterdam, The Netherlands.

Abstract: Hypophosphatasia (HPP) is a rare inborn error of metabolism caused by pathogenic variants in *ALPL*, coding for tissue non-specific alkaline phosphatase. HPP patients suffer from impaired bone mineralization, and in severe cases from vitamin B_6-responsive seizures. To study HPP, we generated *alpl*$^{-/-}$ zebrafish using CRISPR/Cas9 gene-editing technology. At 5 days post fertilization (dpf), no *alpl* mRNA and 89% lower total alkaline phosphatase activity was detected in *alpl*$^{-/-}$ compared to *alpl*$^{+/+}$ embryos. The survival of *alpl*$^{-/-}$ zebrafish was strongly decreased. Alizarin red staining showed decreased bone mineralization in *alpl*$^{-/-}$ embryos. B_6 vitamer analysis revealed depletion of pyridoxal and its degradation product 4-pyridoxic acid in *alpl*$^{-/-}$ embryos. Accumulation of d3-pyridoxal 5′-phosphate (d3-PLP) and reduced formation of d3-pyridoxal in *alpl*$^{-/-}$ embryos incubated with d3-PLP confirmed Alpl involvement in vitamin B_6 metabolism. Locomotion analysis showed pyridoxine treatment-responsive spontaneous seizures in *alpl*$^{-/-}$ embryos. Metabolic profiling of *alpl*$^{-/-}$ larvae using direct-infusion high-resolution mass spectrometry showed abnormalities in polyamine and neurotransmitter metabolism, suggesting dysfunction of vitamin B_6-dependent enzymes. Accumulation of N-methylethanolaminium phosphate indicated abnormalities in phosphoethanolamine metabolism. Taken together, we generated the first zebrafish model of HPP that shows multiple features of human disease and which is suitable for the study of the pathophysiology of HPP and for the testing of novel treatments.

Keywords: vitamin B_6; hypophosphatasia; alpl deficiency; zebrafish; direct-infusion high-resolution mass spectrometry

1. Introduction

Hypophosphatasia (HPP) is a rare inborn error of metabolism that affects the development of bones and teeth and is caused by pathogenic variants in the *ALPL* gene, coding for the tissue-nonspecific isozyme alkaline phosphatase (TNSALP, EC:3.1.3.1). Since the first description by Rathbun [1], more than 400 variants in *ALPL*, predominantly missense, have been identified [2], explaining the highly variable clinical phenotype of HPP. Both autosomal recessive and autosomal dominant modes of inheritance have been reported, with the former often, but not always, corresponding to a more severe clinical phenotype [3]. TNSALP is expressed in the liver, bone (synthesized by the osteoblasts), and kidney, as well as the brain [4], and it functions as a homodimeric ectoenzyme with a broad phospho-substrate specificity [5]. Natural substrates of TNSALP include, but likely are not limited to [5], inorganic pyrophosphate (PP_i) [6,7], pyridoxal 5′-phosphate (PLP, the active form of vitamin B_6) [8–10], and phosphoethanolamine [8,11,12]. The severity of clinical phenotype strongly correlates with residual TNSALP enzyme activity [3,5,13]. Increased levels of plasma PLP and urinary and plasma PEA along with reduced serum unfractionated alkaline phosphatase activity serve as diagnostic markers of HPP [5,14].

Accumulation of TNSALP substrates in HPP patients reflects the physiological role of the enzyme and clarifies the metabolic basis of HPP. Based on the age of diagnosis/onset of symptoms and severity of clinical phenotype, HPP is classified as perinatal (benign), perinatal (severe), infantile, childhood (mild), childhood (severe), adult, or odontohypophosphatasia, with the perinatal (severe) and infantile forms being most severe [14]. The main clinical feature of HPP is abnormal bone mineralization causing premature loss of deciduous teeth, rickets in children, and osteomalacia in adults. The bone phenotype of HPP is explained by accumulation of PP_i, a potent inhibitor of bone mineralization [15]. Paradoxically, in some cases, premature closure of cranial sutures (craniosynostosis) occurs in infantile and childhood HPP, causing intracranial hypertension [5,16]. In addition, blocked entry of minerals into the skeleton may lead to hypercalcemia/hypercalciuria, nephrocalcinosis, and renal impairment [5,17]. In the most severe forms of perinatal and infantile HPP, vitamin B_6-dependent seizures may occur, indicating a lethal prognosis [18–22]. The seizures begin in the first hours after birth and are refractory to standard anticonvulsant drugs, but are responsive to pyridoxine (unphosphorylated form of vitamin B_6) [18–22]. B_6-dependent seizures are explained by the role of TNSALP in the cellular uptake of PLP (phosphorylated form of vitamin B_6) [23], with the decreased TNSALP activity presumably leading to vitamin B_6 deficiency in the central nervous system (CNS) (Figure 1a). The correlation between the response to pyridoxine and the severity of pediatric HPP reinforces TNSALP's role in vitamin B_6 metabolism [24]. Other, less well understood neurological symptoms of HPP may include depression, memory loss, ADHD, anxiety, headache, and sleep disturbance [25].

HPP is an incurable disease. In addition to symptomatic treatment, enzyme replacement therapy with asfotase alfa, a mineral-targeted human recombinant TNSALP, is available for treatment of the bone phenotype of HPP [5,26]. Therefore, understanding TNSALP function in the kidney, liver, brain, and other soft tissues as well as the mechanistic basis of milder neurological symptoms of HPP is becoming more relevant in improving quality of life of HPP patients.

The zebrafish (*Danio rerio*) is a promising model organism in the study of human diseases [27], including HPP [28,29]. Zebrafish have four genes coding for alkaline phosphatases: two for intestinal alkaline phosphatases, *alpi.1* and *alpi.2* (gene duplication), one for alkaline phosphatase 3, *alp3* (also expressed in the intestine), and one for tissue-nonspecific alkaline phosphatase *alpl*, which also shows a high degree of genetic conservation with human *ALPL* [28,29]. In the present study, we generated the first $alpl^{-/-}$ zebrafish line using CRISPR/Cas9 gene-editing technology. Biochemical and behavioral characterization of $alpl^{-/-}$ zebrafish showed that they display multiple features of infantile HPP.

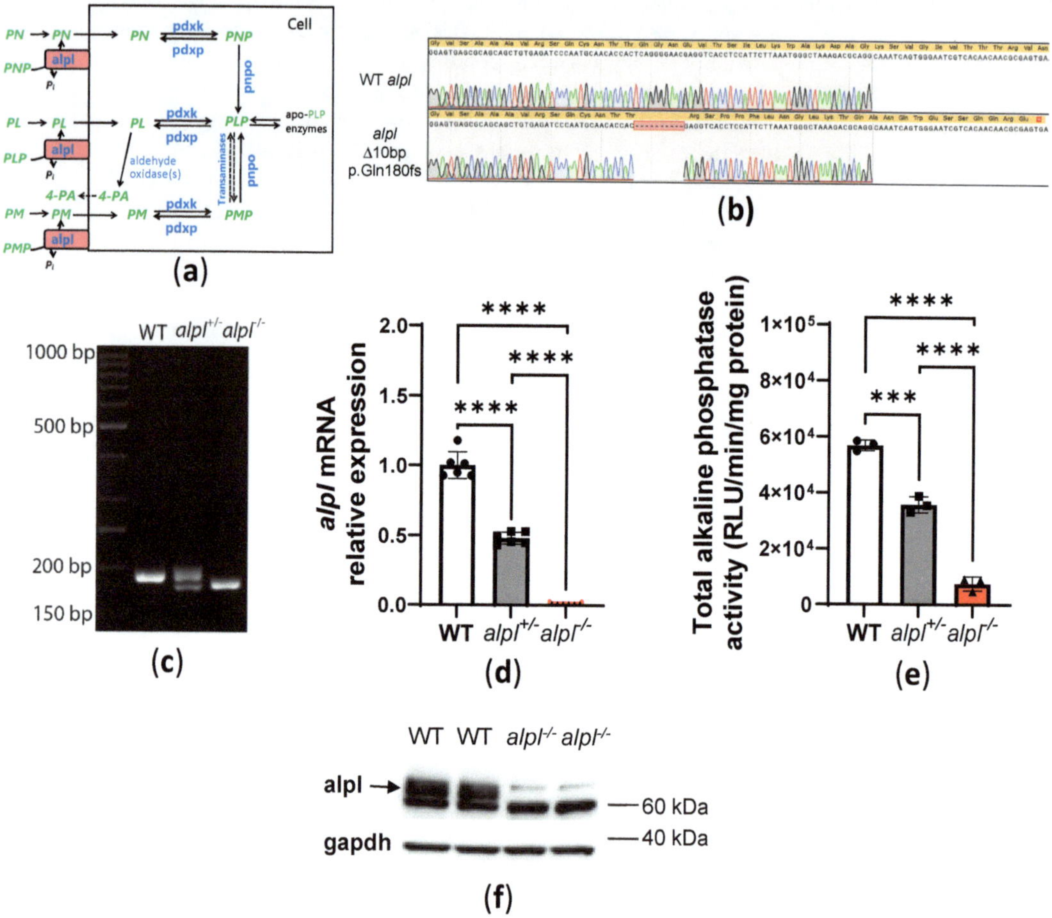

Figure 1. Generation and basic characterization of *alpl*$^{-/-}$ zebrafish line. (**a**) Schematic representation of vitamin B_6 metabolism. (**b**) Sanger sequencing results showing a 10 bp deletion at the CRISPR site in exon 5 of the *alpl* gene (c.623_632del, p.Gln180*fs*), which is predicted to result in a truncated protein. (**c**) Agarose electrophoresis of genotyping PCR products for the CRISPR site in exon 5 of the *alpl* gene in wild type (WT), *alp*$^{+/-}$, *alpl*$^{-/-}$ 5 dpf old zebrafish embryos. (**d**) Relative *alpl* mRNA expression in WT, *alpl*$^{+/-}$, and *alpl*$^{-/-}$ 5 dpf old zebrafish embryos. Data are means from $n = 3$ pools (10 embryos per pool) per genotype measured in duplicate ± SD. **** $p < 0.0001$. (**e**) Total alkaline phosphatase activity in whole-embryo extracts of WT, *alpl*$^{+/-}$, and *alpl*$^{-/-}$ 5 dpf old zebrafish. RLU, relative light unit. Data are means from $n = 3$ pools (10 embryos per pool) per genotype ± SD. **** $p < 0.0001$ and *** $p < 0.001$. (**f**) Alpl protein expression in total WT and *alpl*$^{-/-}$ 5 dpf zebrafish embryo extracts showing lack of alpl protein in *alpl*$^{-/-}$ embryos (predicted molecular weight of alpl protein (NP_957301.2) is 62.5 kDa). Data are from pools of $n = 32$ embryos per genotype. Gapdh protein expression was used as the loading control.

2. Results

2.1. Generation of the alpl Knockout Zebrafish

The *alpl* knockout zebrafish were generated using CRISPR/Cas9 gene editing. Sanger sequencing of F1 embryos derived from outcrossing F0 mosaic zebrafish revealed a 10 base pair out-of-frame deletion (c.623_632del, p.Gln180fs), predicted to result in a truncated protein (Figure 1b,c). Sequencing results also showed an intact reference sequence at the CRISPR target site in exon 6. The genotyping results of all untreated 5 dpf old embryos used in the present study (n = 1752) showed the following genotype distribution: 27% *alpl*$^{+/+}$, 51% *alpl*$^{+/-}$, and 22% *alpl*$^{-/-}$. Upon visual inspection at 5 dpf, *alpl*$^{-/-}$ embryos were morphologically normal, indistinguishable from *alpl*$^{+/+}$ and *alpl*$^{+/-}$ embryos. Compared to the *alpl*$^{+/+}$ embryos, ~50% ($p < 0.0001$) less *alpl* mRNA was detected in *alpl*$^{+/-}$ and virtually no *alpl* mRNA was detected in *alpl*$^{-/-}$ zebrafish embryos at 5 dpf (Figure 1d). Total alkaline phosphatase enzyme activity measured in the whole embryo extracts was 37% ($p < 0.0001$) and 87% ($p < 0.0001$) lower in *alpl*$^{+/-}$ and *alpl*$^{-/-}$ zebrafish embryos, respectively, compared to WT (Figure 1e). The contribution of other alkaline phosphatase isoforms, such as intestinal alkaline phosphatase encoded by *alpi.1* and *alpi.2*, and possibly alkaline phosphatase 3, which is also expressed in the intestine, could explain the residual enzyme activity measured in *alpl*$^{-/-}$ embryos, since intestinal isoforms are insensitive to (-)-tetramisole HCl [30]. In contrast to WT (*alpl*$^{+/+}$) zebrafish, no Alpl protein was detected at 62.5 kDa (the predicted molecular weight of alpl protein (NP_957301.2)) by Western blotting in *alpl*$^{-/-}$ embryos at 5 dpf (Figure 1f).

2.2. *Abnormal Bone Mineralization, Vitamin B$_6$ Metabolism and Locomotion in alpl$^{-/-}$ Embryos*

Decreased bone mineralization due to accumulation of pyrophosphate, one of the substrates of TNSALP, is a key feature of HPP [7,8]). To determine the consequences of *alpl* knockout on bone mineralization, we stained 5 dpf embryos with acid-free alizarin red (bone) and alcian blue (cartilage) double stain. Less alizarin red staining of mineralized structures, such as notochord [31], in *alpl*$^{-/-}$ compared to WT and *alpl*$^{+/-}$ embryos at 5 dpf suggested a negative effect of alpl deficiency on bone mineralization (Figure 2a).

Another important TNSALP substrate is extracellular (circulating) PLP (Figure 1a). We assessed the consequences of *alpl* knockout on PLP metabolism by measuring the utilization of d3-PLP by live 5 dpf old embryos. After 3 h incubation with 100 μM d3-PLP in embryo water, significantly more d3-PLP (Figure 2b) and significantly less d3-PL (Figure 2c) was observed in *alpl*$^{-/-}$ compared to WT and *alpl*$^{+/-}$ embryos, indicating impaired d3-PLP hydrolysis to d3-PL.

The steady-state PLP concentration measured in whole embryo extracts was slightly but significantly lower in *alpl*$^{-/-}$ compared to WT and *alpl*$^{+/-}$ embryos (Figure 2d). Since Alpl deficiency is expected to result in increased PLP concentration in fluid extracellular matrix and decreased intracellular PLP concentration (see Figure 1a), the lower PLP concentration measured in whole *alpl*$^{-/-}$ embryo extracts likely indicates that cell-derived PLP dominated the total measured PLP. Concentrations of PL and its degradation product 4-PA were strongly decreased in *alpl*$^{-/-}$ compared to WT and *alpl*$^{+/-}$ embryos (Figures 2e and 2f, respectively), in agreement with the predicted negative effect of Alpl deficiency on both the extra- and intracellular concentrations of these compounds. PMP concentration was not significantly changed in *alpl*$^{-/-}$ compared to WT and *alpl*$^{+/-}$ embryos (Figure 2g). Negligible concentrations of PM and PNP were measured in whole embryo extracts, which were similar in all three genotypes (Figures 2h and 2i, respectively).

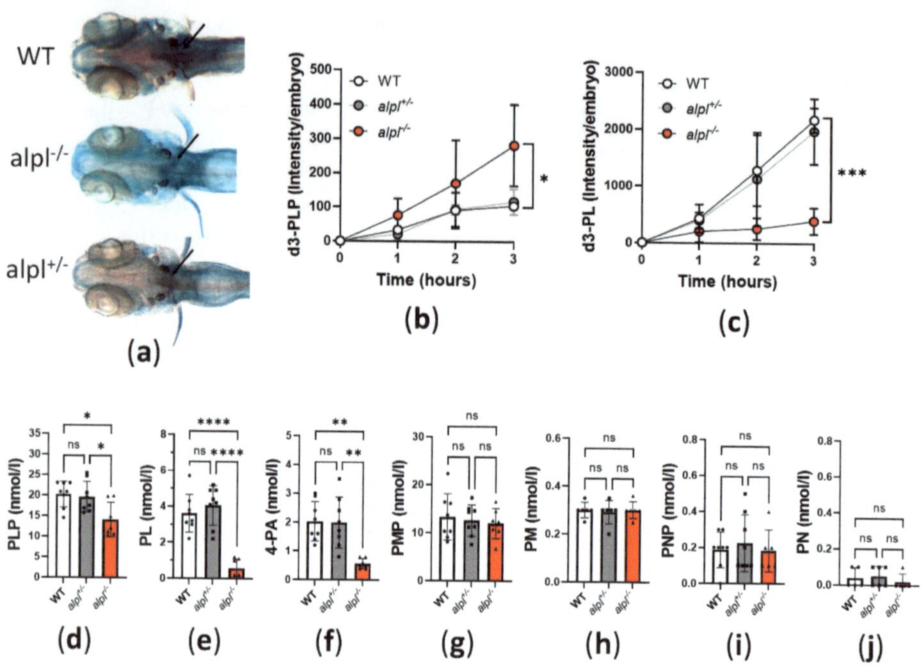

Figure 2. Impaired bone mineralization and abnormal vitamin B_6 metabolism in *alpl* knockout zebrafish. (**a**) Decreased alizarin red staining of mineralized structures (notochord, indicated by arrows) in *alpl*$^{-/-}$ compared to WT and *alpl*$^{+/-}$ embryos at 5 dpf suggests a negative effect of alpl deficiency on bone mineralization. (**b**) Accumulation of d3-PLP and (**c**) decreased production of d3-pyridoxal (d3-PL) in *alpl*$^{-/-}$ compared to WT and *alpl*$^{+/-}$ embryos incubated with 100 μM d3-PLP in embryo water (E3) for 3 h at 28 °C. Data are means from $n = 3$ pools (3 embryos per pool) per time point and genotype ± SD. (**d**) Steady-state PLP, (**e**) pyridoxal (PL), (**f**) 4-pyridoxic acid (4-PA), (**g**) pyridoxamine 5′-phospate (PMP), (**h**) pyridoxamine (PM), (**i**) pyridoxine 5′-phosphate (PNP), and (**j**) pyridoxine (PN) concentrations in whole embryo extracts. Data are means from $n = 8$ pools (3 embryos per pool) per genotype ± SD. **** $p < 0.0001$, *** $p < 0.001$ ** $p < 0.01$, * $p < 0.05$, and ns—not significant ($p > 0.05$).

Intracellular PLP deficiency can have negative effects on neurotransmitter biosynthesis, leading to vitamin B_6-dependent seizures [28]. In zebrafish embryos, seizures cause typical fast, circular swimming bursts [32–35]. Therefore, we recorded the locomotion of 5 dpf old WT, *alpl*$^{+/-}$, and *alpl*$^{-/-}$ embryos with a Zebrabox (Figures 3a and S1). Locomotion analysis revealed that inactivity count (i.e., number of times spent below the inactivity threshold) was significantly lower in *alpl*$^{-/-}$ compared to WT embryos (Figure 3b) without a change in the duration of inactivity (Figure 3c). Furthermore, the count, duration, and distance swum in small movements (speed < 30 mm/s) was significantly lower in *alpl*$^{-/-}$ compared to WT embryos (Figures 3d and 3e, respectively). In contrast, the count, duration, and distance swum in burst movements (speed > 30 mm/s) was significantly increased in *alpl*$^{-/-}$ compared to WT embryos (Figures 3g, 3h and 3i, respectively), indicating spontaneous seizures in the former. No locomotion parameters of *alpl*$^{+/-}$ embryos were significantly different from WT (Figure 3).

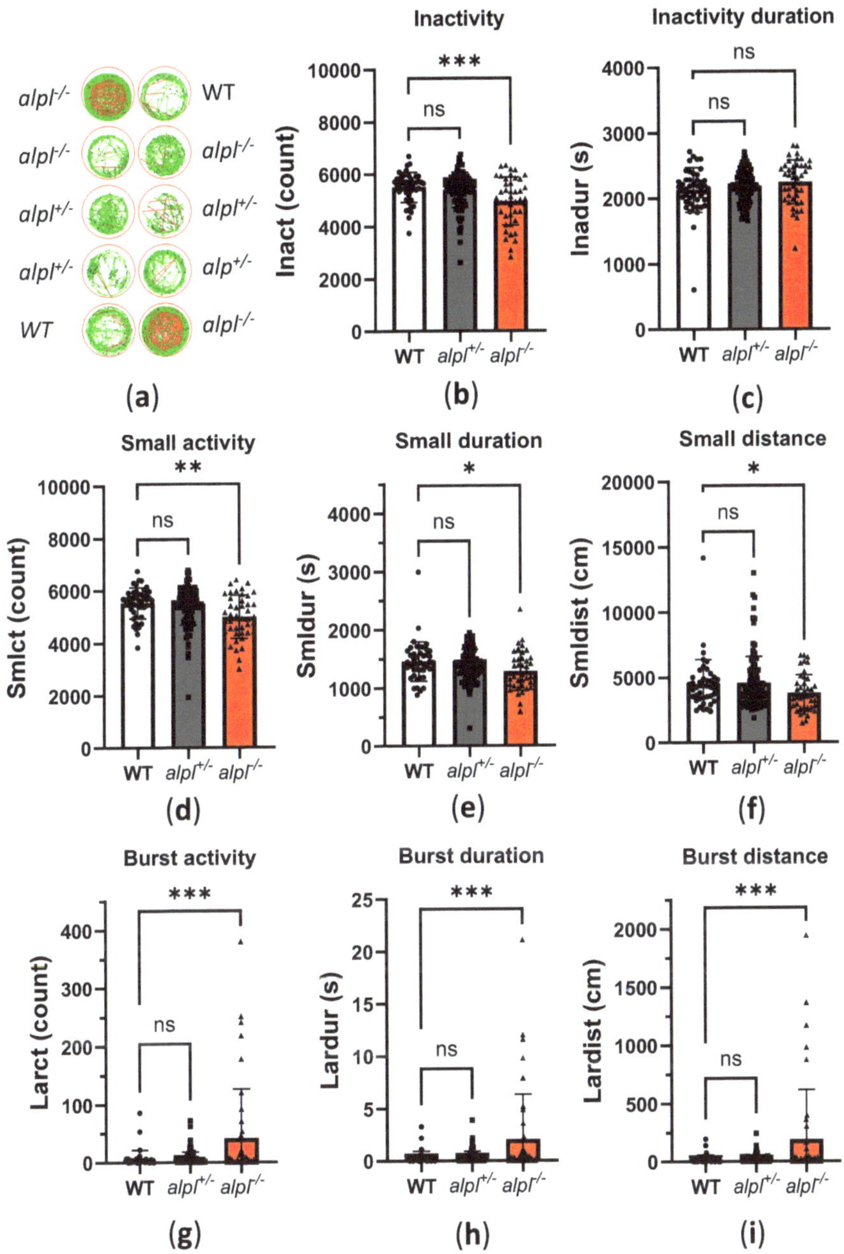

Figure 3. Locomotion analysis of 5 dpf old WT, *alpl*$^{+/-}$, and *alpl*$^{-/-}$ zebrafish embryos. (**a**) Examples of swimming trajectories of 5 dpf old embryos recorded with Zebrabox for 1 h at 28 °C in the dark. Green—movement speed < 30 mm/s (small activity), red—movement speed > 30 mm/s (burst activity), black—no movement (inactivity). (**b**) Inactivity count. (**c**) Duration of inactivity. (**d**) Small activity count. (**e**) Duration of small activity. (**f**) Distance swum during small activity. (**g**) Burst activity count. (**h**) Duration of burst activity. (**i**) Distance swum during burst activity. Data are means from $n = 45$ (WT), $n = 105$ (*alpl*$^{+/-}$), and $n = 42$ (*alpl*$^{-/-}$) embryos ± SD. *** $p < 0.001$, ** $p < 0.01$, * $p < 0.05$, and ns—not significant ($p > 0.05$).

2.3. *Pyridoxine Treatment Normalizes Locomotion and Some Metabolic Abnormalities in alpl$^{-/-}$ Zebrafish Embryos*

Next, we investigated how vitamin B_6 treatment affects locomotion and biochemical abnormalities in 5 dpf old *alpl*$^{-/-}$ zebrafish embryos. Initial experiments showed no significant effects of a single, 3 h treatment with 100 μM pyridoxine (PN) on B_6 vitamers and locomotion parameters in WT and *alpl*$^{-/-}$ embryos. Increasing the duration of the treatment with 100 μM PN to 72 h resulted in a strong increase in PN concentrations of a similar magnitude in WT and *alpl*$^{-/-}$ embryos (Figure 4a). PN treatment had no significant effect on the PLP concentration in WT embryos (Figure 4b). In contrast, in *alpl*$^{-/-}$ embryos, PN treatment led to an increase in PLP concentration to a level that was significantly higher than in WT (Figure 4b). Concentrations of PL and its degradation product 4-PA increased in both WT and *alpl*$^{-/-}$ embryos in response to PN treatment (Figures 4c and 4d, respectively). However, accumulation of both PL and 4-PA was less pronounced in *alpl*$^{-/-}$ embryos (Figures 4c and 4d, respectively). Concentrations of PM, PMP, and PNP were not significantly affected by PN treatment (Supplemental Figure S2a–c, respectively). Locomotion analysis showed that 72 h of PN treatment resulted in a significant reduction in parameters indicating spontaneous seizures in *alpl*$^{-/-}$ embryos. Specifically, burst activity count (Figure 4e), duration of burst activity (Figure 4f), and distance swum during burst activity (Figure 4g) were significantly lower in PN-treated compared to untreated *alpl*$^{-/-}$ embryos, and were not significantly different from WT. Furthermore, the concentration of the inhibitory neurotransmitter γ-aminobutyric acid (GABA), which was significantly lower in untreated *alpl*$^{-/-}$ embryos, was restored after PN treatment to the WT level (Figure 4h). Since PLP is a cofactor in many enzymes involved in amino acid metabolism, we analyzed amino acids. Glutamate, glutamine, and asparagine concentrations were significantly lower in untreated *alpl*$^{-/-}$ compared to WT embryos. These amino acids were normalized after 72 h of PN treatment (Figures 4i and S3A). Methionine, the only amino acid that significantly accumulated in untreated *alpl*$^{-/-}$ embryos, was normalized to WT level after PN treatment (Figures 4i and S3A). Furthermore, concentrations of several amino acids significantly increased, independent of genotype, in response to PN treatment (proline, alanine, valine, leucine, serine, tryptophan, and histidine), and the concentration of only ornithine decreased, independent of genotype, in response to PN treatment (Figure S3B). The concentrations of the remaining quantified amino acids were not affected either by genotype nor PN treatment (Supplemental Figure S3C).

To explore the global metabolic response to PN treatment we analyzed the metabolomes of untreated and 100 μM PN-treated WT and *alpl*$^{-/-}$ embryo extracts by direct-infusion high-resolution mass spectrometry (DI-HRMS). Using an in-house developed peak calling pipeline [36], 1919 mass peaks were annotated with 3941 metabolites that could occur endogenously. Partial least squares-discriminant analysis (PLS-DA) showed a clear discrimination of *alpl*$^{-/-}$ and WT metabolomes, which became less pronounced upon PN treatment, suggesting normalization of *alpl*$^{-/-}$ zebrafish metabolome after 72 h treatment with 100 μM PN (Figure 5a). Among the metabolites of component 1, identified by PLS-DA as contributing the most to the separation of the data, were several phosphorylated compounds (adenosine monophosphate (AMP), glycerol 3-phosphate, uridine 5′-monophosphate (UMP)), the vitamin B_6 degradation product 4-pyridoxic acid (4-PA), as well as several amino acids and their derivatives (glutamine, GABA, methionine, methionine sulfoxide, and cystathionine) (Figure 5b). Statistical analysis of the data using one-way ANOVA with Tukey's post-hoc test identified 284 significantly altered metabolites (Supplemental Table S2). The heatmap overview of the 50 highest-ranking metabolites based on one-way ANOVA results is shown in Figure 5c. Closer inspection of Tukey's post-hoc test results showed that 144 metabolites identified by one-way ANOVA were significantly changed in *alpl*$^{-/-}$ embryos compared to WT, of which 64 metabolites were subsequently significantly affected by PN treatment

(Supplemental Table S2 and Figure S4A). Among phospho compounds that were increased in *alpl*$^{-/-}$ embryos compared to WT, but did not respond to PN treatment and therefore could be directly related to the alpl phosphatase function, were: inositol cyclic phosphate, cytidine monophosphate, phosphoguanidinoacetate, (S)-5-diphosphomevalonic acid, phosphodimethylethanolamine, and glycerylphosphorylethanolamine (Supplemental Table S2). In contrast, the level of O-phosphoethanolamine was normalized by PN treatment, indicating that its accumulation was caused by a different, vitamin B_6-dependent, mechanism (Supplemental Table S2). Figure 5d shows a selection of highest-ranking metabolites that were responsive to PN treatment. The localization of these metabolites within the metabolic pathways is marked based on the KEGG pathway database information in Figure 5e. Abnormalities in methionine, its oxidation product methionine sulfoxide, and cystathionine levels in *alpl*$^{-/-}$ embryos (Figure 5d) suggested impaired functioning of methionine and/or folate cycles and the transsulfuration pathway, which in turn is important for synthesis of glutathione (Figure 5e). These pathways contain vitamin B_6-dependent enzymes (such as CBS and CGL; see Figure 5e), and the fact that all three metabolites were normalized after PN-treatment indicated that the effects were likely mediated by these enzymes. While glutathione levels were similar in WT and *alpl*$^{-/-}$ embryos, they significantly increased in both genotypes upon PN treatment (Supplemental Figure S4B), indicating the sensitivity of glutathione synthesis to vitamin B_6 availability. Increased levels of 2-hydroxybutyric acid and propionylcarnitine in *alpl*$^{-/-}$ embryos suggested that surplus cystathionine was diverted from the transsulfuration pathway, possibly due to impaired activity of vitamin B_6-dependent cystathionine-β-synthase (Figure 5d,e). Moreover, the impairment of transsulfuration pathway activity could underlie the elevation of N1-acetylspermidine levels in *alpl*$^{-/-}$ embryos (Supplemental Figure S4C) by pushing the methionine cycle intermediates into the polyamine pathway (Figure 5e). While none of the measured canonic folate cycle intermediates were altered in *alpl*$^{-/-}$ embryos, we observed accumulation of 10-formyldihydrofolate (Supplemental Figure S4D), which has been shown to participate in purine synthesis in mammalian cells [37]. Consequently, we found increased levels of AMP and inosine in *alpl*$^{-/-}$ embryos, which decreased after PN treatment (Figure 5d,e). Additionally, PN treatment-responsive abnormalities in the pyrimidine synthesis pathway intermediates UMP (Figure 5d,e) and orotidine (Supplemental Figure S4E) in *alpl*$^{-/-}$ embryos may have resulted from abnormal folate cycle activity (Figure 5e).

2.4. *Progression of Metabolic Abnormalities and Decreased Survival in alpl*$^{-/-}$ *Zebrafish Larvae*

To assess the progression of metabolic abnormalities, we analyzed 10 dpf old WT and *alpl*$^{-/-}$ zebrafish larvae. Only 24% of *alpl*$^{-/-}$ larvae were alive at 10 dpf (Figure 6a). Treatment with 100 μM PN for 5 days (from 5 dpf to 10 dpf, 30 min/day) led to a significant improvement in the survival of *alpl*$^{-/-}$ zebrafish larvae (Figure 6a); however, it did not restore the survival to WT levels. While this treatment regimen resulted in increases in PN and PLP concentrations in *alpl*$^{-/-}$ zebrafish larvae, the concentrations of PL and its degradation product 4-PA remained low (Figure 6b). Locomotion analysis using Zebrabox did not indicate spontaneous seizures during 1 h of measurement in the surviving larvae, as indicated by similar burst activity parameters in WT and *alpl*$^{-/-}$ larvae (Figures 6c and S5A). However, there was a decrease in all small activity parameters, i.e., distance swum in small movements (Figure 6c), as well as the count and duration of small movements in *alpl*$^{-/-}$ larvae (Supplemental Figure S5B), indicating decreased mobility of the mutants, which was normalized by PN treatment. No differences were observed in the inactivity count and duration (Supplemental Figure S5C). The concentration of GABA was lower in untreated *alpl*$^{-/-}$ larvae, but it was normalized by PN treatment (Figure 6d).

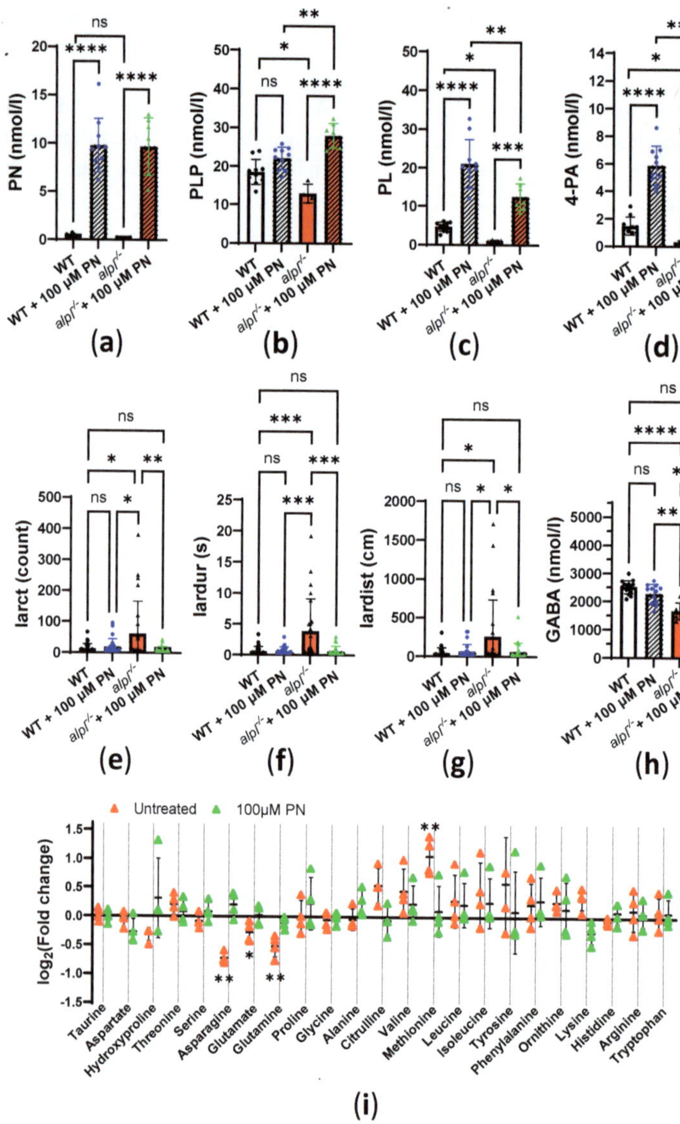

Figure 4. Normalization of biochemical and locomotion abnormalities in 5 dpf old *alpl*$^{-/-}$ zebrafish embryos after 72 h continuous treatment with 100 µM pyridoxine (PN). Concentrations of (**a**) pyridoxine (PN), (**b**) PLP, (**c**) pyridoxal (PL), and (**d**) 4-pyridoxic acid (4-PA) measured in whole 5 dpf old embryo extracts. Data are means from n = 4–10 pools (3 embryos per pool) per genotype ± SD. Zebrabox analysis of (**e**) burst activity count, (**f**) duration of burst activity, and (**g**) distance swum during burst activity in 5 dpf old embryos measured for 1 h at 28 °C in the dark. Data are means from n = 23–25 embryos per genotype and treatment ± SD. (**h**) γ-Aminobutyric acid (GABA) concentrations. Data are means from n = 7–15 pools (3 embryos per pool) per genotype and treatment ± SD. (**i**) Fold-change analysis of amino acid concentrations in untreated *alpl*$^{-/-}$ and 100 µM PN-treated *alpl*$^{-/-}$ zebrafish embryos compared to the corresponding WT group. Data are from n = 4 pools (3 embryos per pool). **** $p < 0.0001$, *** $p < 0.001$, ** $p < 0.01$, * $p < 0.05$, and ns—not significant ($p > 0.05$).

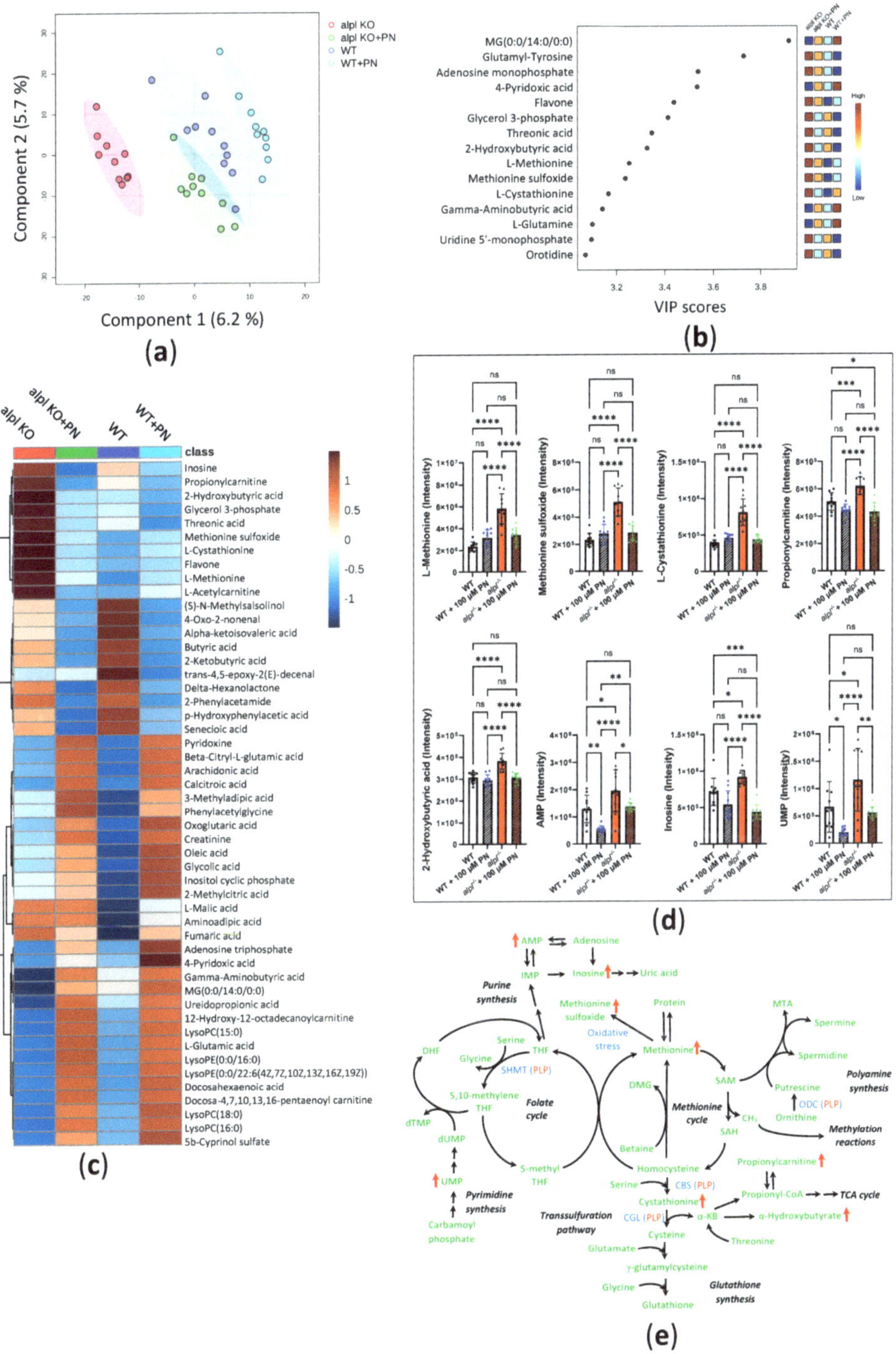

Figure 5. Global metabolic response to pyridoxine (PN) treatment in 5 dpf old WT and *alpl*$^{-/-}$ zebrafish embryos. (**a**) Partial least squares-discriminant analysis (PLS-DA) scores plot (principal component 1 (*x*-axis) and component 2 (*Y*-axis)). The explained variances are shown in brackets. 95% confidence

intervals are shown for each group. (**b**) Important metabolites identified by PLS-DA for principal component 1. Metabolites with the highest variable importance in projection (VIP) scores are shown. The colored boxes on the right indicate the relative intensity of the corresponding metabolite in each group. (**c**) Heatmap visualization of group-average intensities of the 50 highest-ranking metabolites based on one-way ANOVA results. Euclidean distance and Ward's clustering algorithm were used for the hierarchical clustering of metabolites. (**d**) Intensities of a selection of the highest-ranking metabolites based on one-way ANOVA results. Data are means from $n = 10$ pools (3 embryos per pool) per genotype and treatment group ± SD. **** $p < 0.0001$, *** $p < 0.001$, ** $p < 0.01$, * $p < 0.05$, and ns—not significant ($p > 0.05$); comparisons as indicated in the graphs. (**e**) Schematic visualization of metabolic pathways where strongest effects of Alpl deficiency were observed followed by normalization after pyridoxine treatment. Metabolites are shown in green; red arrows indicate the effect of Alpl deficiency. Key PLP-dependent enzymes are shown in blue: SHMT, serine hydroxymethyltransferase (EC 2.1.2.1); CBS, cystathionine-β-synthase (EC 4.2.1.22); CGL, cystathionine γ-lyase (EC 4.4.1.1); ODC, ornithine decarboxylase (EC 4.1.1.17).

Next, we analyzed the metabolomes of 10 dpf old WT and *alpl*$^{-/-}$ larvae with DI-HRMS. Consequently, 1899 mass peaks were annotated with 3891 metabolites that could occur endogenously [36]. Statistical analysis using a *t*-test identified 183 metabolites significantly ($p < 0.05$) altered in *alpl*$^{-/-}$ compared to WT larvae (Supplemental Table S3). Further analysis using a volcano plot identified several metabolites that were strongly and significantly changed in *alpl*$^{-/-}$ larvae (Figure 6e, Supplemental Table S4). Statistical analysis revealed abnormalities in several neurotransmitters, including decreased levels of GABA (Supplemental Table S3) and increased levels of L-DOPA in *alpl*$^{-/-}$ larvae (Figure 6f, Supplemental Table S3). Accumulation of L-DOPA in conjunction with the elevation of vanillactic acid (Figure 6f) pointed towards abnormal activity of the PLP-dependent aromatic L-amino acid decarboxylase (AADC, EC: 4.1.1.28) [38]. Consequently, the downstream products of AADC activity, dopamine and epinephrine, were negatively affected by Alpl deficiency (Figure 6f). 2-(3-Carboxy-3-(methylammonio)propyl)-L-histidine, a post-translationally modified histidine that serves as a substrate for diphthine synthase (EC:2.1.1.98), a methyltransferase involved in transfer of one-carbon groups, was strongly decreased in *alpl*$^{-/-}$ larvae (Figure 6e, Supplemental Tables S3 and S4). Furthermore, changed levels of several polyamine species (N-acetylcadaverine, N-acetylputrescine, dehydrospermidine, and N1-acetylspermidine (all decreased) and norspermidine (increased)) in *alpl*$^{-/-}$ zebrafish larvae (Figure 6e, Supplemental Tables S3 and S4) indicated altered activity of a polyamine synthesis pathway. An increase in vitamin A (retinol) and decrease in retinal (oxidized form of retinol, component of visual pigment [39]) (Figure 6g, Supplemental Tables S3 and S4) pointed towards abnormalities in vitamin A metabolism in *alpl*$^{-/-}$ larvae. Moreover, accumulation of N-methylethanolaminium phosphate in *alpl*$^{-/-}$ larvae (Figure 6h) suggested abnormal metabolism of phosphoethanolamines.

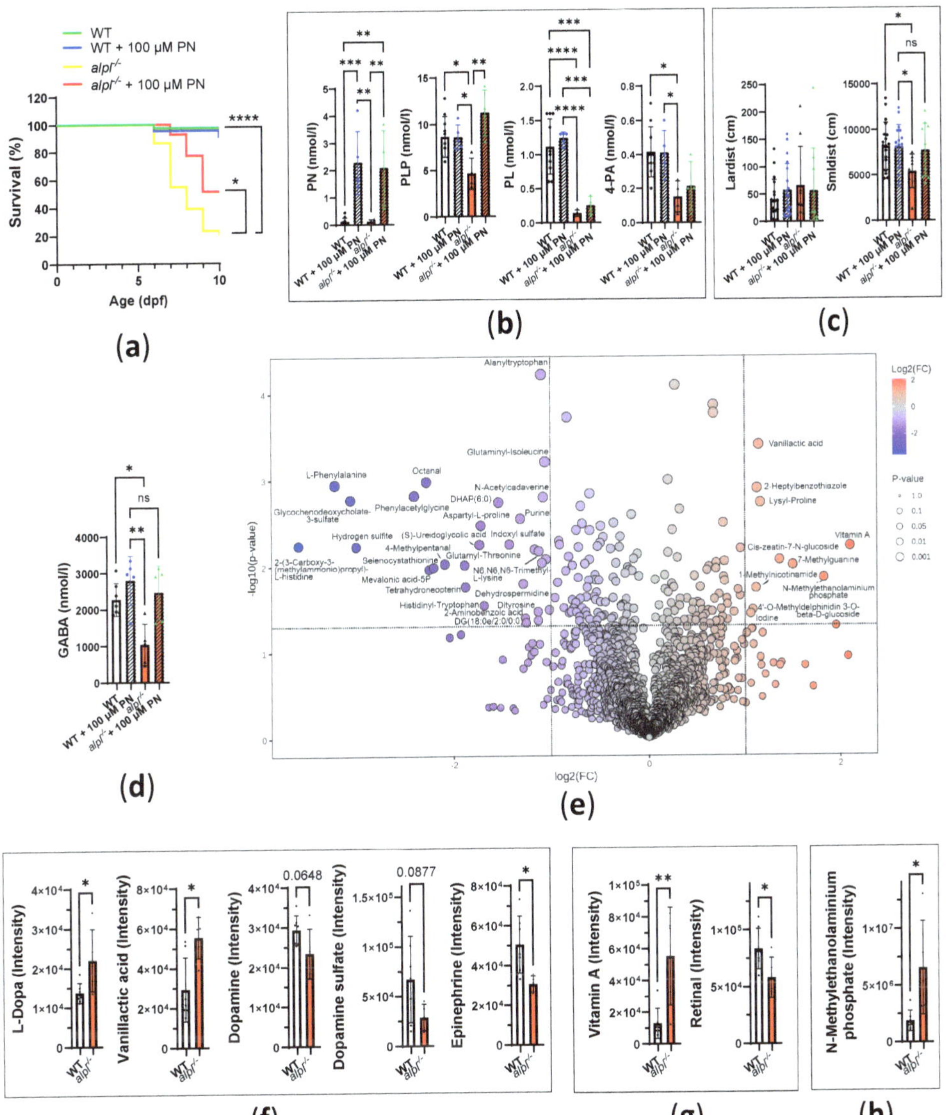

Figure 6. Progression of metabolic abnormalities in 10 dpf old *alpl*$^{-/-}$ zebrafish larvae. (**a**) Kaplan–Meier survival curves of untreated and 100 µM PN-treated WT and *alpl*$^{-/-}$ zebrafish until 10 dpf. Data are from n = 18–42 zebrafish per genotype and treatment condition. (**b**) Concentrations of pyridoxine (PN), PLP, pyridoxal (PL) and its degradation product 4-pyridoxic acid (4-PA) in untreated and 100 µM PN-treated WT and *alpl*$^{-/-}$ larvae. Data are means from n = 4–10 larvae per genotype and treatment condition ± SD. (**c**) Distance swum during burst activity (lardist, movement speed > 30 mm/s) and during small activity (smldist, movement speed < 30 mm/s) during 1 h measurement using Zebrabox at 28 °C in the dark. Data are means from n = 9–19 larvae per genotype and treatment condition ± SD.

(**d**) Concentrations of GABA in untreated and 100 μM PN-treated WT and *alpl*$^{-/-}$ larvae. Data are means from n = 5–7 larvae per genotype and treatment condition ± SD. (**e**) Important features in metabolome of *alpl*$^{-/-}$ larvae selected by volcano plot with fold change (FC) threshold equal to 2 (x-axis) and t-tests threshold of $p < 0.05$ (y-axis). (**f**) Accumulation of L-Dopa and vanillactic acid, and decreased level of dopamine, dopamine sulfate (dopamine 4-sulfate and dopamine 3-O-sulfate) and epinephrine in *alpl*$^{-/-}$ larvae suggest impaired activity of aromatic L-amino acid decarboxylase. (**g**) Accumulation of vitamin A (retinol) and decreased level of retinal in *alpl*$^{-/-}$ larvae. (**h**) Accumulation of N-methylethanolaminium phosphate in *alpl*$^{-/-}$ larvae. Data are means from n = 5–7 larvae per genotype ± SD (panels E-H). **** $p < 0.0001$, *** $p < 0.001$, ** $p < 0.01$, and * $p < 0.05$.

3. Discussion

While the mechanistic basis of the bone phenotype of HPP is well understood, several unexplained clinical features remain, including craniosynostosis and neurological manifestations [22,25]. Moreover, the understanding of TNSALP function in the soft tissues, such as the kidney, liver, and brain, is very limited, underscoring the need for further fundamental research. In the present study we described the first genetic model of tissue non-specific alkaline phosphatase deficiency in zebrafish that displayed many key features of human HPP. Our data showed that like TNSALP in humans, Alpl in zebrafish has a function in vitamin B_6 metabolism, as illustrated by strongly impaired ability of *alpl*$^{-/-}$ embryos to hydrolyze d3-PLP to d3-PL. The deficiency in Alpl activity led to lower total PLP, PL, and 4-PA levels, and to vitamin B_6 (pyridoxine)-responsive seizures. Moreover, multiple metabolic abnormalities were identified through untargeted metabolomics that were linked to decreased cellular/tissue PLP levels and impaired activity of PLP-dependent enzymes. However, pyridoxine treatment improved, but did not fully restore to WT levels, the survival of *alpl*$^{-/-}$ zebrafish. This suggests that a deficiency of functions other than those involved in vitamin B_6 metabolism (e.g., bone mineralization) plays an essential role in the lethality of Alpl deficiency in zebrafish.

There is a high degree of evolutionary conservation of the gene coding for TNSALP in vertebrates [40,41], indicating that animal models of TNSALP deficiency can yield valuable insights into the physiological functions of TNSALP and pathophysiology of HPP [29]. Murine models of TNSALP deficiency have already proven essential for the development of enzyme replacement therapy [42], which is currently the only available treatment option effective for the bone phenotype of human HPP [26]. The zebrafish is increasingly used as an alternative model organism to study human diseases due to its ease of breeding, large number of offspring, and short generation time. The conservation of zebrafish Alpl function in the skeleton and nervous system was previously postulated based on the comparison of tissue-specific gene expression patterns in the zebrafish, mouse, and human [41]. In the present study we describe the first genetic model of TNSALP deficiency in zebrafish generated using CRISPR/Cas gene editing. Although there was virtually no mRNA nor protein detectable, there was residual alkaline phosphatase activity in 5 dpf *alpl*$^{-/-}$ embryos. Since enzyme activity measurements were performed in the total embryo extracts, the most plausible explanation of the residual activity is the contribution of intestinal alkaline phosphatase (encoded by *alpi.1* and *alpi.2*) and alkaline phosphatase 3, also expressed in the intestine, to the total measured enzyme activity. Residual alkaline phosphatase activity was also measured in the serum of the first genetic murine model of TNSALP deficiency (*Akp2*$^{-/-}$), which was explained by the contribution of genetically distinct intestinal alkaline phosphatase activity [43]. Furthermore, the biochemical and behavioral characteristics of *alpl*$^{-/-}$ zebrafish described in the present study were comparable to the phenotypic features of *Akp2*$^{-/-}$ mice, which recapitulate lethal infantile HPP extremely well, including bone abnormalities and vitamin B_6-responsive seizures, with untreated seizing animals dying before weaning [43–45]. Data from available murine models of TNSALP deficiency show that in contrast to human TNSALP, murine TNSALP appears to not be essential in the initial events of bone mineralization during intrauterine development (i.e., no severe skeletal abnormalities typical to perinatal HPP), but it becomes important for this process after birth [43–45]. In the present study, we showed that zebrafish Alpl functions in bone mineralization, as indicated by decreased alizarin staining of notochord, one of the earliest mineralizing structures in zebrafish embryos [31]. This effect was comparable to the effect of chemical Alpl inhibition on bone mineralization in wild-type zebrafish embryos [41].

Vitamin B_6 (pyridoxine)-responsive seizures are a rare clinical feature of HPP, observed only in the most severe forms of perinatal and infantile HPP [18–22]. They are presumably caused by decreased PLP and PL availability in the cells of the central nervous system due to decreased/absent hydrolysis of extracellular PLP to PL (Figure 1a). Impaired activity of PLP-dependent enzymes involved in amino acid and neurotransmitter synthesis, e.g., decreased synthesis of the inhibitory neurotransmitter GABA due to lower activity of PLP-dependent glutamate decarboxylase (EC: 4.1.1.15), and a resulting imbalance in the levels of inhibitory and excitatory neurotransmitters, underlies seizures [18,46,47]. Indeed, by using stably labeled PLP, we could show that the hydrolysis of PLP and production of PL is strongly impaired in *alpl*$^{-/-}$ embryos, leading to lower concentrations of PLP and PL, as well as 4-PA (degradation product of PL) in total embryo extracts (dominated by tissue derived PLP and PL). Unfortunately, due to the small size of the embryos, we were unable to separately quantify PLP in the circulation and in the individual tissues. However, PLP deficiency in tissues including the brain were suggested by the observation that GABA concentrations in *alpl*$^{-/-}$ 5 dpf embryos and 10 dpf larvae were lower than in WT zebrafish, and they were normalized by PN treatment. The abnormalities in B_6 vitamer and GABA concentrations in *alpl*$^{-/-}$ zebrafish were in line with the findings in *Akp2*$^{-/-}$ mice, which have high serum PLP and low PL concentrations, as well as low PLP and PL concentrations in various tissues including the brain [43,44], and low brain GABA concentrations [43]. In HPP patients, circulating PLP concentration is elevated [9], making it a good biomarker of HPP [14], while PL concentration is less often reported, ranging from normal [18] to low in severe cases [10]. A single report in various post-mortem tissues of two patients with perinatal HPP showed no alterations in PLP and PL concentrations [10]. Lastly, similar to *Akp2*$^{-/-}$ mice [43–45] and severe forms of perinatal and infantile HPP [18–22], *alpl*$^{-/-}$ zebrafish embryos also developed spontaneous seizures that were responsive to PN treatment. However, we detected no spontaneous seizures in 10 dpf *alpl*$^{-/-}$ larvae, which may possibly be attributed to the fact that the measurements in larvae were conducted in a small number of the surviving larvae. Only a subpopulation of *Akp2*$^{-/-}$ mice experience seizures [43]. The underlying cause of phenotypic heterogeneity despite genetic homogeneity is not clear. We showed that, similarly to *Akp2*$^{-/-}$ mice [43–45] and HPP patients [18–22], spontaneous seizures in *alpl*$^{-/-}$ zebrafish embryos were responsive to PN treatment, leading to improved survival of the mutants. The lack of complete rescue of survival by PN treatment is in line with the multiple non-overlapping TNSALP functions, as demonstrated in *Akp2*$^{-/-}$ mice, where PN treatment prevents seizures without a beneficial effect on the skeletal phenotype [48].

The untargeted metabolomics analysis of the broader consequences of Alpl deficiency showed abnormalities in several neurotransmitter levels attributable to decreased cellular PLP availability. Next to lower GABA levels, increased levels of L-DOPA and vanillactic acid along with lower levels of dopamine and epinephrine pointed towards decreased activity of PLP-dependent AADC in *alpl*$^{-/-}$ larvae. Impaired AADC activity was also implied in

HPP patients based on elevated 3-ortho-methyldopa in CSF [18,47] and increased vanillactic acid in urine [47]. Moreover, we observed that accumulation of N-methylethanolaminium phosphate (PEA) is detectable in *alpl*$^{-/-}$ larvae, but not yet in 5 dpf embryos, indicating that in zebrafish accumulation of PEA develops gradually. In *Akp2*$^{-/-}$ mice, elevated serum PEA concentrations were shown in 8–10 day old pups; however, age dependence was not investigated [43]. Furthermore, we found increased methionine and cystathionine levels in *alpl*$^{-/-}$ embryos, suggesting altered activity of the methionine cycle and transsulfuration pathway, likely caused by the impaired activity of PLP-dependent enzymes, as underscored by the normalization of these metabolites in response to PN treatment (Figure 5e). Interestingly, an elevation in methionine and cystathionine levels was also reported in the brains of 1-week-old *Akp2*$^{-/-}$ mice [49], suggesting common underlying mechanisms in zebrafish and mice. Furthermore, elevated AMP and inosine levels in *alpl*$^{-/-}$ embryos suggest abnormalities in purine metabolism that could be linked to the deficiency in Alpl ectophosphatase function and to the mechanisms of chronic pain via an effect on circulating adenosine levels [50]. Our observation that PN treatment led to normalization of AMP and inosine levels in *alpl*$^{-/-}$ embryos suggests that changes in these compounds were caused by reduced vitamin B_6 availability rather than by the ectophosphatase activity of Alpl. However, empirical pyridoxine therapy for chronic fatigue and pain in four adult-onset HPP patients did not provide symptomatic relief [22], suggesting that the underlying mechanisms are vitamin B_6-independent. It must be noted that due to the contribution of isobaric compounds to the levels of AMP, inosine, and adenosine (not changed in *alpl*$^{-/-}$ embryos determined with DI-HRMS, future follow up research using targeted methods is required to clarify the involvement of Alpl in the regulation of purine metabolism. This could contribute to better understanding of the mechanisms of chronic pain and help to develop new treatments to improve the quality of life of HPP patients.

Lastly, we observed accumulation of vitamin A (retinol) and decreased levels of retinal in *alpl*$^{-/-}$ larvae. Interestingly, high *alpl* expression and alpl enzyme activity were observed in the eyes (especially lens and retina) of zebrafish embryos [41], as well as retina of other vertebrates [51], suggesting that TNSALP has a function in vision. However, no eye-specific phenotype was reported in HPP patients nor TNSALP deficient mice. Abnormalities in vitamin A metabolism have been implicated in the development of craniosynostosis and skeletal abnormalities in humas and zebrafish [52]. The mechanistic basis of how Alpl deficiency leads to retinol accumulation needs further investigation. Possibly, impaired Ca^{2+} homeostasis caused by Alpl deficiency could affect retinol transport into the cell, which is regulated by Ca^{2+}/calmodulin [53], leading to accumulation of circulating retinol and decreased intracellular retinal production.

4. Materials and Methods

4.1. Zebrafish Maintenance and Treatment Protocols

Zebrafish (*Danio rerio*) were raised and maintained under standard laboratory conditions [31]. Animal experiments were approved by and performed according to the guidelines of the Animal Welfare Body Utrecht, Utrecht University (protocol code 1444WP2B2).

Pyridoxine-treatment experiments were carried out in either 5 dpf embryos and 10 dpf larvae. For embryo treatment, batches of 2 dpf embryos (not genotyped), generated by incrossing *alpl*$^{+/-}$ parents, were randomly assigned to untreated or pyridoxine-treated groups. Next, 0 µM (untreated) or 100 µM (treated) pyridoxine (Sigma-Aldrich, Zwijndrecht, The Netherlands) was added to E3 medium in a petri dish and embryos were raised to 5 dpf (3 days continuous treatment). Media were refreshed every 24 h. At 5 dpf, embryos were anesthetized with tricaine, a sample of caudal fin was dissected for DNA isolation and genotyping, and embryos were instantly frozen on dry ice and stored at −80 °C until analysis. For larvae treatment, zebrafish embryos were genotyped at 3 dpf as described in [34]. Only *alpl*$^{+/+}$ and *alpl*$^{-/-}$ embryos were raised to 5 dpf. Starting from 5 dpf, zebrafish were treated for 5 consecutive days (between 9:00 and 10:30 am) with 0 µM or 100 µM pyridoxine for 30 min (Supplemental Figure S6). Treatment included placing zebrafish larvae (n = 18–22 per tank) in plastic tanks containing 500 mL of system water without pyridoxine (untreated) or 500 mL of system water containing 100 µM pyridoxine (treated). After 30 min, zebrafish larvae were rinsed and placed in the home tank. At 10 dpf, zebrafish larvae were anesthetized with tricaine and terminated by instant freezing on dry ice (1 larva per Eppendorf cup) and stored at −80 °C until analysis.

To assess the utilization of stably labeled pyridoxal 5′-phospate, 5 dpf old embryos (~60 embryos/petri dish) were incubated with 100 µM pyridoxal-5′-phosphate (methyl-D3) (d3-PLP) (Buchem, Minden, The Netherlands) in E3 for 0, 1, 2, and 3 h. At specified time-points, embryos were washed with E3, anesthetized with tricaine, dissected for genotyping, snap-frozen on dry ice and stored at −80 °C until analysis.

4.2. sgRNA and Cas9 mRNA Design and Synthesis

CRISPR/Cas9 gene-specific regions for *alpl* were designed by the Sanger Institute (Hinxton, Cambridge, UK) using a modified version of CHOPCHOP (http://chopchop.cbu.uib.no (accessed on 11 January 2015)). Target sites were selected in exon 5 and exon 6 (Supplemental Table S1) [54,55]. The gene-specific oligonucleotides contained the T7 promotor sequence (5′-TAATACGACTCACTATA-3′), the GGN20 target site without the Protospacer Adjacent Motif (PAM), and the constant complementary region 5′-GTTTTAGAGCTAGAAATAGCAAG-3′. Oligonucleotides were ordered at IDT (Integrated DNA Technologies, Coralville, IA, USA) and the zebrafish specific pCS2-nCas9n plasmid was obtained from Addgene (Cambridge, MA, USA). Cas9 mRNA transcription and sgRNA synthesis were performed as described before [56].

4.3. Generation of alpl Knockout Zebrafish

Wild type Tupfel longfin (WT TL) one-cell stage zebrafish embryos were microinjected in the yolk with approximately 1 nl sgRNA mixture (sgRNA targeting exon 5 and 6, each 30 ng/µL) and Cas9 mRNA (250 ng/µL). CRISPR efficiency was determined in a subpopulation of healthy microinjected larvae at 4 dpf. The rest of the healthy microinjected larvae were raised till adulthood. Heterozygous variation was assessed in DNA extracted from healthy embryonal offspring (F1) at 24 dpf. Offspring from a mosaic founder that contained a 10 bp out-of-frame deletion was raised till adulthood and was fin-clipped for genotyping at 9 weeks of age. The mutant zebrafish line was maintained in the heterozygous form by crossing *alpl*$^{+/-}$ zebrafish with WT TL. In this study, F6 zebrafish (*alpl*$^{+/+}$, *alpl*$^{+/-}$ and *alpl*$^{-/-}$) were used, obtained from incrossing F5 *alpl*$^{+/-}$ zebrafish.

4.4. DNA Extraction and Genotyping

Depending on the type of experiment, genotyping was performed on the caudal fin dissections at 3 or 5 dpf (overall experiments), whole 5 dpf embryos (Zebrabox and staining experiments), or adult zebrafish caudal fin dissections (line maintenance) as described in detail in [34]. Briefly, tissue was lysed in single embryo lysis buffer (SEL) containing 10 mM Tris pH 8.2, 10 mM EDTA, 200 mM NaCl, 0.5% sodium dodecyl sulfate (SDS), and 12 U/mL proteinase K (freshly added, Thermo Scientific, Waltham, MA, USA, cat. # EO0491). DNA was isolated using the following thermocycler program: 60 min 60 °C, 15 min 95 °C, 15 min 4 °C, ∞ 12 °C. Genomic regions flanking the CRISPR target sites were amplified with CRISPR site-specific PCR primers (Table S1), using AmpliTaq Gold 360 DNA polymerase (Applied Biosystems, Waltham, MA, USA, cat. # 4398823) in combination with a touch down PCR program as previously described [57]. Amplicons were visualized on a 3% agarose gel and mutations were confirmed by Sanger sequencing.

4.5. RNA Isolation and Real-Time PCR

Zebrafish embryos (5 dpf) were placed in sterile Eppendorf tubes on ice (10 embryos/tube per genotype, 3 tubes/genotype). Sterile, RNAse-free zirconium oxide beads (0.5 mm) and cold 0.5 mL TRI reagent (Sigma-Aldrich, cat. # T9424) were added to each tube. Embryos were homogenized using a bullet blender tissue homogenizer (Next Advance, Troy, NY, USA) for 10 min in stand 8 at 4 °C. Total mRNA was isolated from the embryo homogenates following the manufacturer's recommendations. The quantity and purity of the total RNA was quantified using a NanoDrop spectrophotometer (Thermo Scientific). One µg of total RNA was reverse transcribed to cDNA using M-MLV reverse transcriptase (Sigma-Aldrich, cat. # M1302) according to the manufacturer's protocol. Real-time PCR was performed with a StepOne Real-Time PCR System (Applied Biosystems, Waltham, MA, USA) using the SYBR Select Master Mix (Applied Biosystems, cat. # 4472908) and the primers listed in Table S1. The *alpl* (ZDB-GENE-040420-1, RefSeq:NM_201007.2) mRNA levels were normalized to the mRNA level of β-actin (ZDB-GENE-000329-1, RefSeq:NM_131031.2) and expressed relative to the wild type (calculated according to the ΔΔCt method).

4.6. Alizarin Red and Alcian Blue Staining

Mineralized bone and cartilage were stained in whole 5 dpf embryos with acid-free alizarin red and alcian blue double stain as described in [58]. Briefly, 5 dpf embryos were anesthetized with tricaine and up to 20 embryos were collected per 1.5 mL Eppendorf tube. After removing the medium, 1 mL of 4% paraformaldehyde in phosphate buffered saline was added per tube and embryos were fixed for 2 h with agitation at 500 rpm in an Eppendorf thermomixer at room temperature (RT), followed by washing and dehydration with 1 mL 50% ethanol for 10 min at RT. Embryos were stained overnight with 0.0005% alizarin red and 0.4% alcian blue working solution with agitation at RT. Stained embryos were washed and bleached with 1.5% H_2O_2 containing 1% KOH for 20 min at RT. After removing the bleach solution, 1 mL 20% glycerol containing 0.25% KOH was added and embryos were incubated for 2 h, followed by overnight incubation with 1 mL 50% glycerol containing 0.25% KOH at RT. Next, medium was replaced with 50% glycerol containing 0.1% KOH and embryos were stored at 4 °C. Images were captured with a Leica DFC420C digital microscope camera (Leica Microsystems, Wetzlar, Germany) mounted on a Zeiss Axioplan brightfield microscope (Carl Zeiss AG, Oberkochen, Germany). After the imaging, DNA was extracted from stained embryos and genotype was determined as described in Section 2.4.

4.7. Alkaline Phosphatase Enzyme Activity

Total alkaline phosphatase enzyme activity was determined in the whole embryo homogenates using assay described in [59]. Briefly, 5 dpf embryos (n = 30 per genotype) were homogenized in 200 µL of Dulbecco's Phosphate Buffered Saline (DPBS, Sigma-Aldrich, cat. # D8537) containing 0.1% Triton X100 using a bullet blender tissue homogenizer (Next Advance) for 5 min in stand 8 at 4 °C. Homogenates were centrifuged at 600× g for 5 min at 4 °C and sonicated using an ultrasonic disintegrator (Soniprep 150 Plus, MSE, Cholet, France) for 30 s in the pulse mode (1 s on 1 s off, amplitude 10 µm) on ice. The assay mix contained

20 μL of embryo extract (5× diluted in DPBS, final protein concentration in the assay 0.065 mg/mL), 80 μL DPBS, and 100 μL of CSPD ready-to-use reagent (0.25 mM solution; Roche GmbH, Mannheim, Germany, cat. # CSPD-RO) without Alpl inhibitor (-)-tetramisole HCl (Sigma-Aldrich, cat. # L9756) or with 20 mM (-)-tetramisole HCl. Alpl activity was measured by following the chemiluminescence for 5 min at 37 °C using Clariostar microplate reader (BMG Labtech, Ortenberg, Germany). Alpl activity was expressed as RLU/min/mg protein. Protein concentration in the embryo homogenates was determined using a Pierce BCA protein assay kit according to the manufacturer's protocol (Thermo Scientific, cat. # 23225).

4.8. Western Blotting

Zebrafish embryos (5 dpf, n = 32 per genotype) were placed in Eppendorf tubes on ice. Then, 150 μL RIPA lysis and extraction buffer (Thermo Scientific, cat. #89900) containing 2 mM NaF (Sigma-Aldrich) and protease inhibitor cocktail (1:200, Roche) was added, followed by zirconium oxide beads (0.5 mm). Embryos were homogenized using a bullet blender tissue homogenizer (Next Advance) for 10 min in stand 8 at 4 °C. Tissue homogenates were solubilized with agitation for 2 h at 4 °C, followed by centrifugation at 16,200× g for 10 min at 4 °C. Supernatants were mixed with LDS sample buffer (NuPage, Invitrogen, Waltham, MA, USA, cat. # NP0007) and dithiothreitol (final concentration 50 mM), and denatured at 98 °C for 5 min with agitation. Proteins were resolved on NuPAGE 4–12% Bis-Tris gels (Invitrogen) and transferred to polyvinylidene difluoride (PVDF) membranes (Immobilon-P) with a semi-dry blotting system (Novex, Invitrogen), following the manufacturer's recommendations. Membranes were blocked with tris-buffered saline (TBS) containing 0.1% Tween 20 (TBS-T) and 50 g/L bovine serum albumin (BSA, Sigma-Aldrich) for 1 h at RT. Next, the membranes were incubated overnight at 4 °C with primary rabbit polyclonal anti-ALPL antibody (1:1000, Sigma-Aldrich, cat. # HPA008765) or mouse monoclonal anti-glyceraldehyde-3-phosphate dehydrogenase (GAPDH, 1:5000, Santa Cruz Biotechnology, Dallas, TX, USA, cat. # sc-365062) in TBS-T containing 10 g/L BSA. After washing three times for 10 min each with TBS-T, membranes were incubated with a corresponding horseradish peroxidase-conjugated secondary antibody in TBS-T containing 5 g/L BSA for 1 h at RT. After the final wash of 3 × 10 min with TBS-T, the immunocomplexes were detected using SuperSignal™ West Atto Ultimate Sensitivity Substrate (Thermo Scientific, cat. # A38554) and images were captured with the ChemiDoc MP imaging system (Bio-Rad Laboratories, Hercules, CA, USA).

4.9. B_6 Vitamer Analysis

Frozen zebrafish 5 dpf embryos (3 embryos/100 μL TCA) or 10 dpf larvae (1 larva/100 μL TCA) were homogenized in ice-cold trichloroacetic acid (TCA; 50 g/L) with zirconium oxide beads (0.5 mm) using a bullet blender tissue homogenizer (Next Advance) at a speed of 8 for 10 min at 4 °C. Homogenates were centrifuged at 16,200× g for 5 min at 4 °C. Next, 80 μL of the supernatant was mixed with 80 μL of a solution containing stable isotope-labeled internal standards, vortexed, incubated for 15 min in the dark and centrifuged at 16,200× g for 5 min at 4 °C. B_6 vitamers were quantified using ultra-performance liquid chromatography tandem mass spectrometry (UPLC-MS/MS) as previously described [60], except for using 10 times lower concentrations of the calibration samples. For the analysis of pyridoxal 5′-phosphate-(methyl-d3) (d3-PLP) utilization and pyridoxal-(methyl-d3) (d3-PL) formation, zebrafish embryos were processed and analyzed using the same protocol, except that no stable isotope-labeled internal standards were added during UPLC-MS/MS measurement. During all steps, samples were protected from light as much as possible.

4.10. Non-Quantitative Direct-Infusion High-Resolution Mass Spectrometry (DI-HRMS)

Metabolite profiling was performed in 5 dpf embryos and 10 dpf larvae using a non-quantitative DI-HRMS method described in [36]. For extraction of metabolites, three embryos or a single 10 dpf larvae were homogenized in 100 μL of ice-cold 100% methanol with zirconium oxide beads (0.5 mm) using a bullet blender tissue homogenizer (Next Advance Inc., Averill Park, NY, USA) at a speed of 8 for 10 min at 4 °C. Homogenates were centrifuged at 16,200× g for 5 min at 4 °C. The supernatants (70 μL) were mixed with 60 μL of 0.3% formic acid (Emsure, Darmstadt, Germany) and 70 μL of internal standard working solution described in [36], and filtered using a methanol-preconditioned 96-well filter plate (Pall Corporation, Ann Arbor, MI, USA) loaded onto a vacuum manifold into an Armadillo high-performance 96-well PCR plate (Thermo Fisher Scientific). Samples were analyzed using a TriVersa NanoMate system (Advion, Ithaca, NY, USA) controlled by Chipsoft software (version 8.3.3, Advion). Data were acquired using Xcalibur software (version 3.0, Thermo Scientific, Waltham, MA, USA). Raw mass spectrometry data were analyzed using an in-house developed peak calling pipeline written in R programming language (source code available at https://github.com/UMCUGenetics/DIMS (accessed on 11 January 2015) that utilizes Human Metabolome DataBase (HMDB) for peak annotation with an accuracy of 5 ppm with respect to the theoretical m/z value, as described in detail in [36]. The web-based analysis tool MetaboAnalyst v.6.0 was used for statistical analysis (one factor) [61]. Metabolites with multiple possible annotations (isobaric compounds) were processed as single metabolite for statistical purposes.

4.11. Amino Acid and γ-Aminobutyric Acid (GABA) Analysis

Amino acid analysis was performed in 40 μL of zebrafish embryo (3 embryos/100 μL) extracts in 100% methanol (see Section 4.10 for preparation details) using an UPLC-MS/MS method described in [62]. GABA was quantified in 10 μL of zebrafish embryo (3 embryos/100 μL) or larvae (1 larva/100 μL) extracts in 100% methanol using an UPLC-MS/MS method described in [34].

4.12. Locomotion Analysis

ZebraBox system (ViewPoint Behavior Technology, Lyon, France) was used to track and quantify the locomotion of zebrafish embryos. Populations of 5 dpf old embryos or 10 dpf old larvae (1 embryo/well) were transferred to a 48-well flat-bottom plate (Greiner Bio-one CELLSTAR) containing 0.5 mL embryo E3 medium. Zebrafish embryos/larvae were allowed to acclimatize in the measurement chamber in the dark for 15 min prior to the measurement. Locomotion was assessed in the tracking mode using the following settings: background 15, inactivity threshold <1 mm/s, and burst activity threshold >30 mm/s. Temperature was maintained at 28 ± 1 °C. Locomotion was tracked in the dark without any intervention for 1 h. Movement trajectories were recorded and locomotion parameters were quantified with ZebraLab software (Viewpoint Behavior Technology, Lyon, France, https://www.viewpoint.fr/ (accessed on 23 May 2022).

4.13. Statistical Analysis

Data are presented as means ± SD. The number of zebrafish used for a specific experiment is indicated in the figure legends. Statistical analysis was performed using GraphPad Prism v.10 (GraphPad Software, San Diego, CA, USA). For comparison of two groups, Student's t-test was used. For comparison of three or more groups, one-way ANOVA followed by Tukey's post hoc test was used. The level of significance was set at $p < 0.05$.

5. Conclusions

In conclusion, we generated the first zebrafish model of HPP that shows multiple features of human disease and is suitable for studying the pathophysiology of HPP and testing novel treatments. We showed that Alpl has a function in vitamin B_6 metabolism and bone mineralization in zebrafish. Untargeted metabolomics revealed a multitude of metabolic alterations occurring in response to Alpl deficiency, including but not limited to phosphoetanolamines, neurotransmitters, nucleotides, polyamines, and retinoids, suggesting potential interesting directions for follow-up research on the mechanisms of HPP in zebrafish. Furthermore, the normalization of multiple metabolic abnormalities in response to PN treatment suggests that vitamin B_6 supplementation could be beneficial to HPP patients even in the absence of seizures. This study also revealed the limitations of performing metabolic research in zebrafish embryos, particularly related to the small embryo size that constrained the ability to analyze individual tissues/organs. Nevertheless, the data presented in this study clearly showed that this zebrafish model can serve as a valuable tool for investigating poorly understood aspects of TNSALP function and for developing improved therapies in the future.

Supplementary Materials: The following supporting information can be downloaded at: https://www.mdpi.com/article/10.3390/ijms26073270/s1.

Author Contributions: Conceptualization, J.C., J.J.J. and N.M.V.-D.; methodology, J.C., S.M.C.S., N.W.F.M. and F.T.; validation, J.C., S.M.C.S. and M.A.; formal analysis, J.C., S.M.C.S., N.W.F.M. and M.B.; investigation, J.C., S.M.C.S., N.W.F.M. and M.B.; resources, J.J.J., N.M.V.-D., J.P.W.B. and G.v.H.; data curation, J.C.; writing—original draft preparation, J.C.; writing—review and editing, M.A., S.M.C.S., N.W.F.M., F.T., J.P.W.B., G.v.H., J.J.J. and N.M.V.-D.; visualization, J.C.; supervision, G.v.H., J.J.J. and N.M.V.-D.; project administration, J.C.; funding acquisition, M.A., J.J.J. and N.M.V.-D. All authors have read and agreed to the published version of the manuscript.

Funding: This work was supported by a Wilhelmina Children's Hospital Research Fund grant to dr. M. Albersen (OZF 2014-2015).

Institutional Review Board Statement: The animal study protocol was approved by the Animal Welfare Body Utrecht, Utrecht University (protocol code 1444WP2B2, 6 April 2022).

Informed Consent Statement: Not applicable.

Data Availability Statement: Data are available within this manuscript, figures, Supplemental Figures and Supplemental Tables. Raw DI-HRMS data are available upon request.

Conflicts of Interest: The authors declare no conflicts of interest.

References

1. Rathbun, J.C. Hypophosphatasia; a new developmental anomaly. *Am. J. Dis. Child.* **1948**, *75*, 822–831. [CrossRef]
2. Farman, M.R.; Rehder, C.; Malli, T.; Rockman-Greenberg, C.; Dahir, K.; Martos-Moreno, G.A.; Linglart, A.; Ozono, K.; Seefried, L.; Del Angel, G.; et al. The Global ALPL gene variant classification project: Dedicated to deciphering variants. *Bone* **2024**, *178*, 116947. [CrossRef]
3. Zurutuza, L.; Muller, F.; Gibrat, J.F.; Taillandier, A.; Simon-Bouy, B.; Serre, J.L.; Mornet, E. Correlations of genotype and phenotype in hypophosphatasia. *Hum. Mol. Genet.* **1999**, *8*, 1039–1046. [CrossRef]
4. Brun-Heath, I.; Ermonval, M.; Chabrol, E.; Xiao, J.; Palkovits, M.; Lyck, R.; Miller, F.; Couraud, P.O.; Mornet, E.; Fonta, C. Differential expression of the bone and the liver tissue non-specific alkaline phosphatase isoforms in brain tissues. *Cell Tissue Res.* **2011**, *343*, 521–536. [CrossRef]
5. Millan, J.L.; Whyte, M.P. Alkaline Phosphatase and Hypophosphatasia. *Calcif. Tissue Int.* **2016**, *98*, 398–416. [CrossRef]
6. Russell, R.G. Excretion of Inorganic Pyrophosphate in Hypophosphatasia. *Lancet* **1965**, *2*, 461–464. [CrossRef]

7. Russell, R.G.; Bisaz, S.; Donath, A.; Morgan, D.B.; Fleisch, H. Inorganic pyrophosphate in plasma in normal persons and in patients with hypophosphatasia, osteogenesis imperfecta, and other disorders of bone. *J. Clin. Investig.* **1971**, *50*, 961–969. [CrossRef]
8. Fedde, K.N.; Whyte, M.P. Alkaline phosphatase (tissue-nonspecific isoenzyme) is a phosphoethanolamine and pyridoxal-5′-phosphate ectophosphatase: Normal and hypophosphatasia fibroblast study. *Am. J. Hum. Genet.* **1990**, *47*, 767–775.
9. Whyte, M.P.; Mahuren, J.D.; Vrabel, L.A.; Coburn, S.P. Markedly increased circulating pyridoxal-5′-phosphate levels in hypophosphatasia. Alkaline phosphatase acts in vitamin B6 metabolism. *J. Clin. Investig.* **1985**, *76*, 752–756. [CrossRef]
10. Whyte, M.P.; Mahuren, J.D.; Fedde, K.N.; Cole, F.S.; McCabe, E.R.; Coburn, S.P. Perinatal hypophosphatasia: Tissue levels of vitamin B6 are unremarkable despite markedly increased circulating concentrations of pyridoxal-5′-phosphate. Evidence for an ectoenzyme role for tissue-nonspecific alkaline phosphatase. *J. Clin. Investig.* **1988**, *81*, 1234–1239. [CrossRef]
11. McCance, R.A.; Morrison, A.B.; Dent, C.E. The excretion of phosphoethanolamine and hypophosphatasia. *Lancet* **1955**, *268*, 131. [CrossRef]
12. Fraser, D.; Yendt, E.R.; Christie, F.H. Metabolic abnormalities in hypophosphatasia. *Lancet* **1955**, *268*, 286. [CrossRef]
13. Orimo, H.; Girschick, H.J.; Goseki-Sone, M.; Ito, M.; Oda, K.; Shimada, T. Mutational analysis and functional correlation with phenotype in German patients with childhood-type hypophosphatasia. *J. Bone Miner. Res.* **2001**, *16*, 2313–2319. [CrossRef]
14. Nunes, M.E. Hypophosphatasia. In *GeneReviews((R))*; Adam, M.P., Feldman, J., Mirzaa, G.M., Pagon, R.A., Wallace, S.E., Amemiya, A., Eds.; University of Washington: Seattle, WA, USA, 1993.
15. Meyer, J.L. Can biological calcification occur in the presence of pyrophosphate? *Arch. Biochem. Biophys.* **1984**, *231*, 1–8. [CrossRef]
16. Collmann, H.; Mornet, E.; Gattenlohner, S.; Beck, C.; Girschick, H. Neurosurgical aspects of childhood hypophosphatasia. *Childs Nerv. Syst.* **2009**, *25*, 217–223. [CrossRef]
17. Bacchetta, J. Renal impairment in hypophosphatasia. *Arch. Pediatr.* **2017**, *24*, 5S93–5S95. [CrossRef]
18. Baumgartner-Sigl, S.; Haberlandt, E.; Mumm, S.; Scholl-Burgi, S.; Sergi, C.; Ryan, L.; Ericson, K.L.; Whyte, M.P.; Hogler, W. Pyridoxine-responsive seizures as the first symptom of infantile hypophosphatasia caused by two novel missense mutations (c.677T>C, p.M226T; c.1112C>T, p.T371I) of the tissue-nonspecific alkaline phosphatase gene. *Bone* **2007**, *40*, 1655–1661. [CrossRef]
19. Basura, G.J.; Hagland, S.P.; Wiltse, A.M.; Gospe, S.M. Clinical features and the management of pyridoxine-dependent and pyridoxine-responsive seizures: Review of 63 North American cases submitted to a patient registry. *Eur. J. Pediatr.* **2009**, *168*, 697–704. [CrossRef]
20. Taketani, T. Neurological Symptoms of Hypophosphatasia. *Subcell. Biochem.* **2015**, *76*, 309–322. [CrossRef]
21. Demirbilek, H.; Alanay, Y.; Alikasifoglu, A.; Topcu, M.; Mornet, E.; Gonc, N.; Ozon, A.; Kandemir, N. Hypophosphatasia presenting with pyridoxine-responsive seizures, hypercalcemia, and pseudotumor cerebri: Case report. *J. Clin. Res. Pediatr. Endocrinol.* **2012**, *4*, 34–38. [CrossRef]
22. Lefever, E.; Witters, P.; Gielen, E.; Vanclooster, A.; Meersseman, W.; Morava, E.; Cassiman, D.; Laurent, M.R. Hypophosphatasia in Adults: Clinical Spectrum and Its Association With Genetics and Metabolic Substrates. *J. Clin. Densitom.* **2020**, *23*, 340–348. [CrossRef]
23. di Salvo, M.L.; Contestabile, R.; Safo, M.K. Vitamin B(6) salvage enzymes: Mechanism, structure and regulation. *Biochim. Biophys. Acta* **2011**, *1814*, 1597–1608. [CrossRef]
24. Whyte, M.P.; Zhang, F.; Mack, K.E.; Wenkert, D.; Gottesman, G.S.; Ericson, K.L.; Cole, J.T.; Coburn, S.P. Pyridoxine challenge reflects pediatric hypophosphatasia severity and thereby examines tissue-nonspecific alkaline phosphatase's role in vitamin B(6) metabolism. *Bone* **2024**, *181*, 117033. [CrossRef]
25. Colazo, J.M.; Hu, J.R.; Dahir, K.M.; Simmons, J.H. Correction to: Neurological Symptoms in Hypophosphatasia. *Osteoporos. Int.* **2019**, *30*, 535. [CrossRef]
26. Whyte, M.P.; Greenberg, C.R.; Salman, N.J.; Bober, M.B.; McAlister, W.H.; Wenkert, D.; Van Sickle, B.J.; Simmons, J.H.; Edgar, T.S.; Bauer, M.L.; et al. Enzyme-replacement therapy in life-threatening hypophosphatasia. *N. Engl. J. Med.* **2012**, *366*, 904–913. [CrossRef]
27. Lieschke, G.J.; Currie, P.D. Animal models of human disease: Zebrafish swim into view. *Nat. Rev. Genet.* **2007**, *8*, 353–367. [CrossRef]
28. Abe, M.; Matsuda, M. A correlation between changes in gamma-aminobutyric acid metabolism and seizures induced by antivitamin B6. *J. Biochem.* **1976**, *80*, 1165–1171. [CrossRef]
29. Liedtke, D.; Hofmann, C.; Jakob, F.; Klopocki, E.; Graser, S. Tissue-Nonspecific Alkaline Phosphatase-A Gatekeeper of Physiological Conditions in Health and a Modulator of Biological Environments in Disease. *Biomolecules* **2020**, *10*, 1648. [CrossRef]
30. Van Belle, H. Alkaline phosphatase. I. Kinetics and inhibition by levamisole of purified isoenzymes from humans. *Clin. Chem.* **1976**, *22*, 972–976. [CrossRef]

31. Bensimon-Brito, A.; Cardeira, J.; Cancela, M.L.; Huysseune, A.; Witten, P.E. Distinct patterns of notochord mineralization in zebrafish coincide with the localization of Osteocalcin isoform 1 during early vertebral centra formation. *BMC Dev. Biol.* **2012**, *12*, 28. [CrossRef]
32. Baraban, S.C.; Taylor, M.R.; Castro, P.A.; Baier, H. Pentylenetetrazole induced changes in zebrafish behavior, neural activity and c-fos expression. *Neuroscience* **2005**, *131*, 759–768. [CrossRef]
33. Pena, I.A.; Roussel, Y.; Daniel, K.; Mongeon, K.; Johnstone, D.; Weinschutz Mendes, H.; Bosma, M.; Saxena, V.; Lepage, N.; Chakraborty, P.; et al. Pyridoxine-Dependent Epilepsy in Zebrafish Caused by Aldh7a1 Deficiency. *Genetics* **2017**, *207*, 1501–1518. [CrossRef] [PubMed]
34. Ciapaite, J.; Albersen, M.; Savelberg, S.M.C.; Bosma, M.; Tessadori, F.; Gerrits, J.; Lansu, N.; Zwakenberg, S.; Bakkers, J.P.W.; Zwartkruis, F.J.T.; et al. Pyridox(am)ine 5′-phosphate oxidase (PNPO) deficiency in zebrafish results in fatal seizures and metabolic aberrations. *Biochim. Biophys. Acta Mol. Basis Dis.* **2020**, *1866*, 165607. [CrossRef]
35. Afrikanova, T.; Serruys, A.S.; Buenafe, O.E.; Clinckers, R.; Smolders, I.; de Witte, P.A.; Crawford, A.D.; Esguerra, C.V. Validation of the zebrafish pentylenetetrazol seizure model: Locomotor versus electrographic responses to antiepileptic drugs. *PLoS ONE* **2013**, *8*, e54166. [CrossRef]
36. Haijes, H.A.; Willemsen, M.; Van der Ham, M.; Gerrits, J.; Pras-Raves, M.L.; Prinsen, H.; Van Hasselt, P.M.; De Sain-van der Velden, M.G.M.; Verhoeven-Duif, N.M.; Jans, J.J.M. Direct Infusion Based Metabolomics Identifies Metabolic Disease in Patients' Dried Blood Spots and Plasma. *Metabolites* **2019**, *9*, 12. [CrossRef]
37. Baggott, J.E.; Tamura, T. Metabolism of 10-formyldihydrofolate in humans. *Biomed. Pharmacother.* **2001**, *55*, 454–457. [CrossRef]
38. Brun, L.; Ngu, L.H.; Keng, W.T.; Ch'ng, G.S.; Choy, Y.S.; Hwu, W.L.; Lee, W.T.; Willemsen, M.A.; Verbeek, M.M.; Wassenberg, T.; et al. Clinical and biochemical features of aromatic L-amino acid decarboxylase deficiency. *Neurology* **2010**, *75*, 64–71. [CrossRef]
39. Nawrocki, L.; BreMiller, R.; Streisinger, G.; Kaplan, M. Larval and adult visual pigments of the zebrafish, *Brachydanio rerio*. *Vision. Res.* **1985**, *25*, 1569–1576. [CrossRef]
40. Yang, Y.; Wandler, A.M.; Postlethwait, J.H.; Guillemin, K. Dynamic Evolution of the LPS-Detoxifying Enzyme Intestinal Alkaline Phosphatase in Zebrafish and Other Vertebrates. *Front. Immunol.* **2012**, *3*, 314. [CrossRef]
41. Ohlebusch, B.; Borst, A.; Frankenbach, T.; Klopocki, E.; Jakob, F.; Liedtke, D.; Graser, S. Investigation of alpl expression and Tnap-activity in zebrafish implies conserved functions during skeletal and neuronal development. *Sci. Rep.* **2020**, *10*, 13321. [CrossRef]
42. Millan, J.L.; Narisawa, S.; Lemire, I.; Loisel, T.P.; Boileau, G.; Leonard, P.; Gramatikova, S.; Terkeltaub, R.; Camacho, N.P.; McKee, M.D.; et al. Enzyme replacement therapy for murine hypophosphatasia. *J. Bone Miner. Res.* **2008**, *23*, 777–787. [CrossRef] [PubMed]
43. Waymire, K.G.; Mahuren, J.D.; Jaje, J.M.; Guilarte, T.R.; Coburn, S.P.; MacGregor, G.R. Mice lacking tissue non-specific alkaline phosphatase die from seizures due to defective metabolism of vitamin B-6. *Nat. Genet.* **1995**, *11*, 45–51. [CrossRef]
44. Fedde, K.N.; Blair, L.; Silverstein, J.; Coburn, S.P.; Ryan, L.M.; Weinstein, R.S.; Waymire, K.; Narisawa, S.; Millan, J.L.; MacGregor, G.R.; et al. Alkaline phosphatase knock-out mice recapitulate the metabolic and skeletal defects of infantile hypophosphatasia. *J. Bone Miner. Res.* **1999**, *14*, 2015–2026. [CrossRef]
45. Narisawa, S.; Frohlander, N.; Millan, J.L. Inactivation of two mouse alkaline phosphatase genes and establishment of a model of infantile hypophosphatasia. *Dev. Dyn.* **1997**, *208*, 432–446. [CrossRef]
46. Belachew, D.; Kazmerski, T.; Libman, I.; Goldstein, A.C.; Stevens, S.T.; Deward, S.; Vockley, J.; Sperling, M.A.; Balest, A.L. Infantile hypophosphatasia secondary to a novel compound heterozygous mutation presenting with pyridoxine-responsive seizures. *JIMD Rep.* **2013**, *11*, 17–24. [CrossRef]
47. Balasubramaniam, S.; Bowling, F.; Carpenter, K.; Earl, J.; Chaitow, J.; Pitt, J.; Mornet, E.; Sillence, D.; Ellaway, C. Perinatal hypophosphatasia presenting as neonatal epileptic encephalopathy with abnormal neurotransmitter metabolism secondary to reduced co-factor pyridoxal-5′-phosphate availability. *J. Inherit. Metab. Dis.* **2010**, *33* (Suppl. 3), S25–S33. [CrossRef]
48. Narisawa, S.; Wennberg, C.; Millan, J.L. Abnormal vitamin B6 metabolism in alkaline phosphatase knock-out mice causes multiple abnormalities, but not the impaired bone mineralization. *J. Pathol.* **2001**, *193*, 125–133. [CrossRef]
49. Cruz, T.; Gleizes, M.; Balayssac, S.; Mornet, E.; Marsal, G.; Millan, J.L.; Malet-Martino, M.; Nowak, L.G.; Gilard, V.; Fonta, C. Identification of altered brain metabolites associated with TNAP activity in a mouse model of hypophosphatasia using untargeted NMR-based metabolomics analysis. *J. Neurochem.* **2017**, *140*, 919–940. [CrossRef]
50. Street, S.E.; Kramer, N.J.; Walsh, P.L.; Taylor-Blake, B.; Yadav, M.C.; King, I.F.; Vihko, P.; Wightman, R.M.; Millan, J.L.; Zylka, M.J. Tissue-nonspecific alkaline phosphatase acts redundantly with PAP and NT5E to generate adenosine in the dorsal spinal cord. *J. Neurosci.* **2013**, *33*, 11314–11322. [CrossRef]

51. Kantor, O.; Varga, A.; Kovacs-Oller, T.; Enzsoly, A.; Balogh, L.; Baksa, G.; Szepessy, Z.; Fonta, C.; Roe, A.W.; Nitschke, R.; et al. TNAP activity is localized at critical sites of retinal neurotransmission across various vertebrate species. *Cell Tissue Res.* **2014**, *358*, 85–98. [CrossRef]
52. Laue, K.; Pogoda, H.M.; Daniel, P.B.; van Haeringen, A.; Alanay, Y.; von Ameln, S.; Rachwalski, M.; Morgan, T.; Gray, M.J.; Breuning, M.H.; et al. Craniosynostosis and multiple skeletal anomalies in humans and zebrafish result from a defect in the localized degradation of retinoic acid. *Am. J. Hum. Genet.* **2011**, *89*, 595–606. [CrossRef]
53. Zhong, M.; Kawaguchi, R.; Costabile, B.; Tang, Y.; Hu, J.; Cheng, G.; Kassai, M.; Ribalet, B.; Mancia, F.; Bok, D.; et al. Regulatory mechanism for the transmembrane receptor that mediates bidirectional vitamin A transport. *Proc. Natl. Acad. Sci. USA* **2020**, *117*, 9857–9864. [CrossRef]
54. Cho, S.W.; Kim, S.; Kim, Y.; Kweon, J.; Kim, H.S.; Bae, S.; Kim, J.S. Analysis of off-target effects of CRISPR/Cas-derived RNA-guided endonucleases and nickases. *Genome Res.* **2014**, *24*, 132–141. [CrossRef] [PubMed]
55. Brocal, I.; White, R.J.; Dooley, C.M.; Carruthers, S.N.; Clark, R.; Hall, A.; Busch-Nentwich, E.M.; Stemple, D.L.; Kettleborough, R.N. Efficient identification of CRISPR/Cas9-induced insertions/deletions by direct germline screening in zebrafish. *BMC Genom.* **2016**, *17*, 259. [CrossRef]
56. Gagnon, J.A.; Valen, E.; Thyme, S.B.; Huang, P.; Akhmetova, L.; Pauli, A.; Montague, T.G.; Zimmerman, S.; Richter, C.; Schier, A.F. Efficient mutagenesis by Cas9 protein-mediated oligonucleotide insertion and large-scale assessment of single-guide RNAs. *PLoS ONE* **2014**, *9*, e98186. [CrossRef]
57. Tessadori, F.; Roessler, H.I.; Savelberg, S.M.C.; Chocron, S.; Kamel, S.M.; Duran, K.J.; van Haelst, M.M.; van Haaften, G.; Bakkers, J. Effective CRISPR/Cas9-based nucleotide editing in zebrafish to model human genetic cardiovascular disorders. *Dis. Model. Mech.* **2018**, *11*, dmm035469. [CrossRef]
58. Walker, M.B.; Kimmel, C.B. A two-color acid-free cartilage and bone stain for zebrafish larvae. *Biotech. Histochem.* **2007**, *82*, 23–28. [CrossRef]
59. Graser, S.; Mentrup, B.; Schneider, D.; Klein-Hitpass, L.; Jakob, F.; Hofmann, C. Overexpression of tissue-nonspecific alkaline phosphatase increases the expression of neurogenic differentiation markers in the human SH-SY5Y neuroblastoma cell line. *Bone* **2015**, *79*, 150–161. [CrossRef]
60. van der Ham, M.; Albersen, M.; de Koning, T.J.; Visser, G.; Middendorp, A.; Bosma, M.; Verhoeven-Duif, N.M.; de Sain-van der Velden, M.G. Quantification of vitamin B6 vitamers in human cerebrospinal fluid by ultra performance liquid chromatography-tandem mass spectrometry. *Anal. Chim. Acta* **2012**, *712*, 108–114. [CrossRef]
61. Pang, Z.; Lu, Y.; Zhou, G.; Hui, F.; Xu, L.; Viau, C.; Spigelman, A.F.; MacDonald, P.E.; Wishart, D.S.; Li, S.; et al. MetaboAnalyst 6.0: Towards a unified platform for metabolomics data processing, analysis and interpretation. *Nucleic Acids Res.* **2024**, *52*, W398–W406. [CrossRef]
62. Prinsen, H.; Schiebergen-Bronkhorst, B.G.M.; Roeleveld, M.W.; Jans, J.J.M.; de Sain-van der Velden, M.G.M.; Visser, G.; van Hasselt, P.M.; Verhoeven-Duif, N.M. Rapid quantification of underivatized amino acids in plasma by hydrophilic interaction liquid chromatography (HILIC) coupled with tandem mass-spectrometry. *J. Inherit. Metab. Dis.* **2016**, *39*, 651–660. [CrossRef] [PubMed]

International Journal of *Molecular Sciences*

MDPI

Article

The Impact of Oxytocin on Stimulus Discrimination of Zebrafish Albino and Non-Albino Models

Ana-Maria Danila [1,†], Alexandra Savuca [1,*], Alin Stelian Ciobica [1,2,3,4,†], Irina Luciana Gurzu [5], Mircea Nicusor Nicoara [1] and Bogdan Gurzu [6]

1 Department of Biology, Faculty of Biology, Alexandru Ioan Cuza University of Iasi, Bd. Carol I No. 20A, 700505 Iasi, Romania; danilamariuca@yahoo.com (A.-M.D.); alin.ciobica@uaic.ro (A.S.C.); mirmag@uaic.ro (M.N.N.)
2 Centre of Biomedical Research, Romanian Academy, Bd. Carol I, No. 8, 700506 Iasi, Romania
3 Academy of Romanian Scientists, Str. Splaiul Independentei No. 54, Sector 5, 050094 Bucharest, Romania
4 "Ion Haulica" Institute, Apollonia University, Păcurari Street 11, 700511 Iasi, Romania
5 Department of Preventive Medicine and Interdisciplinarity, Faculty of Medicine, "Grigore T. Popa" University of Medicine and Pharmacy, 700115 Iasi, Romania; irina-luciana.gurzu@umfiasi.ro
6 Department of Morfofunctional Sciences, Faculty of Medicine, "Grigore T. Popa" University of Medicine and Pharmacy, 16th Universitatii Street, 700115 Iasi, Romania; bgurzu@yahoo.com
* Correspondence: alexandra.savuca@student.uaic.ro
† These authors contributed equally to this work.

Abstract: Zebrafish have the ability, to a certain extent, to distinguish between different types of stimuli, including distinguishing between videos of conspecifics and non-conspecifics, a skill known as stimulus discrimination. In this study, we investigated the effects of oxytocin on this ability in albino and non-albino zebrafish models, focusing on the correlations between albinism, sensory deficiencies, and socio-emotional behaviors. Our hypothesis is based on the premise that oxytocin influences socio-emotional behaviors in zebrafish, with varying effects depending on phenotype (albino vs. non-albino), social context, and treatment duration. Studies have shown that albino zebrafish have more pronounced sensory deficiencies, meaning they may benefit more from oxytocin in terms of increased social comfort and interactions with conspecifics, while non-albino zebrafish would experience a reduction in defensive behaviors and anxiety. To test this, two experiments were conducted: one assessing the responses to video predator stimuli and the other comparing social interactions with real and video conspecifics. The results showed significant differences between the two groups: non-albino zebrafish exhibited stronger long-term reductions in anxiety-related behaviors, such as reaction speed and freezing, suggesting that oxytocin regulates defensive responses and aggression. Meanwhile, albino zebrafish showed greater improvements in social interactions, reflecting the nuanced, phenotype-dependent effects of oxytocin. These results not only confirm existing research but also highlight the therapeutic potential of oxytocin in treating socio-emotional deficiencies.

Keywords: social discrimination; albino; oxytocin; zebrafish; predator

1. Introduction

Albinism is an autosomal recessive disorder characterized by reduced melanin biosynthesis in melanocytes in the epidermis and hair follicles; the absence or reduction in pigmentation can occur in the skin, hair, and eyes, known as oculocutaneous albinism, or it can affect only the pigmentation of the eyes, known as ocular albinism [1,2]. The

World Health Organization (WHO) recognizes albinism as a genetic condition that affects individuals from all social classes and from all countries around the world [3]. The global incidence of albinism is 1 in 17,000–20,000 people in Europe and the United States but higher among populations in Africa (1 in 1000), particularly within the Tonga tribe in Zimbabwe, an isolated rural community [4]. Consanguinity and limited geographical mobility are relevant factors in assessing the current and future prevalence of albinism [5]. There are also much rarer forms of albinism, such as Hermansky–Pudlak syndrome and Chediak–Higashi syndrome, characterized by more severe phenotypes that affect a wide range of cell types beyond the pigmentary cells. Studies indicate that albino individuals in sub-Saharan Africa have approximately 1000 times higher risk of developing skin cancer compared to the general population, being extremely vulnerable to the harmful effects of ultraviolet (UV) radiation. This increased susceptibility significantly contributes to the premature mortality of individuals with albinism, with many dying from skin cancer before the age of 30 and, frequently, between the ages of 30 and 40 [6–8]. Albinism human patients often suffer from reduced visual acuity, refractive errors, translucent iris, nystagmus, foveal hypoplasia, hypopigmentation of the fundus, and abnormal decussation of the optic nerve fibers at the level of the optic chiasm, which can lead to strabismus and deterioration of stereoscopic vision [9]. Recently, new genes associated with albinism have been discovered, both in oculocutaneous and syndromic forms. Specifically, two new genes associated with oculocutaneous albinism (*OCA*) have been identified: *SLC24A5* and *C10orf11*, referred to as *OCA6* and *OCA7*, respectively. These discoveries contribute to understanding the genetic complexity of albinism and may facilitate more accurate diagnosis and management of this condition [10,11]. Thus, albinism and its numerous forms have been extensively investigated in genetic association studies [12], but qualitative studies regarding the life experiences of individuals with albinism show that, aside from medical concerns, individuals with albinism also face psychological and social challenges, such as depression, anxiety, social discrimination, abuse, and stigmatization affecting self-esteem and educational and professional opportunities [13]. Although albinism affects a small number of people compared to other major issues, its extensive implications make it a public health problem that requires greater attention, especially for increasing awareness and knowledge in the field [14–16].

Oxytocin, produced in the paraventricular nuclei of the hypothalamus, regulates social behaviors in animals and promotes prosocial behaviors by encouraging closeness and reducing avoidance [17,18]. In humans, social bonds are often reflected through emotional experiences towards others. Therefore, the effects of oxytocin in humans should influence behavior and our emotional perception of others [19]. Regarding the connections between oxytocin and affective disorders, it is important to note that oxytocin acts as a modulator of the neuroendocrine axis, specifically regulating cortisol levels, especially under stress conditions [20]. Oxytocin is crucial in the mesocorticolimbic dopaminergic circuit (reward system), with extensive implications in addictive behaviors, motivation, survival, reproduction, feeding, drinking, and sexuality [21]. These processes are significantly disrupted in depression, manifested by decreased motivation, reduced appetite, low libido, and suicidal behavior. At the same time, some clinical studies have clearly indicated the involvement of oxytocin in other psychiatric disorders and the possible beneficial effects of oxytocin administration in depression and anxiety [22,23].

Oxytocin is a neuropeptide synthesized in the hypothalamus and released by the pituitary gland, playing an essential role in regulating social behavior, emotional responses, and stress levels. Olff et al. (2013) suggest that oxytocin plays a significant role in the regulation of social behavior, emotional responses, and stress, highlighting its influence on brain regions associated with these processes, such as the amygdala and prefrontal cortex [24]. For instance, Parker et al. (2019) suggest that oxytocin interacts with the amygdala and dopaminergic system to mediate social rewards. This aligns with the findings of our study, where we observed that albino zebrafish showed an increased preference for social interactions after oxytocin treatment, indicating a more intense activation of the reward system [25]. It is the mechanism of action that involves the activation of oxytocin receptors in various brain regions, including the amygdala and prefrontal cortex, structures involved in processing emotions and social interactions. Studies have shown that oxytocin modulates stress responses by regulating the hypothalamic–pituitary–adrenal axis, reducing cortisol levels, and promoting prosocial behaviors such as empathy and cooperation.

Individuals with albinism often face psychosocial difficulties, including stigmatization, social isolation, and increased anxiety, factors that can affect the development and maintenance of interpersonal relationships. In this context, oxytocin may play an important role in improving social adaptation and reducing perceived stress. Evidence from the literature, such as studies by Jin et al. (2023), suggests that the administration of oxytocin can facilitate social interactions and reduce anxiety in vulnerable groups, supporting the idea that this neuropeptide could have a beneficial effect on individuals with albinism who face challenges related to social integration [26].

Recent studies on albinism and its social and psychological impact create the opportunity to investigate the role of oxytocin in this context. Oxytocin is known for its role in facilitating positive social behaviors, including increasing trust and empathy. In the case of individuals with albinism, who can face social ostracism and the psychological impact of stigmatization, investigating the effects and how this hormone can influence social adaptability and emotional resilience becomes essential [27,28]. By examining the interaction between oxytocin and the psychosocial response of individuals with albinism, research can contribute to the development of intervention strategies aimed at improving the quality of life for these vulnerable individuals. Furthermore, a deeper understanding of how oxytocin influences social perception and adaptive behaviors could have practical applications in managing and alleviating the negative effects of social and psychological discrimination associated with albinism [29].

The zebrafish (*Danio rerio*) is considered an ideal research model for the study of albinism for several fundamental reasons. Firstly, zebrafish exhibit significant genetic similarity to humans. Many of the genes that control pigment development and melanocyte function in humans and mice are conserved in zebrafish as well as in other chordates [30–32]. Some examples of conserved melanocyte genes that control the analogous cell type, melanophores, in zebrafish, include the transcription factor associated with microphthalmia (referred to as *MITF* in humans, *Mitf* in mice, and *mitf* in zebrafish), dopachrome tautomerase (*DCT*), tyrosinase (*TYR*), tyrosinase-related protein 1 (*TYRP1*), and oculocutaneous albinism 2 (*OCA2*) [33]. In addition to this aspect, zebrafish learn voluntary behaviors through operant conditioning when the same stimuli serve as consequences of behavior. Albino zebrafish models exhibit more pronounced behavioral responses to social and environmental stimuli [34]. Additionally, the lack of pigmentation facilitates detailed analysis of brain structures and neural pathways using imaging techniques, providing an in-depth perspective on the effects of oxytocin on neural networks involved in anxiety and social behavior. Thus, zebrafish have achieved recognition in the neuroscience area and have a high potential to serve as a connecting bridge between biomedical and behavioral sciences, including the study of social behavior [35,36]. Using modern technology, it is possible to monitor the behavioral responses of albino zebrafish treated with oxytocin in various stimulation contexts.

Some studies, such as those presented at the Virtual Reality Symposium of the International Ethology Congress held in 2015 in Cairns, Australia, ref. [37], use robotic fish in the context of anti-predator behavior in zebrafish [38], ultimately addressing the construct validity of paradigms involving zebrafish as biological models of human brain function [39]. Zebrafish have innate abilities to respond to various stimuli, they can hear [13], smell [40], and perceive low-frequency vibrations through their lateral line (structure equivalent to the tactile stimulus perception in terrestrial species). In the present study, we focused on visual stimuli.

In this study, we aimed to investigate the effects of oxytocin treatment on social behavior in albino zebrafish, considering their visual impairments and socio-emotional deficiencies, by using real and virtual stimuli to test social preferences, anxiety levels, adaptability, and responses to risk exposure.

2. Results

2.1. *Behavioral Response of Zebrafish to Predator Video Stimuli—Oxytocin Influences Defensive and Escape Responses in Albino and Non-Albino Zebrafish*

Our study investigated whether zebrafish can distinguish between a video stimulus and a real stimulus, considering their reality recognition abilities suggested by the previous literature [41,42]. Our hypothesis was that, if the fish perceived the video stimulus as a real predator, they would exhibit significant changes in behavioral parameters, such as adjusting their distance from the stimulus, freezing duration, turning angle, and swimming velocity. However, if they recognized it as a screen, these variables would remain relatively stable after the initial acclimatization period. To test this hypothesis, the aquarium was divided into two zones: the "screen zone" (where the video stimulus was displayed) and the "safe zone" (a refuge area). The fish were exposed to a video depicting a natural predator (African leaf fish), and behavioral parameters were analyzed at regular intervals. The results indicate significant differences in behavioral responses depending on zebrafish phenotype and the timing of oxytocin administration, supporting the idea that this neuropeptide can influence defensive reactions, but with variable effects depending on the individual characteristics of the fish.

In the predator video stimulus test, we obtained several interesting findings. As presented in Figure 1, no significant differences were observed between the albino and non-albino groups in terms of velocity. However, in the non-albino group, significant differences were noted between the groups treated 1 h (N1h) and 48 h (N48h) after oxytocin administration. Specifically, differences were observed in the first 20 s ($p = 0.048$) and at the 60 s interval (when the stimulus appeared) ($p = 0.031$), suggesting that the behavioral response to oxytocin varies over time. For the distance to the stimulus, no significant differences were observed in albino fish, while, in non-albino fish, the control group (CTR N) showed a significant increase in this parameter between 20 and 60 s ($p = 0.044$) and between 20 and 80 s ($p = 0.019$), as well as for 120 s, suggesting the fact that there is a possibility of recognizing the visual stimulus as a real one. Additionally, a significant difference was observed between the CTR N and N48h groups during the first 20 s ($p = 0.004$), indicating that oxytocin treatment at 48 h influences approach behavior toward the stimulus, with higher values in the N48h group compared to the CTR N group. Regarding the turn angle, only non-albino fish displayed differences between the CTR N and N1h groups during the first 20 s ($p = 0.009$), with a lower angle in the N1h group compared to CTR N, and between 20 and 40 s in the CTR N group ($p = 0.028$), suggesting rapid behavioral adaptation to the predator stimulus. As shown in Figure 2, for freezing duration, only non-albino fish exhibited significant changes. A notable difference was observed between the CTR N and

N48h groups at 20 s ($p = 0.004$), with a longer freezing duration in the N48h group compared to CTR N, and between the N1h and N48h groups ($p = 0.01$), as well as between the N24h and N48h groups ($p = 0.005$), indicating that the effect of oxytocin on freezing behavior increases significantly, with higher values as time passes after administration. Finally, while the anxiety index did not show significant differences between groups, visual observations suggested a possible anxiogenic effect of oxytocin in albino fish, particularly one hour after administration, when anxiety values were higher in the N1h group compared to CTR N.

Freezing behavior, observed in the predator stimulus test, is clearly defined as an anxiety-related response. It is a defensive behavior specifically associated with fear and anxiety reactions when facing a potentially threatening stimulus. The results showed that, in non-albino fish, oxytocin treatments significantly influenced the duration of freezing, indicating a modification of the defensive response depending on the timing of oxytocin administration ($p = 0.004$ at 20 s and $p = 0.005$ between the groups treated at 24 h and 48 h). This behavior is linked to anxiety levels, and the observed changes suggest that oxytocin may influence this anxiety-related response, particularly when the administration of oxytocin is longer.

Our findings show that non-albino zebrafish exhibit significant behavioral responses to a predator video, while albino zebrafish do not. Oxytocin influenced defensive behavior in non-albino fish, especially 48 h post-administration, affecting distance to the stimulus ($p = 0.004$), turning angle ($p = 0.009$), and freezing duration ($p = 0.004$). These findings suggest that zebrafish can recognize virtual stimuli, with oxytocin modulating their responses based on phenotype and administration timing.

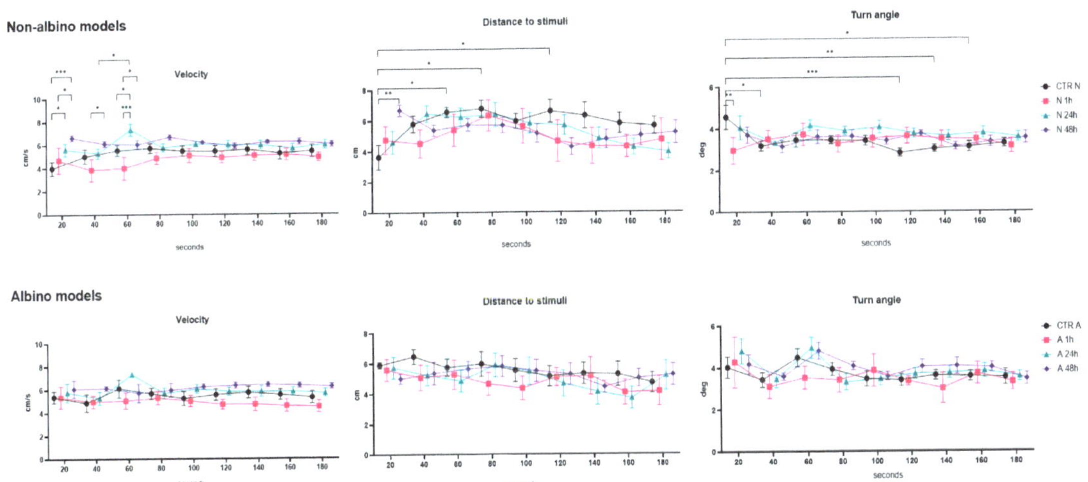

Figure 1. Graphical representation of the velocity (cm/s), distance to stimuli (cm), and turn angle (deg) results for both albino (A) and non-albino (N) zebrafish models in the predator video stimuli test. "1 h/24 h/48 h" represents the exposure time to oxytocin. The blue mark represents the moment when the stimulus appeared during the test. The data are expressed as the mean ± SEM, and a $p < 0.05$ was considered to be statistically significant (* $p < 0.05$; ** $p < 0.01$; *** $p < 0.001$).

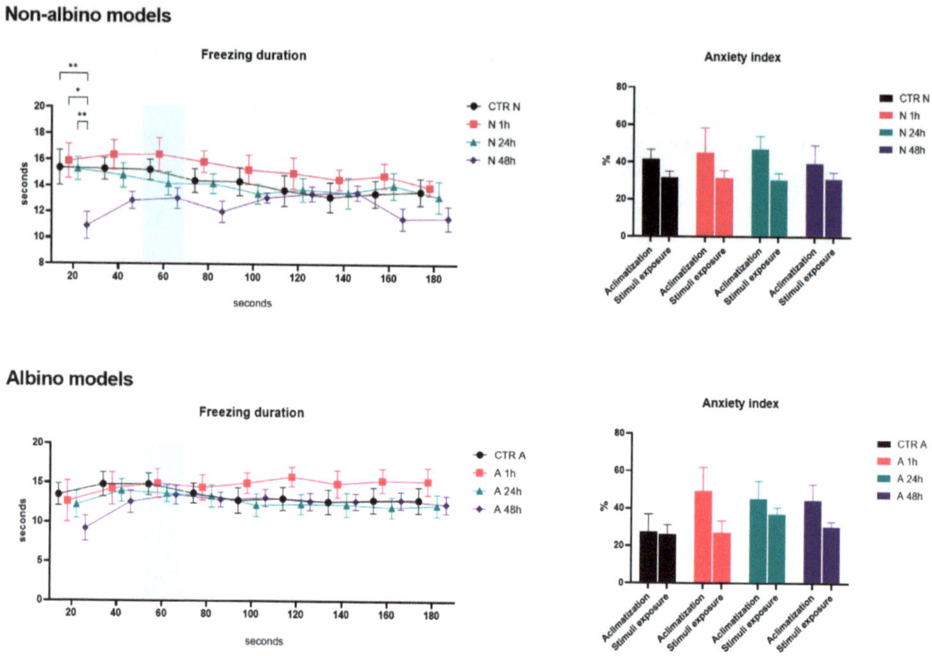

Figure 2. Graphical representation of the freezing duration (seconds) and anxiety index (%) results for both albino (A) and non-albino (N) zebrafish models in the predator video stimuli test. "1 h/24 h/48 h" represents the exposure time to oxytocin. The blue mark represents the moment when the stimulus appeared during the test. The data are expressed as the mean ± SEM, and a $p < 0.05$ was considered to be statistically significant (* $p < 0.05$; ** $p < 0.01$).

2.2. *Influence of Video vs. Real Conspecifics on Zebrafish Behavior—Oxytocin Affects Anxiety and Social Behaviors in Albino and Non-Albino Zebrafish*

Our hypothesis was that zebrafish can distinguish between a video stimulus and a real fish, and that oxytocin can influence this ability. Therefore, we aimed to test their preference between a real conspecific and a virtual one. If the fish perceive the video as a real conspecific, we expect them to spend a similar amount of time near the screen and the live fish. If they recognize the difference, they should prefer interacting with the real fish. To test this hypothesis, we used the conspecific video vs. real fish preference test, comparing the behavior of albino and non-albino zebrafish before and after oxytocin administration. To test this, we used a simple maze setup, with a two-option environment: a screen displaying a zebrafish video and a compartment containing a real fish. Fish activity was recorded by an overhead camera. Each fish was placed individually at the center of the arena and allowed to explore freely, choosing between the two options.

In the preference test between the video stimuli and real fish, the behaviors of albino and non-albino zebrafish were evaluated regarding the time spent in the screen vicinity, frequency of interactions with it, and distance from the screen. Statistical analysis revealed several significant results suggesting the influence of oxytocin on the social behavior of these models (Figure 3). The time spent in the screen zone was significantly higher in the group treated with oxytocin for 48 h (A48h) compared to the albino control group (CTR A) ($p = 0.048$). Similarly, in the non-albino fish, the time spent in the screen zone was significantly higher in the group treated with oxytocin for 1 h (N1h) compared to the non-albino control group (CTR N) ($p = 0.048$). This suggests an effect of oxytocin on social behavior. The frequency of interactions with the screen was significantly higher in the albino group treated with oxytocin for 48 h (A48h) compared to the albino control group (CTR A) ($p = 0.023$). Similarly, in non-albino fish, the frequency of interactions was significantly higher in the group treated with oxytocin for 1 h (N1h) compared to the control group ($p = 0.018$), indicating a positive effect of oxytocin on social interactions. In albino fish, the distance from the screen was significantly smaller in the group treated with oxytocin for 48 h (A48h) compared to the control group (CTR A) ($p = 0.011$), suggesting a closer approach to the visual stimulus. In contrast, no significant differences in distance from the screen were observed in non-albino fish.

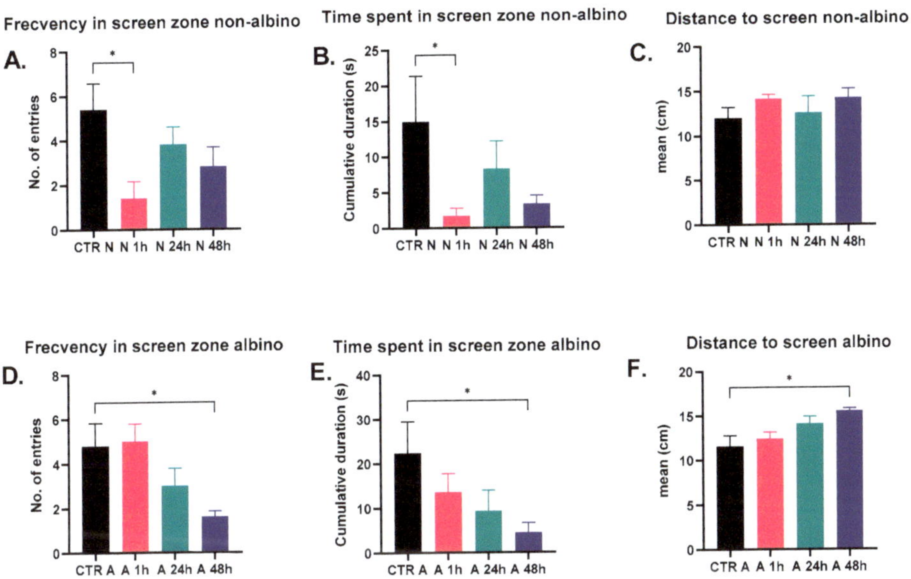

Figure 3. Graphical representation of the results for both albino and non-albino zebrafish models in the conspecific video vs. real fish preference test ((**A**) Frecvency in screen zone non-albino; (**B**) Time spent in screen zone non-albino; (**C**) Distance to screen non-albino; (**D**) Frecvency in screen zone albino; (**E**) Time spent in screen zone albino; (**F**) Distance to screen albino). The data are expressed as the mean ± SEM, and a $p < 0.05$ was considered to be statistically significant (* $p < 0.05$).

In short, in albino zebrafish, the time spent near the screen and the number of interactions significantly increased after 48 h of oxytocin treatment. In non-albino zebrafish, oxytocin administered for 1 h also increased the number of interactions, but there were no significant changes in the distance from the screen. These results suggest that oxytocin affects social behavior, with the effects varying depending on the zebrafish phenotype (albino or non-albino) and the timing of administration.

3. Discussion

According to research conducted by Onarheim and colleagues (2022), anxiety behaviors in zebrafish are characterized by parameters such as freezing duration, reaction speed to the predator stimulus, and the anxiety index, which reflects defensive responses and high stress levels. These behaviors are associated with avoiding open areas and reduced activity in light conditions, indicating a state of distress [43]. In contrast, social behaviors are expressed through time spent near conspecifics, the frequency of social interactions, and the distance from the screen—factors that reflect the desire for social affiliation and exploratory behaviors. According to the research by Fontana et al. (2021), social behaviors are related to interactions between individuals and the desire to approach conspecifics, being correlated with increased social activity, especially in a stimulating environment [44].

The previous research complemented by our study provides a comprehensive understanding of the effects of oxytocin on social behavior and anxiety in zebrafish, highlighting both similarities and differences between various phenotypes and the duration of the treatment. Existing studies suggest that oxytocin can influence social and anxiety-related behaviors in zebrafish, with its effects being dependent on the social context and type of treatment administered [45]. As suggested by studies in the literature, Ricci et al. (2013) indicate that oxytocin receptors may respond differently depending on the fish's age or social context, and, in our study, significant differences were observed between albino and non-albino fish [46]. Additionally, Chuang et al. (2021) suggest that oxytocin may modulate not only social behaviors but also defensive reactions, being more effective in reducing anxiety and behaviors related to speed and freezing, particularly in non-albino fish [47].

The anxiogenic effect observed in the group treated with oxytocin at 48 h could be explained by the influence of oxytocin on the hypothalamic–pituitary–adrenal (HPA) axis. Although oxytocin is well-known for its anxiolytic effects, previous studies suggest that it may have bidirectional effects on emotional stress, depending on the context and individual factors [48,49]. Specifically, oxytocin can amplify stress responses when individuals are exposed to unfamiliar environments or threatening stimuli [46,47]. This hypothesis is supported by studies showing that oxytocin administration can increase amygdala reactivity to negative stimuli, potentially leading to a paradoxical increase in anxiety under certain conditions [50,51]. Another possible mechanism for the observed anxiogenic effect could be the influence of oxytocin on amygdala activity [52]. While oxytocin may reduce amygdala activity in positive social contexts, some studies have shown that it can have the opposite effect in stressful situations, increasing neuronal reactivity and amplifying defensive responses. The study by Labuschagne et al. (2010) explores the effects of oxytocin on amygdala reactivity in stressful contexts, highlighting how this neuropeptide can influence emotional responses [52]. This reaction may explain why, in our predator stimulus test, the oxytocin-treated fish exhibited a longer freezing behavior, indicating an increase in anxiety in the presence of a threatening stimulus.

Regarding anxiety-like behaviors and defensive responses, our research highlighted an anxiogenic effect of oxytocin, particularly in the group treated for 48 h. This group exhibited a longer duration of freezing, suggesting a suppression of defensive reactions, an effect also observed in previous studies. For example, Gemmer et al. [53] and evidence from other studies suggest that oxytocin regulates aggression and defensive behaviors in zebrafish. In our study, long exposure time effects were more pronounced in the non-albino fish groups. Thus, it is confirmed that oxytocin can have a significant impact on defensive behavior and anxiety, but its effects may vary depending on phenotype and treatment duration.

Furthermore, we observed a significant increase in the preference for social interactions in albino fish, especially after 48 h of oxytocin treatment, which aligns with the existing literature suggesting that oxytocin contributes to the development and maintenance of social behavior [54]. Prior studies have shown that exposure to oxytocin activates the social system in zebrafish, with social behaviors being more prominently regulated by oxytocin, particularly in the context of interactions with conspecifics. Social interaction behaviors were clearly associated with social preference rather than anxiety reduction. For example, albino fish treated with oxytocin at 48 h (A48h) showed a significant increase in the frequency of interactions with the screen ($p = 0.023$) and a greater interest in real conspecifics ($p = 0.011$ compared to the albino control group). These results suggest that oxytocin influences social behaviors by enhancing social interaction rather than by reducing anxiety. Thus, we have clarified that oxytocin reduces anxiety (by influencing freezing behavior) and enhances social behaviors (by increasing interaction frequency and interest in real conspecifics). In addition, non-albino fish maintained a consistent behavior towards stimuli, while albino fish became more attracted to real conspecifics after oxytocin treatment. These results suggest that oxytocin influences social behaviors differently, depending on the zebrafish phenotype, contributing more to the social comfort of albino fish. These observations are supported by prior studies, which show that the effects of oxytocin depend on the type of stimuli and social context [55,56].

Prior studies also support our results, such as those of Ricci et al. (2013) on the effect of interdependent movements on the social preferences of zebrafish. Their results showed a significant preference for interacting with socially interdependent stimuli animated in 3D [46]. Similarly, the study by Chuang et al. (2021) explored an innovative system for the simultaneous observation of the behavior of eight adult zebrafish to analyze their responses to these visual stimuli. They found that zebrafish exhibit specific reactions to visual stimuli, including increased exploratory abilities that elicit a negative optomotor reaction [47]. Our results show that the responses to virtual stimuli vary depending on the zebrafish phenotype (albino or non-albino) and the timing of oxytocin administration, suggesting the high level of impact of oxytocin exposure modulations on this aspect. For example, the time spent near the screen zone in the social preference test decreases with increasing duration of oxytocin treatment in albino fish.

Regarding the visual stimuli, Gerlai R. (2017) used, in his study, animated images to examine zebrafish behavior, allowing for the simulation of visual stimuli and observation of their reactions in a controlled environment (e.g., exploration, avoidance reactions, freezing behavior, and social interaction). This method aids in understanding both instinctive and learned behaviors as well as social interactions [37]. In our study, we investigated the relevance of oxytocin, which regulates the behavior and emotional responses of zebrafish in the context of visual and social stimuli.

A connection between oxytocin and the dopaminergic system could be observed, which regulates social behaviors and emotional processes. Singh and Ousby (2021) demonstrated that oxytocin modulates the social preferences of zebrafish, with its effects varying based on their phenotype—a finding confirmed also in our research, where albino fish exhibited a positive adaptation to social stimuli compared to non-albino fish [57]. Moreover, the study by Umeoka et al. (2020) deepens the understanding of the interaction between oxytocin and the dopaminergic system, indicating that oxytocin influences dopamine release in the nucleus accumbent, having a differential impact on social behavior and direct interactions. This observation is in line with our results, where oxytocin had a more significant effect on social behavior in albino zebrafish; this could suggest a direct link between the dopaminergic system and social responses to stimuli [58]. Additionally, Silva et al. (2019) outlined the anxiolytic effects of oxytocin in zebrafish models by exploring the role of oxytocin receptor subtypes. These findings, along with ours, which showed a reduction in anxiety-like behavior and defensive responses in oxytocin-treated groups, indicate that oxytocin not only regulates social behaviors but also emotional and defensive responses [59].

Additional knowledge is provided within our findings in the field of confirming that oxytocin plays an important role in the modulation of social behavior and emotional responses in zebrafish, particularly with regard to social interactions and anxiety behaviors.

Consistent with the existing evidence that oxytocin promotes positive social behaviors and reduces aggressive and defensive behaviors, albino fish benefited more from oxytocin in terms of social comfort and desire to interact with conspecifics [60].

Our study suggests a potential role of oxytocin in improving social interactions in albino zebrafish models, adding knowledge to the evidence that many of the fundamental mechanisms of oxytocin are evolutionarily conserved, making these results relevant for further research in mammals and humans, especially regarding the time of oxytocin administration. Studies in mice and primates have shown that oxytocin influences social behaviors and reduces stress, while, in humans, oxytocin administration has been associated with improved social interactions [61,62]. However, further research is needed, and this study's limitations should be considered to validate these results in more complex contexts.

4. Materials and Methods

4.1. Ethical Note

All animals were treated and maintained in accordance with the EU Commission Recommendation (2007), Directive 2010/63/EU of the European Parliament, and the Council of 22 September 2010 guidelines for the accommodation, care, and protection of animals used for experimental and other scientific purposes. The protocol we followed received approval from the Ethics Committee of the Faculty of Biology, "Alexandru Ioan Cuza" University, Iasi, with registration no. 1349/20 March 2024.

4.2. Animal Maintenance

For our study, we used 70 adult zebrafish (*Danio rerio*), 35 normal fish (non-albino) (N), and 35 albino fish (A) from an authorized breeder. The zebrafish used in this experiment had an acclimatization period under experimental laboratory conditions for three weeks prior to the experiments in 10 L aquariums equipped with oxygen pumps and with the water changed daily. After this period, the zebrafish were randomly distributed into experimental groups (n = 10/5).

4.3. Experimental Design

The zebrafish were randomly selected and acclimatized within the corresponding groups under the conditions mentioned above. We created eight experimental groups to be exposed to oxytocin treatment at a concentration of 33.2 ng/mL for 15 min. According to previous studies, oxytocin remains active at room temperature in water for 21 days with normal function. Oxytocin was purchased in pharmaceutical liquid form (Pasteur Company, Bucharest, Romania, 10 I.U./mL). The dose of oxytocin was determined in accordance with the existing literature [63–65]. Behavioral responses were analyzed 24 h after the first dose (groups A 24 h, N 24 h) (n = 10) and 48 h after the administration of two doses (groups A 48 h, N 48 h) (n = 10), along with two control groups for each specific model (CTR A, CTR N) (Table 1). In addition, two groups with n = 5 received the same experimental dose for 15 min and were tested 1 h after exposure (A 1 h, N 1 h). The following test was assessed:

Table 1. Experimental group assignment.

Experimental Group	No. of Individuals	Oxytocin Concentration	Time of Exposure (hours)
CN	10	33.2 ng/mL	-
CA	10	33.2 ng/mL	-
A24h	10	33.2 ng/mL	24
N24h	10	33.2 ng/mL	24
A48h	10	33.2 ng/mL	48
N48h	10	33.2 ng/mL	48
A1h	5	33.2 ng/mL	1
N1h	5	33.2 ng/mL	1

4.3.1. Predator Video Stimulus Test

The experimental apparatus consisted of a rectangular tank measuring 30 cm × 20 cm × 20 cm, filled with 6 L of water, equipped with a camera above to record the fish's activity during the test. A monitor covering the entire right side of the tank was placed on the right side (Figure 4). The fish were placed in the apparatus for a 1 min acclimatization period without visual stimuli, after which a video featuring an African leaf fish (its natural predator [66]) was played for 2 min. The distance to the visual stimuli, speed, turning angle, and freezing behavior were recorded using the EthoVision XT 16 Software, Noldus, Netherlands, every 20 s as the mean of the sample once every 0.4 s.

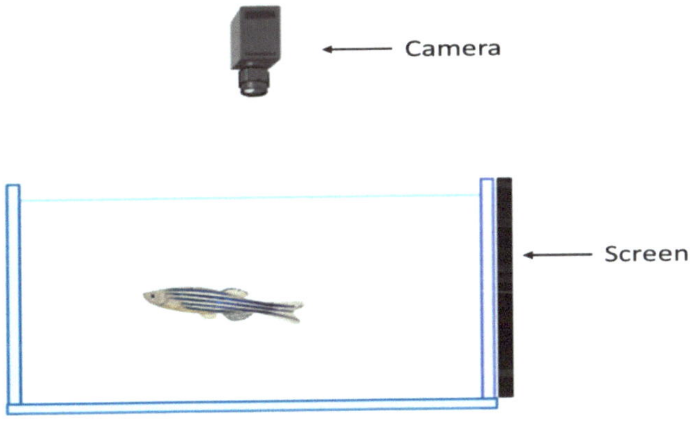

Figure 4. Experimental setup of the predator video stimulus test. Image created in Biorender.com www.biorender.com Accessed in 2 April 2024.

Our hypothesis, based on previous observations made by us, was that the distance to the visual stimuli, freezing, turning angle, and velocity (due to high-speed escape episodes) would increase upon the presentation of stimuli if they recognized it as a real predator; otherwise, all these parameters would remain normal without specific variations every 10 s, especially after the 1 min acclimatization period [67].

In addition, we evaluated the anxiety index based on suggestions from the literature and previous studies for both acclimatization and stimulus exposure periods. The anxiety index was calculated as the time spent in the peripheral zone (peripheral areas) divided by the total testing time (60 s for acclimatization and 120 s for stimulus exposure) multiplied by 100 [68,69].

4.3.2. Video Conspecific vs. Real Fish Preference Test

The experimental configuration was carried out using a cross maze, closing one arm with lids. The testing arena consisted of a rectangular tank measuring 20 cm × 10 cm × 10 cm that was made up of two arms which were filled with water and equipped with a camera above to record the fish's activity. In the left arm, a monitor displayed a video stimulus featuring a zebrafish (Figure 5); for the albino models, we used a video of the same fish, and vice versa. A real zebrafish was placed in the right arm. To initiate the test, the fish were placed one by one in the center of the testing arena and allowed to explore the environment, reacting quickly and concisely while choosing between the options. The time spent in the video stimulus zone and the number of entries into it were recorded by the EthoVision XT 16 Software, Noldus, Netherlands, for 2 min. Additionally, the distance to the visual stimuli was recorded every 20 s as the mean of the sample once every 0.1 s.

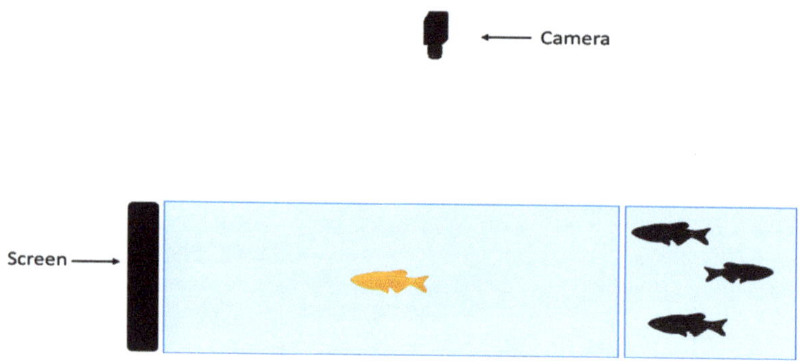

Figure 5. Experimental setup for the video conspecific vs. real fish preference test. Image created in Biorender.com www.biorender.com Accessed in 2 April 2024.

4.4. Statistical Analysis

The normality and distribution of the data were determined by the Shapiro–Wilk test using the Graph Pad Prism 9 software (San Diego, CA, USA). Outliers were removed by ROUT analysis. Multiple comparisons between the groups (by time and group) and post hoc analysis were then performed using Two-way ANOVA followed by Šídák's multiple comparisons test for the predator video stimuli test and One-way ANOVA followed by Dunnet's multiple comparison test for the conspecific video vs. real fish preference test. The data are expressed as the mean ± SEM, and a $p < 0.05$ was considered to be statistically significant.

5. Conclusions

Our study suggests that oxytocin plays an important role in enhancing social interactions in albino zebrafish, indicating a potential neurochemical mechanism that may also be relevant to other species, including mammals. Additionally, our findings highlight the zebrafish's ability to recognize and differentiate between real and virtual stimuli, suggesting a stimulus discrimination process that can be influenced by oxytocin administration. However, it is important to note that extrapolating these results to higher species, such as mammals or humans, should be performed cautiously, given the complexity of their neurobiological systems. Additionally, clinical studies investigating the effects of oxytocin on social behavior in humans have shown mixed results, indicating that therapeutic interventions with oxytocin need further studies and should be approached with caution and personalized according to each patient's needs.

Based on our findings, we conclude that the preference test between video conspecifics and real fish revealed that oxytocin enhances social behaviors in zebrafish, particularly in albino models. Albino fish showed increased interactions and spent more time near real conspecifics, especially 48 h after oxytocin treatment, indicating that oxytocin promotes social engagement and reduces social anxiety. In contrast, non-albino fish exhibited stable interactions with both stimuli, suggesting that oxytocin's effects are more pronounced in fish with sensory or social deficits, such as albinos. This highlights the potential of oxytocin in improving social cognition and adaptation.

Additionally, the conspecific video vs. real fish preference test showed that oxytocin's impact on predator stimulus responses varied in non-albino fish, depending on the time elapsed since administration. While albino fish did not show significant differences in most parameters, the effects observed in non-albino groups suggest that oxytocin plays a role in regulating defensive and anxiety-related behaviors. These findings underscore the need for further research on oxytocin's influence on social interactions and threat responses both in albino and non-albino zebrafish models.

Regarding the theoretical foundation, previous research on mammals and humans has demonstrated that oxytocin plays a role in forming and maintaining social bonds, as well as in cognitive processes related to recognizing and interpreting social signals. However, the existing literature also highlights limitations in using oxytocin for therapeutic purposes, such as the variable effects between individuals and the risks associated with its irregular or high-dose administration. Therefore, while this study provides a promising starting point for future research, it is essential to consider these limitations and continue investigating the impact of oxytocin on social behaviors in a clinical context. Future studies should ideally employ more rigorous approaches and closely assess how different oxytocin levels influence social and therapeutic behaviors in individuals with social behavior disorders.

Author Contributions: Conceptualization, A.-M.D., A.S., B.G. and A.S.C.; methodology, A.S., A.-M.D. and B.G.; software, A.S. and I.L.G.; validation, A.S.C., M.N.N. and B.G.; formal analysis, A.-M.D., A.S. and I.L.G.; investigation, A.-M.D. and A.S.; resources, M.N.N.; data curation, A.S. and A.S.C.; writing—original draft preparation, A.-M.D., A.S. and A.S.C.; writing—review and editing, B.G. and M.N.N.; visualization, A.-M.D., A.S. and A.S.C.; supervision, A.S.C. and B.G. All authors have read and agreed to the published version of the manuscript.

Funding: This research received no external funding.

Institutional Review Board Statement: The animal study protocol was approved by the Ethics Committee of the Faculty of Biology, "Alexandru Ioan Cuza" University, Iasi, with registration no. 1349/20 March 2024.

Informed Consent Statement: Not applicable.

Data Availability Statement: Data is contained within the article.

Conflicts of Interest: The authors declare no conflicts of interest.

References

1. Summers, C.G. Albinism: Classification, clinical characteristics, and recent findings. *Optom. Vis. Sci.* **2009**, *86*, 659–662. [CrossRef]
2. Bolognia, J.; Jorizzo, J.; Schaffer, J. *Dermatology*, 3rd ed.; Saunders Elsevier: Philadelphia, PA, USA, 2012.
3. Lund, P.M. Oculocutaneous albinism in southern Africa: Population structure, health, and genetic care. *Ann. Hum. Biol.* **2005**, *32*, 168–173. [CrossRef]
4. Lund, P.M.; Puri, N.; Durham-Pierre, D.; King, R.A.; Brilliant, M.H. Oculocutaneous albinism in an isolated Tonga community in Zimbabwe. *J. Med. Genet.* **1997**, *34*, 733–735. [CrossRef]
5. Kromberg, J.G.; Jenkins, T. Prevalence of albinism in the South African negro. *S. Afr. Med. J.* **1982**, *61*, 383–386. [PubMed]

6. Grønskov, K.; Ek, J.; Brondum-Nielsen, K. Oculocutaneous albinism. *Orphanet J. Rare Dis.* **2007**, *2*, 43. [CrossRef] [PubMed]
7. Castori, M.; Morrone, A.; Kanitakis, J.; Grammatico, P. Genetic skin diseases predisposing to basal cell carcinoma. *Eur. J. Dermatol.* **2012**, *22*, 299–309. [CrossRef]
8. Dessinioti, C.; Stratigos, A.J.; Rigopoulos, D.; Katsambas, A.D. A review of genetic disorders of hypopigmentation: Lessons learned from the biology of melanocytes. *Exp. Dermatol.* **2009**, *18*, 741–749. [CrossRef]
9. Kirkwood, B.J. Albinism and its implications with vision. *Insight* **2009**, *34*, 13–16. [PubMed]
10. Pennamen, P.; Tingaud-Sequeira, A.; Gazova, I.; Keighren, M.; McKie, L.; Marlin, S.; Gherbi Halem, S.; Kaplan, J.; Delevoye, C.; Lacombe, D.; et al. Dopachrome tautomerase variants in patients with oculocutaneous albinism. *Genet. Med.* **2021**, *23*, 479–487. [CrossRef]
11. Wei, A.H.; Yang, X. Oculocutaneous albinism: New insights and new genes. *Curr. Opin. Genet. Dev.* **2014**, *29*, 82–89. [CrossRef]
12. Lasseaux, E.; Plaisant, C.; Michaud, V.; Pennamen, P.; Trimouille, A.; Gaston, L.; Monfermé, S.; Lacombe, D.; Rooryck, C.; Morice-Picard, F.; et al. Molecular characterization of a series of 990 index patients with albinism. *Pigment. Cell Melanoma Res.* **2018**, *31*, 466–474. [CrossRef]
13. Ladich, F. Fish bioacoustics. *Curr. Opin. Neurobiol.* **2014**, *28*, 121–127. [CrossRef] [PubMed]
14. McBride, S.R.; Leppard, B.J. Attitudes and beliefs of an albino population toward sun avoidance: Advice and services provided by an outreach albino clinic in Tanzania. *Arch. Dermatol.* **2002**, *138*, 629–632. [CrossRef] [PubMed]
15. Lund, P.M.; Gaigher, R. A health intervention programme for children with albinism at a special school in South Africa. *Health Educ. Res.* **2002**, *17*, 365–372. [CrossRef]
16. Ezeilo, B.N. Psychological aspects of albinism: An exploratory study with Nigerian (Igbo) albino subjects. *Soc. Sci. Med.* **1989**, *29*, 1129–1131. [CrossRef] [PubMed]
17. Insel, T.R.; Gingrich, B.S.; Young, L.J. Oxytocin: Who needs it? *Prog. Brain Res.* **2001**, *133*, 59–66.
18. Young, L.J. The neurobiology of social recognition, approach, and avoidance. *Biol. Psychiatry* **2002**, *51*, 18–26. [CrossRef] [PubMed]
19. Kosfeld, M.; Heinrichs, M.; Zak, P.J.; Fischbacher, U.; Fehr, E. Oxytocin increases trust in humans. *Nature* **2005**, *435*, 673–676. [CrossRef] [PubMed]
20. Turner, R.A.; Altemus, M.; Enos, T.; Cooper, B.; McGuinness, T. Preliminary research on plasma oxytocin in normal cycling women: Investigating emotion and interpersonal distress. *Psychiatry* **1999**, *62*, 97–113. [CrossRef]
21. Kelley, A.E.; Berridge, K.C. The neuroscience of natural rewards: Relevance to addictive drugs. *J. Neurosci.* **2002**, *22*, 3306–3311. [CrossRef]
22. Blume, A.; Bosch, O.J.; Miklos, S.; Torner, L.; Wales, L.; Waldherr, M.; Neumann, I.D. Oxytocin reduces anxiety via ERK1/2 activation: Local effect within the rat hypothalamic paraventricular nucleus. *Eur. J. Neurosci.* **2008**, *27*, 1947–1956. [CrossRef]
23. Kozak, M.A.; Kato, H.; Borthwick, T. Oxytocin as a potential therapeutic agent in anxiety and depression. *Front. Psychiatry* **2018**, *9*, 15–27. [CrossRef]
24. Olff, M.; Frijling, J.L.; Kubzansky, L.D.; Bradley, B.; Ellenbogen, M.A.; Cardoso, C.; Bartz, J.A.; Yee, J.R.; van Zuiden, M. The role of oxytocin in social bonding, stress regulation and mental health: An update on the moderating effects of context and interindividual differences. *Psychoneuroendocrinology* **2013**, *38*, 1883–1894. [CrossRef]
25. Parker, M.O.; Gaviria, J.; Haigh, A.; Millington, M.E.; Brown, V.J. Rewarding outcomes of social interactions are regulated by the amygdala and oxytocin system. *Sci. Rep.* **2019**, *9*, 12410.
26. Jin, Y.; Song, D.; Yan, Y.; Quan, Z.; Qing, H. The Role of Oxytocin in Early-Life-Stress-Related Neuropsychiatric Disorders. *Int. J. Mol. Sci.* **2023**, *24*, 10430. [CrossRef] [PubMed]
27. Carlson, L.E.; Speca, M.; Patel, K.D.; Goody, P. Mindfulness based stress reduction in relation to quality of life, mood, symptoms outpatients. *Psychosom. Med.* **2003**, *65*, 571–581. [CrossRef] [PubMed]
28. Vander Kolk, C.J.; Bright, B.C. Albinism: A survey attitudes and behavior. *J. Vis. Impair. Blind.* **1983**, *77*, 49–51. [CrossRef]
29. Ciobica, A.; Balmus, I.M.; Padurariu, M. Is Oxytocin relevant for the affective disorders? *Acta Endocrinol.* **2016**, *12*, 65–71. [CrossRef] [PubMed]
30. Braasch, I.; Schartl, M.; Volff, J.N. Evolution of pigment synthesis pathways by gene and genome duplication in fish. *BMC Evol. Biol.* **2007**, *7*, 74. [CrossRef]
31. Howe, K.; Clark, M.D.; Torroja, C.F.; Torrance, J.; Berthelot, C.; Muffato, M.; Collins, J.E.; Humphray, S.; McLaren, K.; Matthews, L.; et al. The zebrafish reference genome sequence and its relationship to the human genome. *Nature* **2013**, *496*, 498–503. [CrossRef] [PubMed]
32. Beirl, A.J.; Linbo, T.H.; Cobb, M.J.; Cooper, C.D. Oca2 regulation of chromatophore differentiation and number is cell type specific in zebrafish. *Pigment. Cell Melanoma Res.* **2014**, *27*, 178–189. [CrossRef] [PubMed]
33. Oetting, W.S.; King, R.A. Molecular basis of albinism: Mutations and polymorphisms of pigmentation genes associated with albinism. *Hum. Mutat.* **1999**, *13*, 99–115. [CrossRef]

34. Goodson, J.L.; Thompson, R.R. Oxytocin receptors regulate social preference in zebrafish. *PLoS ONE* **2020**, *15*, e0228886. [CrossRef]
35. Braubach, O.R.; Wood, H.D.; Gadbois, S.; Fine, A.; Croll, R.P. Olfactory conditioning in the zebrafish (*Danio rerio*). *Behav. Brain Res.* **2009**, *198*, 190–198. [CrossRef]
36. Cerutti, D.T.; Jozefowiez, J.; Staddon, J.E. Rapid, accurate time estimation in zebrafish (*Danio rerio*). *Behav. Process* **2013**, *99*, 21–25. [CrossRef]
37. Gerlai, R. Animated images in the analysis of zebrafish behavior. *Curr. Zool.* **2017**, *63*, 35–44. [CrossRef]
38. Ladu, F.; Bartolini, T.; Panitz, S.G.; Chiarotti, F.; Butail, S.; Macrì, S.; Porfiri, M. Live predators, robots, and computer-animated images elicit differential avoidance responses in zebrafish. *Zebrafish* **2015**, *12*, 205–214. [CrossRef]
39. Neuhauss, S.C. Behavioral genetic approaches to visual system development and function in zebrafish. *J. Neurobiol.* **2003**, *54*, 148–160. [CrossRef]
40. Bleckmann, H.; Zelick, R. Lateral line system of fish. *Integr. Zool.* **2009**, *4*, 13–25. [CrossRef]
41. Qin, M.; Wong, A.; Seguin, D.; Gerlai, R. Induction of social behavior in zebrafish: Live versus computer animated fish as stimuli. *Zebrafish* **2014**, *11*, 185–197. [CrossRef] [PubMed]
42. Wee, C.L.; Nikitchenko, M.; Wang, W.C.; Luks-Morgan, S.; Song, E.; Gagnon, J.; Randlett, O.; Bianco, I.H.; Lacoste, A.M.B.; Glushenkova, E.; et al. Zebrafish oxytocin neurons drive nocifensive behavior via brainstem premotor targets. *Nat. Neurosci.* **2019**, *22*, 1477–1492. [CrossRef] [PubMed]
43. Onarheim, T.; Janczak, A.M.; Nordgreen, J. The effects of social vs. individual housing of zebrafish on whole-body cortisol and behavior in two tests of anxiety. *Psychopharmacology* **2020**, *239*, 859848. [CrossRef] [PubMed]
44. Fontana, B.; Alnassar, N.; Parker, M.O. The zebrafish (*Danio rerio*) anxiety test battery: Comparison of behavioral responses in the novel tank diving and light-dark tasks following exposure to anxiogenic and anxiolytic compounds. *Sci. Rep.* **2021**, *11*, 23170. [CrossRef] [PubMed]
45. Saszik, S.M.; Smith, C.M. The impact of stress on social behavior in adult zebrafish (*Danio rerio*). *Behav. Pharmacol.* **2018**, *29*, 53–59. [CrossRef]
46. Ricci, L.; Summers, C.H.; Larson, E.T.; O'Malley, D.; Melloni, R.H. Development of aggressive phenotypes in zebrafish: Interactions of age, experience and social status. *Anim. Behav.* **2013**, *86*, 245–252. [CrossRef]
47. Chuang, H.J.; Chang, C.Y.; Ho, H.P.; Chou, M.Y. Oxytocin signaling acts as a marker for environmental stressors in Zebrafish. *Int. J. Mol. Sci.* **2021**, *22*, 7459. [CrossRef]
48. Takayanagi, Y.; Tatsushi, O. Roles of Oxytocin in Stress Responses, Allostasis and Resilience. *Int. J. Mol. Sci.* **2021**, *23*, 150. [CrossRef] [PubMed]
49. Buisman-Pijlman, F.T.; Sumracki, N.M.; Gordon, J.J.; Hull, P.R.; Carter, C.S.; Tops, M. Individual Differences Underlying Susceptibility to Addiction: Role for the Endogenous Oxytocin System. *Pharmacol. Biochem. Behav.* **2014**, *119*, 23–38. [CrossRef]
50. Neumann, I. Brain oxytocin: A key regulator of emotional and social behaviours in both females and males. *J. Neuroendocrinol.* **2008**, *20*, 858–865. [CrossRef] [PubMed]
51. Bakermans-Kranenburg, M.J.; Van IJzendoorn, M.H. Sniffing around oxytocin: Review and meta-analyses of trials in healthy and clinical groups with implications for pharmacotherapy. *Transl. Psychiatry* **2023**, *3*, e258. [CrossRef]
52. Labuschagne, I.; Phan, K.L.; Wood, A.; Angstadt, M.; Chua, P.; Heinrichs, M.; Stout, J.C.; Nathan, P.J. Oxytocin Attenuates Amygdala Reactivity to Fear in Generalized Social Anxiety Disorder. *Neuropsychopharmacology* **2010**, *35*, 2403–2413. [CrossRef]
53. Gemmer, A.; Mirkes, K.; Anneser, L.; Eilers, T.; Kibat, C.; Mathuru, A.; Ryu, S.; Schuman, E. Oxytocin receptors influence the development and maintenance of social behavior in zebrafish (*Danio rerio*). *Sci. Rep.* **2022**, *12*, 4322. [CrossRef]
54. Jones, C.; Barrera, I.; Brothers, S.; Ring, R.; Wahlestedt, C. Oxytocin and social functioning. *Dialogues Clin. Neurosci.* **2017**, *19*, 193–201. [CrossRef] [PubMed]
55. Ghazy, A.A.; Soliman, O.A.; Elbahnasi, A.I.; Alawy, A.Y.; Mansour, A.M.; Gowayed, M.A. Role of Oxytocin in different neuropsychiatric, neurodegenerative, and neurodevelopmental disorders. *Rev. Physiol. Biochem. Pharmacol.* **2023**, *186*, 95–134.
56. Landin, J.; Hovey, D.; Xu, B.; Lagman, D.; Zettergren, A.; Larhammar, D.; Kettunen, P.; Westberg, L. Oxytocin receptors regulate social preference in Zebrafish. *Sci. Rep.* **2020**, *10*, 5435. [CrossRef] [PubMed]
57. Singh, C.; Ousby, L. Modulation of social preference by oxytocin in the zebrafish *Danio rerio* behavior. *Brain Res.* **2021**, *399*, 113004.
58. Umeoka, E.H.; García-Chequer, A.J.; Oliveira, R.F. Oxytocin modulates dopamine release in the nucleus accumbens shell of zebrafish: Differential effect in social approach and focal visiting. *Front. Behav. Neurosci.* **2020**, *14*, 27.
59. Silva, P.F.; Freitas, E.G.; Azeredo, R.; Oliveira, R.F.; Silva, R. Anxiolytic effects of oxytocin in zebrafish: Investigating the role of oxytocin receptor subtypes and the receptor antagonist atosiban. *Peptides* **2019**, *113*, 14–21.
60. Nunes, A.R.; Gliksberg, M.; Varela, S.A.M.; Teles, M.; Wircer, E.; Blechman, J.; Petri, G.; Levkowitz, G.; Oliveira, R.F. Developmental Effects of Oxytocin Neurons on Social Affiliation and Processing of Social Information. *J. Neurosci.* **2021**, *41*, 8742–8760. [CrossRef] [PubMed]

61. Kirsch, P.; Esslinger, C.; Chen, Q.; Mier, D.; Lis, S.; Siddhanti, S.; Gruppe, H.; Mattay, V.S.; Gallhofer, B.; Meyer-Lindenberg, A. Oxytocin modulates neural circuitry for social cognition and fear in humans. *J. Neurosci.* **2005**, *25*, 11489–11493. [CrossRef] [PubMed]
62. Guzmán, Y.F.; Tronson, N.C.; Sato, K.; Mesic, I.; Guedea, A.L.; Nishimori, K.; Radulovic, J. Role of Oxytocin receptors in modulation of fear by social memory. *Psychopharmacology* **2013**, *231*, 2097–2105. [CrossRef]
63. Balmus, I.; Strungaru, Ș.A.; Nicoara, M.; Plavan, G.; Cojocaru, S.; Simion, L. Preliminary data regarding the effects of Oxytocin administration on the oxidative stress status of Zebrafish (*Danio rerio*). *Rev. Chim.* **2017**, *68*, 1640–1643. [CrossRef]
64. Robea, M.A.; Ciobica, A.; Curpan, A.S.; Plavan, G.; Strungaru, S.; Lefter, R.; Nicoara, M. Preliminary results regarding sleep in a zebrafish model of autism spectrum disorder. *Brain Sci.* **2021**, *11*, 556. [CrossRef] [PubMed]
65. Kaushal, G.; Sayre, B.E.; Prettyman, T. Stability-indicating HPLC method for the determination of the stability of oxytocin parenteral solutions prepared in polyolefin bags. *Drug Discov. Ther.* **2012**, *6*, 49–54. [CrossRef]
66. Kassi, G.B.; Zeré, M.G.; Stanislas, S.Y.; N'douba, V. Régime Alimentaire de Ctenopoma Petherici (Perciformes, Anabantidae) Dans La Rivière Agnéby et Dans Le Lac de Barrage Hydroélectrique d'Ayamé 2 (Côte d'Ivoire). Available online: https://popups.uliege.be/2295-8010 (accessed on 13 June 2023).
67. Ahmed, O.; Seguin, D.; Gerlai, R. An automated predator avoidance task in zebrafish. *Behav. Brain Res.* **2011**, *216*, 166–171. [CrossRef] [PubMed]
68. Haghani, S.; Karia, M.; Cheng, R.-K.; Mathuru, A.S. An automated assay system to study novel tank induced anxiety. *Front. Behav. Neurosci.* **2019**, *13*, 180. [CrossRef] [PubMed]
69. Savuca, A.; Chelaru, I.A.; Balmus, I.M.; Curpan, A.S.; Nicoara, M.N.; Ciobica, A.S. Toxicological response of zebrafish exposed to polymeric material cocktails and valproic acid. *Sustainability* **2024**, *16*, 2057. [CrossRef]

International Journal of
Molecular Sciences

MDPI

Article

Deletion of Slc1a4 Suppresses Single Mauthner Cell Axon Regeneration In Vivo through Growth-Associated Protein 43

Keqiang Li [1,†], Dinggang Fan [1,†], Junhui Zhou [1], Ziang Zhao [1], Along Han [1], Zheng Song [1], Xiahui Tang [1] and Bing Hu [1,2,*]

1 Center for Advanced Interdisciplinary Science and Biomedicine of IHM, Division of Life Sciences and Medicine, University of Science and Technology of China, Hefei 230026, China; likeqiang@mail.ustc.edu.cn (K.L.); fdg@mail.ustc.edu.cn (D.F.)
2 Division of Life Sciences and Medicine, University of Science and Technology of China, Hefei 230001, China
* Correspondence: bhu@ustc.edu.cn; Tel.: +86-551-63602489
† These authors contributed equally to this work.

Abstract: Spinal cord injury (SCI) is a debilitating central nervous system (CNS) disorder that leads to significant motor and sensory impairments. Given the limited regenerative capacity of adult mammalian neurons, this study presents an innovative strategy to enhance axonal regeneration and functional recovery by identifying a novel factor that markedly promotes axonal regeneration. Employing a zebrafish model with targeted single axon injury in Mauthner cells (M-cells) and utilizing the Tg (Tol056: EGFP) transgenic line for in vivo monitoring, we investigate the intrinsic mechanisms underlying axonal regeneration. This research specifically examines the role of amino acid transport, emphasizing the role of the solute carrier 1A4 amino acid transporter in axonal regeneration. Our findings demonstrate that Slc1a4 overexpression significantly enhances axonal regeneration in M-cells, whereas Slc1a4 deficiency impedes this process, which is concomitant with the downregulation of the P53/Gap43 signaling pathway. By elucidating the fundamental role of Slc1a4 in axonal regeneration and uncovering its underlying mechanisms, this study thus provides novel insights into therapeutic strategies for SCI.

Keywords: *slc1a4*; single Mauthner cell; Gap43; CNS axon regeneration; zebrafish

1. Introduction

Spinal cord injury (SCI) is a highly debilitating central nervous system (CNS) disorder, with its incidence rate increasing sharply each year, leading to severe motor and sensory dysfunctions that drastically affect patients' quality of life [1,2]. The primary challenge in spinal cord injury (SCI) repair lies in the inherent difficulty of axonal regeneration within the adult human CNS, as opposed to species that are capable of axonal regeneration within the adult CNS [3,4]. Axons in the adult mammalian CNS face significant obstacles in regenerating and re-innervating target areas due to both intrinsic and extrinsic factors [5,6]. Extrinsic factors include the presence of inhibitory molecules in the glial scar, such as chondroitin sulfate proteoglycans, which impede axonal growth [7]. Additionally, the lack of growth promoting signals in the post-injury environment further complicates regeneration efforts [8]. Intrinsic factors relate to the inability of mature neurons to re-enter a growth state and regenerate after development, due to specific physiological and molecular changes [9].

A substantial body of literature has demonstrated that amino acids play a vital role in the process of axon regeneration, essential for the recovery of nervous system function following injury [5,10–12]. For instance, local translation of amino acids in axons is crucial for protein synthesis necessary for axon growth and repair [10]. Specifically, neutral amino acids have been identified as particularly influential. It has been demonstrated that serine, one of the neutral amino acids, can promote axon growth and nerve regeneration through its role in the PI3K/Akt/mTOR pathway [11]. Amino acids are crucial for neuron growth, metabolism, electrical activity, and nutritional support. Slc1a4 is an amino acid transporter that facilitates the uptake of neutral amino acids, but its role in axon regeneration remains unknown [12]. Zebrafish, as a classical model animal, is highly genetically homologous to humans [13–15]. In recent years, due to its optical transparency, zebrafish has become an emerging model for studying axon regeneration, allowing for in vivo imaging in situ [16–18]. Mauthner cells (M-cells) are a pair of central neurons located in the hindbrain of zebrafish, with large axon diameters that run throughout the spinal cord [19–21].

M-cells exhibit strong regenerative capacity after axonal injury, allowing for the real-time visualization of axon regeneration at the single-cell level [22]. In this study, we will take full advantage of in vivo imaging of zebrafish to find out the roles of slc1a4 in the axonal regeneration. Our research aims to investigate the regulation of intrinsic factors involved in the axonal regeneration process of the CNS. Based on a new analysis, supported by data, we found that Slc1a4, among the seven members of the solute carrier family 1, exhibits the highest expression level and the most significant changes, with an increase of several folds compared to the uninjured group (Figure S1). Based on this, we speculate that Slc1a4 is crucial for axon regeneration. In vivo imaging in *slc1a4* mutant larval zebrafish constructed by CRISPR-Cas9 technology revealed that the mutation of *slc1a4* suppressed the M-cells' axonal regeneration. Using single-cell electroporation technology, we found that the overexpression of Slc1a4 significantly promotes axonal regeneration in M-cells. RNA sequencing (RNA-seq) results indicated that the absence of Slc1a4 may inhibit regeneration through the downregulation of the P53/Gap43 axonal regeneration pathway. Real-time quantitative PCR results demonstrated that the mutation of *slc1a4* led to the downregulation of P53/Gap43 pathway relative genes. Utilizing Western blot experiments, we found that *slc1a4* mutation significantly reduced the Gap43 protein expression level. Ultimately, our study indicates that Slc1a4 plays an important role in the process of axon regeneration and provides new insights into the treatment of SCI.

2. Results

2.1. Slc1a4 Overexpression in Mauthner Cells Enhances Axon Regeneration In Vivo

To investigate the role of Slc1a4 in the regeneration of M-cells within the zebrafish CNS (Figure 1a) and to determine whether Slc1a4 functions as a critical transporter protein during the axonal regeneration of M-cells, we developed an overexpression plasmid system [23]. This setup includes CMV-GAL4-VP16/UAS-mCherry (which served as the control group) and CMV-GAL4-VP16/UAS-mCherry/UAS-*slc1a4*. First, the effectiveness of the plasmid and its expression in embryos needed to be confirmed. We utilized microinjection techniques to introduce the plasmids into embryos [13]. Screening and a quantitative polymerase chain reaction (qPCR) analysis were performed at 2 days post-fertilization (dpf). The findings revealed a significant increase in the expression level of Slc1a4 within the overexpression group. Subsequently, we aimed to further validate the impact of Slc1a4 overexpression on axonal regeneration. We mainly employed the Tg (Tol-056) line (Figure 1b), which labels M-cells, to explore the function of Slc1a4 during the axonal regeneration of these neurons [23]. With the single-cell electroporation platform developed by our lab, we transferred the overexpression plasmid system into the cell bodies of unilateral M-cells in zebrafish larvae at 4 dpf. Two days after electroporation, we carried out two-photon laser ablation right above the lesioned side of the cloacal pore. Axonal regeneration was then observed at 8 dpf using confocal imaging. A statistical analysis of the regenerative lengths indicated that M-cells, after single-cell electroporation with the Slc1a4 overexpression plasmid (Figure 1c–e), showed significantly greater regeneration lengths at 2 dpf compared to the control group. Next, we interfered with the expression of Slc1a4 to explore its impact on axonal regeneration (Figure 1f). We then proceeded to explore the effect of disrupting Slc1a4 expression on the post-injury axonal regeneration of M-cells. The larvae at 4 dpf were subjected to the inhibitor (OH-Pro), the competitive inhibitor of the Slc1a4 protein, followed by two-photon laser ablation at 6 dpf, and the regeneration length was measured at 8 dpf. The statistical analysis revealed that following the suppression of slc1a4 function, the length of axon regeneration was significantly reduced compared to the control group (Figure 1g,h). The inhibitor effect reached a threshold when the inhibitor concentration was 1 mM, and when it was greater than this concentration, the inhibition effect was comparable. This further implies that the inhibition of Slc1a4 protein function can suppress the regenerative capacity of zebrafish M-cells after injury, playing an essential role in the axonal regeneration process of these neurons.

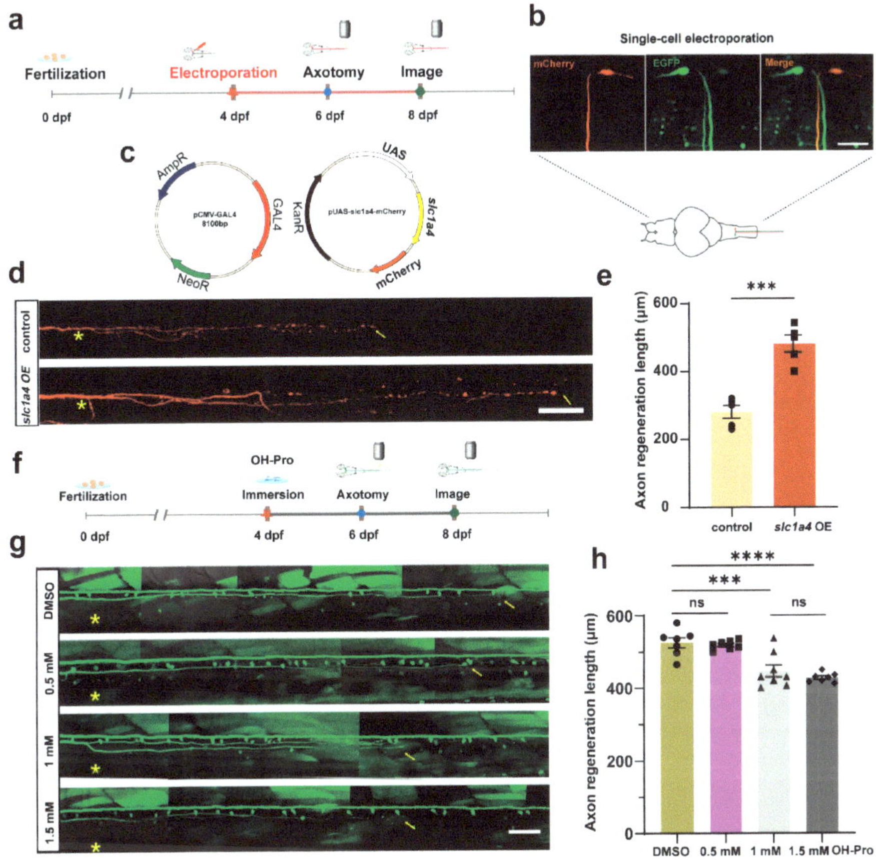

Figure 1. Overexpression of slc1a4 in Mauthner cells promotes axon regeneration in vivo. (**a**) Timeline of time points of electroporation, axotomy, and imaging. (**b**) Pattern diagram of electroporation and confocal images of positive expression in M-cells; three photos represent position of M-cell under 40× magnification. Scale bar, 50 μm (EGFP: labeled M-cells in Tol-056 zebrafish strain; mCherry: fluorescent reporter gene in foreign plasmid). (**c**) Schematic diagram of microinjection using two-plasmid system. (**d**) Representative diagram of confocal imaging of M-cells' axon regeneration. Asterisk, ablation site; Arrow, regeneration endpoint location. scale bar, 50 μm (control: control; *slc1a4* OE: overexpression). (**e**) Statistical quantitative diagram of axon regeneration. Data shown as mean ± sem (control: 280.0 ± 18.8 μm, n = 5; *slc1a4* OE: 452.8 ± 34.0 μm, n = 5). Assessed by unpaired, two-tailed Student's t-test. *** $p < 0.001$. (**f**) Timing of inhibitor processing, laser damage, and imaging. (**g**) Representative diagram of confocal imaging of M-cells' axon regeneration between DMSO and inhibitor (concentration gradient, OH-Pro: 0.5 mM, 1 mM, 1.5 mM). Asterisk: ablation site. Arrowhead: axon regeneration terminal; scale bar, 50 μm. (**h**) Statistical quantitative diagram of axon

regeneration. Data shown as mean ± sem (DMSO: 525.6 ± 13.9 μm, n = 7; 0.5 mM: 519.0 ± 4.6 μm, n = 7; 1 mM: 446.9 ± 17.0 μm, n = 8; 1.5 mM: 427.3 ± 4.879 μm, n = 7); data were analyzed with one-way ANOVA. *** $p < 0.001$; **** $p < 0.0001$; ns, not significant.

2.2. *Identification of slc1a4 Mutant Zebrafish*

To elucidate the role of Slc1a4 in the development of M-cells and their axonal regeneration, we employed clustered regularly interspaced short palindromic repeat CRISPR-Cas9 technology to generate an *slc1a4* gene knockout zebrafish line [24,25]. This genetically modified line was created to facilitate further experimental investigations into the functional implications of Slc1a4 deficiency.

The gene knockout target was strategically located on the first exon of the *slc1a4* gene (Figure 2a,b). Through successive generations of breeding and meticulous tail-clip sequencing for identification, we successfully established a homozygous *slc1a4* knockout line. Sequencing results revealed a deletion of ten bases in the first exon, which induced a frameshift mutation. This mutation resulted in the premature appearance of a stop codon, consequently leading to the early termination of translation (Figure 2c,d).

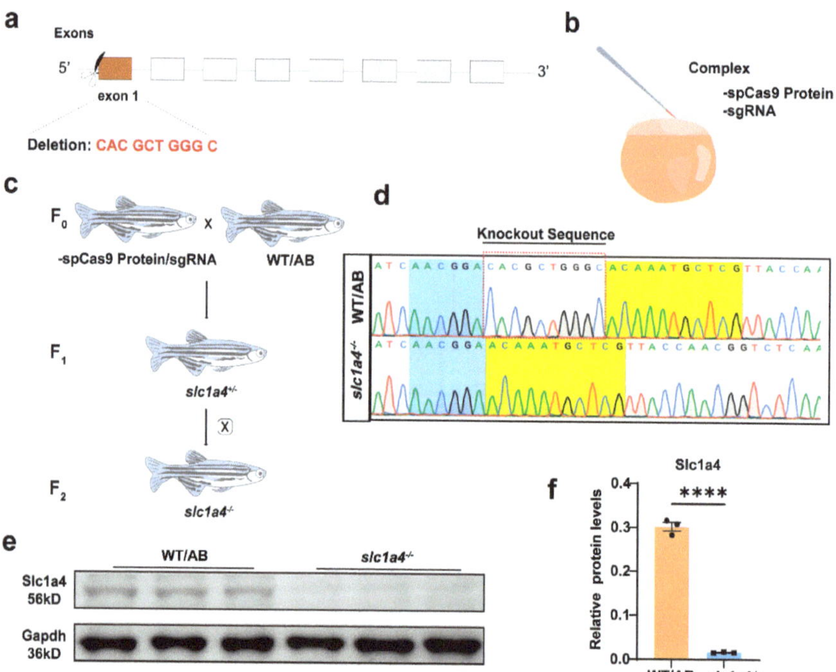

Figure 2. Identification of *slc1a4* mutant zebrafish. (**a**) Schematic of Cas9-sgRNA targeted site located in first exon of *slc1a4*. (**b**) Schematic of complex injected into one-cell embryos. (**c**) Procedure for obtaining *slc1a4*$^{-/-}$ mutant lines. (**d**) Representative sequencing results of wild-type and mutated zebrafish lines. Mutant sequencing results showed that 10 bp were deleted in *slc1a4*$^{-/-}$ zebrafish. (**e**,**f**) Western blotting analysis showed that Slc1a4 protein expression is inhibited in mutant group compared with wild type. Protein expressions were quantified by Image J software version 1.54f. Experiment was repeated three times with three independent samples. **** $p < 0.0001$. Assessed by unpaired t-test.

To confirm whether the translation of the *slc1a4* gene was indeed disrupted at the protein level in this knockout line, we conducted Western blotting (WB) analyses [26]. The WB results unequivocally indicated that the expression of the Slc1a4 protein was significantly impaired in the zebrafish knockout line. This impairment was evident from the absence or drastic reduction in the protein bands corresponding to Slc1a4 in the knockout samples compared to the wild-type controls (Figure 2e,f).

These findings substantiate the successful knockout of the *slc1a4* gene and underscore the resultant deficiency in Slc1a4 protein expression. Consequently, this knockout model provides a robust platform for investigating the physiological and molecular functions of Slc1a4 in M-cell development and axonal regeneration. Future experiments will leverage this model to elucidate the mechanistic pathways through which Slc1a4 influences these critical neurobiological processes.

2.3. *Slc1a4 Deficiency Does Not Impact Mauthner Cell Development or Motor Function*

To facilitate further investigation into the impact of Slc1a4 on axonal regeneration in the *slc1a4* knockout line, we crossed this homozygous knockout line with the Tg (Tol-056) line for two consecutive generations. This breeding strategy resulted in a homozygous zebrafish line with *slc1a4* knockout (Figure 3a), marked by enhanced green fluorescent protein (EGFP) [2].

Given that growth and developmental signals can influence the length of axonal regeneration, we aimed to determine whether the absence of Slc1a4 affects the development of M-cells in juvenile fish. We first measured the entire body length of the zebrafish. There was no significant difference in body length between the control and mutant strains, suggesting that the mutation in *slc1a4* did not result in an overall length difference (Figure 3b,e). We used confocal microscopy to observe the morphology of the M-cell bodies at 6 dpf, using the cell body area as an assessment criterion (Figure 3c,f). The imaging results showed that in the absence of Slc1a4, the size of M-cell bodies at 6 dpf did not change significantly, and morphologically, there was no obvious difference between the *slc1a4* mutant and the wild-type cytosol [19]. Therefore, the deficiency of Slc1a4 did not affect the development of the M-cell bodies. Next, to rule out the impact of Slc1a4 deficiency on axonal development, we measured the length of the M-cell axons uniformly. We chose 6 dpf as the time point to rule out the possibility that the absence of Slc1a4 causes differences in the rate of axonal growth. Using the axon directly above the cloaca as the starting point (which is also the starting point of injury), we measured its extension to the tail end. This length was used for a relative comparison of the total length to determine if there were any differences between the *slc1a4* knockout group and the WT (Figure 3d,g). Imaging results showed that the axon length of M cells was not affected by the absence of Slc1a4, being similar to that of WT, with no significant differences. In addition, juvenile fish treated with an Slc1a4 inhibitor did not exhibit significant changes in spontaneous movements compared to controls (Figure 3h,i). In addition, we treated zebrafish with an inhibitor of *slc1a4* and then examined spontaneous movements after treatment. After the assay, the results showed that the inhibitor treatment did not affect the spontaneous movements of zebrafish (Figure S2a,b). Based on these findings, we conclude that the lack of Slc1a4 does not impact the growth and development of M-axons on the whole.

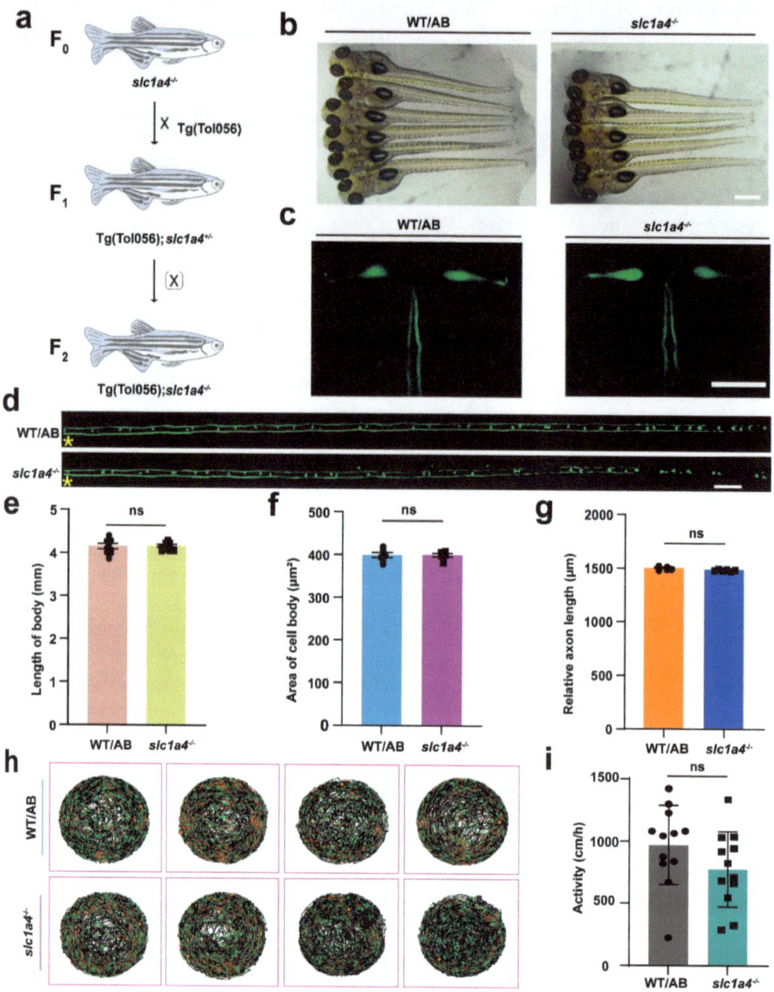

Figure 3. Deficiency in Slc1a4 does not affect development of M-cells and motor function of juvenile fish. (**a**) Hybridization of transgenic line: Tg (Tol 056: EGFP) and *slc1a4* mutants were crossed for two consecutive generations to obtain Tg (Tol 056: EGFP)/*slc1a4*$^{+/-}$ and Tg (T056: EGFP); *slc1a4*$^{-/-}$ lines. (**b**) Representative images of embryos from wild type and mutant at 6 dpf (scale bar, 500 μm). (**c**) Representative images of area of cell body (scale bar, 50 μm). (**d**) Representative images of relative axon length from cloacal pore to end (scale bar, 50 μm). Asterisk, ablation site. (**e**) Statistical quantitative diagram represents length of fish at 6 dpf. Data shown as mean ± sem (WT/AB: 4.1 ± 0.1 mm, n = 9; *slc1a4*$^{-/-}$: 4.2 ± 0.0 mm, n = 10). Assessed by unpaired t-test. ns, not significant. (**f**) Statistical quantitative diagram represents area of cell body. Data shown as mean ± sem (WT/AB: 399.3 ± 5.9 μm^2, n = 6; *slc1a4*$^{-/-}$: 399.7 ± 3.9 μm^2, n = 6). Assessed by unpaired t-test. ns, not significant. (**g**) Statistical quantitative diagram represents relative axon length. Data shown as mean ± sem (WT/AB: 1498.0 ± 6.1 μm, n = 6; *slc1a4*$^{-/-}$: 1485.0 ± 3.0 μm, n = 6). Assessed by unpaired t-test. ns, not significant. (**h**,**i**) Line illustrates 6 dpf zebrafish larvae's swimming trajectory differences from WT and *slc1a4*$^{-/-}$ groups evaluated over 1 h. Data shown as mean ± sem (WT/AB: 1230.0 ± 70.1 cm/h, n = 6; *slc1a4*$^{-/-}$: 844.2 ± 58.0 cm/h, n = 12). Assessed by unpaired t-test. ns, not significant.

2.4. Slc1a4 Deficiency Suppresses Mauthner Cell Axon Regeneration and Associated Functions In Vivo

To explore the effect of Slc1a4 on regeneration after axonal damage, we followed the same timeline as before for processing the *slc1a4* mutant line (6 dpf for laser damage, 8 dpf for imaging statistical length) [27]. After the data analysis, it became evident that the regeneration length of axons significantly decreased in the absence of Slc1a4 compared to WT (Figure 4a–c). In addition, we looked at the regeneration of the $slc1a4^{+/-}$ mutant lines, and after counting, it is clear that even single-stranded mutations result in differences in regeneration length (Figure S3a,b). Previous studies have indicated that M-cells are particularly important for rapid escape responses. The C-Start is a very fast startle or escape reflex where M-cells play a crucial role [28–30]. Under predator pressure, zebrafish exhibit C-Start escape behavior. After receiving a stimulus, M-cells generate a strong electrical signal, which causes a strong and rapid contraction upon reaching the muscles, leading to the fish making a C-shape curve movement. Based on these principles and other studies, we constructed a C-Start escape response device for testing the function of Mauthner neurons (Figure 4d). The main indicators for assessing the recovery of Mauthner neuron function are the maximum deflection angle reached by the zebrafish after stimulation and the time taken to reach this maximum deflection angle (Figure 4e).

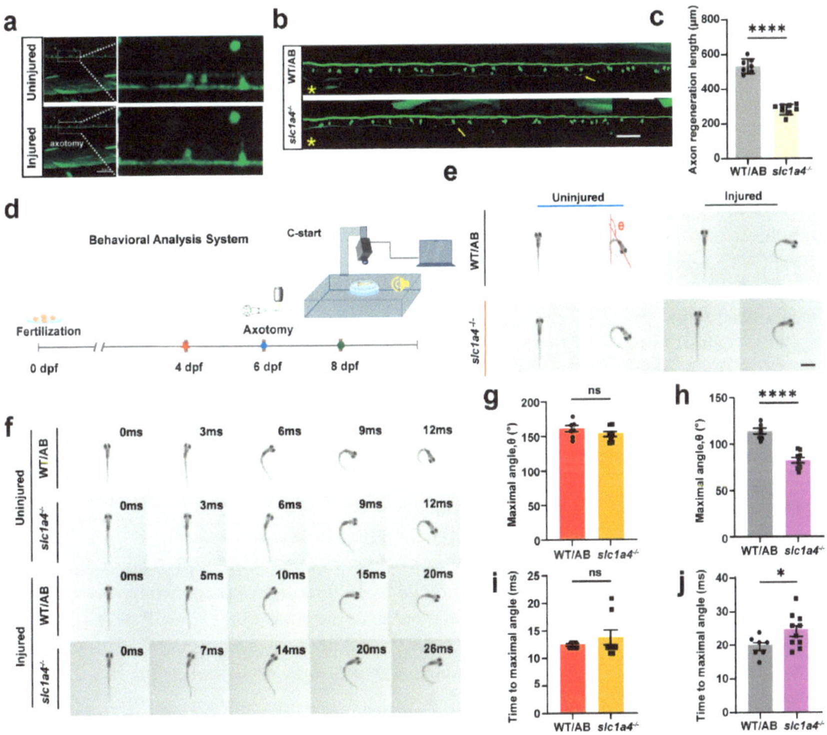

Figure 4. Deficiency in Slc1a4 suppresses Mauthner cell axon regeneration and relative function in vivo. (**a**) Representative images of confocal imaging of M-cell axon before and after ablation by two-photon laser above cloacal pore (scale bar, 50 μm). (**b**,**c**) Representative diagram of confocal imaging of M-cells' axon regeneration between WT and $slc1a4^{-/-}$. Data shown as mean ± sem. Scale bar, 50 μm. Assessed by unpaired *t*-test. **** $p < 0.0001$. Asterisk, ablation site; Arrow, regeneration end-point location. (**d**) Device for testing escape behavior. (**e**) Representative images of original orientation

and maximal turn angle position from WT and *slc1a4*$^{-/-}$ zebrafish larvae in uninjured and injured groups. Red lines indicate heading direction. **(f)** Series of images of movement trajectory from WT and *slc1a4*$^{-/-}$ zebrafish larvae in uninjured and injured groups. **(g,h)** Statistical diagram of maximal turn angle, θ. Data shown as mean ± sem. Uninjured: ns, not significant (WT/AB: 161.2 ± 11.89°, n = 7; *slc1a4*$^{-/-}$: 152.9 ± 10.36°, n = 8); injured: **** $p < 0.0001$ (WT/AB: 113.9 ± 3.1°, n = 7; *slc1a4*$^{-/-}$: 82.36 ± 2.917°, n = 10). ns, not significant. Scale bar, 1 mm. Assessed by unpaired t-test. **(i,j)** Statistical diagram of time to maximal turn angle. Data shown as mean ± sem. Uninjured, ns, not significant (WT/AB: 12.4 ± 0.2 ms, n = 7; *slc1a4*$^{-/-}$: 13.75 ± 1.386 ms, n = 8); injured, * $p < 0.05$ (WT/AB: 19.86 ± 1.143 ms, n = 7; *slc1a4*$^{-/-}$: 24.50 ± 1.586 ms, n = 10); ns, not significant; * $p < 0.05$. Assessed by unpaired t-test.

We performed laser ablation at 6 dpf and conducted behavioral tests at 8 dpf. After the statistical analysis, it was found that in the group where Mauthner neurons were not damaged, there were no significant differences in the time taken to reach the maximum deflection angle or the maximum deflection angle itself after stimulation (Figure 4f). However, in the knockout strains, significant changes were observed in both the time taken to reach the maximum deflection angle and the maximum deflection angle itself after the damage. This further indicates that the knockout of *slc1a4* severely inhibits the axonal regeneration ability and functional recovery of Mauthner neurons (Figure 4g–j).

In summary, the knockout of *slc1a4* did not affect the development of the M-cell bodies and axons, allowing us to exclude development as a factor in the influence of Slc1a4 on axonal regeneration. Moreover, our behavioral results also indicate that Slc1a4 does not affect the activity level of the larvae, thus ruling out spontaneous activity as an influencing factor on axonal regeneration.

2.5. RNA-Seq Revealed That Slc1a4 May Influence Mauthner Axons' Regeneration through P53 Signaling Pathway

To uncover the specific molecular mechanisms affecting regeneration, we performed whole-genome transcriptome sequencing (Figure 5a) on *slc1a4*$^{-/-}$ zebrafish and wild-type (WT) zebrafish at 4 dpf [31]. The bioinformatics analysis of the sequencing data revealed 4372 out of 24,064 core genes with significant expression changes, including 1863 upregulated and 2509 downregulated genes (Figure 5b,c).

The Gene Ontology (GO) enrichment analysis (Figure 5d) and Kyoto Encyclopedia of Genes and Genomes (KEGG) functional enrichment analysis (Figure 5e) revealed that a subset of these differentially expressed genes are primarily enriched in the P53 signaling pathway, amino acid metabolism, and regeneration processes. Research indicates that the P53 signaling pathway is associated with neuronal regeneration [32,33]. Existing research indicates that *tp53* influences the expression levels of known neuronal regeneration-related factors such as *gap43*, *rab13*, and *coronin1a*.

These findings suggest that the P53 signaling pathway may play a crucial role in mediating the effects of Slc1a4 on axonal regeneration. The differential expression of genes involved in this pathway highlights a potential molecular mechanism through which Slc1a4 deficiency impacts the regenerative capacity of Mauthner neuron axons. Further investigation into these specific genes and their interactions within the P53 signaling pathway could provide deeper insights into the molecular underpinnings of axonal regeneration and the role of Slc1a4 in this process.

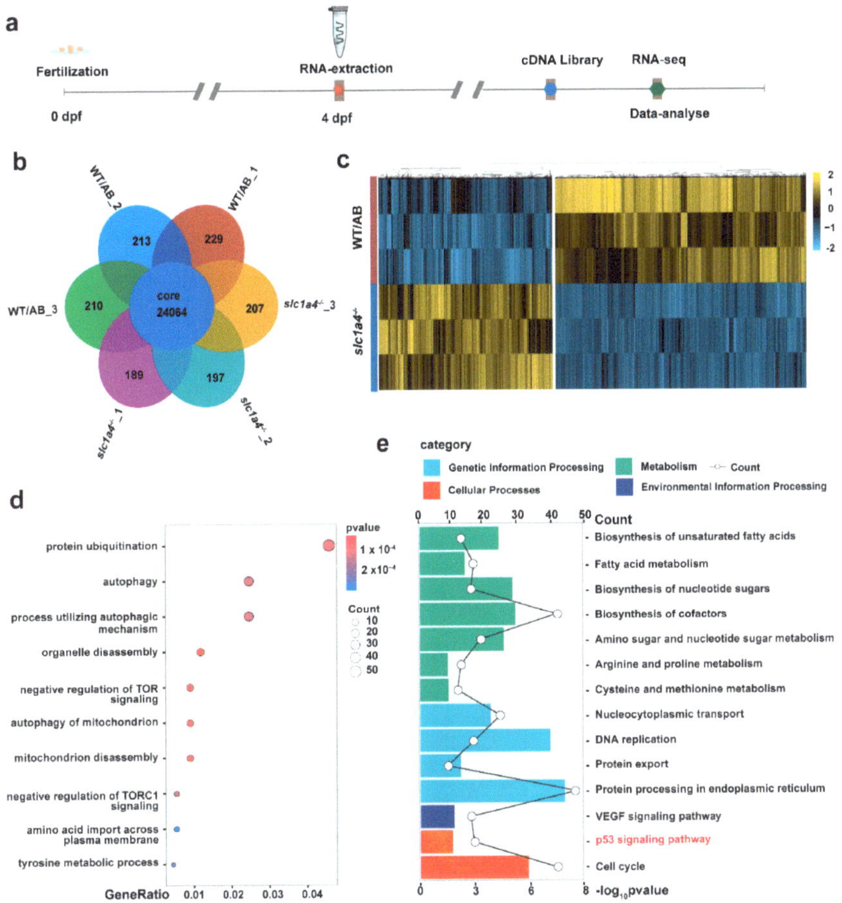

Figure 5. RNA-seq revealed that Slc1a4 may influence regeneration of Mauthner axons through P53 signal pathway. (**a**) Timeline of time points of RNA extraction and RNA-seq. (**b**,**c**) Heatmap shows downregulated and upregulated genes in *slc1a4* mutant zebrafish (downregulated genes: 2509; upregulated genes: 1863). (**d**) GO enrichment analysis of upregulated genes. (**e**) Enrichment for KEGG pathway analysis of downregulated genes.

2.6. Slc1a4 Deletion May Inhibit Mauthner Axons' Regeneration via Gap43 Suppression

The sequencing results showed that genes such as *tp53*, *gap43*, and *rab13* all experienced a significant decrease (Figure 6a). Based on the sequencing results, it is speculated that the lack of Slc1a4 may lead to a deficiency of nutritional factors, resulting in the generation of nutritional stress signals, which in turn causes a decrease in the expression level of *tp53*. This leads to a synchronous decrease in the expression levels of genes such as *gap43* and *rab13*, thereby affecting the ability of axonal regeneration. To prove this hypothesis, we tested the expression levels of these genes. After the qPCR assay, it can be found that the expression levels of *tp53*, *gap43*, and *rab13* in the *slc1a4*$^{-/-}$ mutant lines all underwent a significant reduction relative to the wild type (Figure 6b). After the WB assay, it can be seen that a significant reduction in the protein expression level of *gap43* occurred in the *slc1a4*$^{-/-}$ mutant line (Figure 6c,d). It is tentatively possible to conclude that mutant strains of *slc1a4*$^{-/-}$ may affect axon regeneration through a pathway that affects *tp53* and *gap43*.

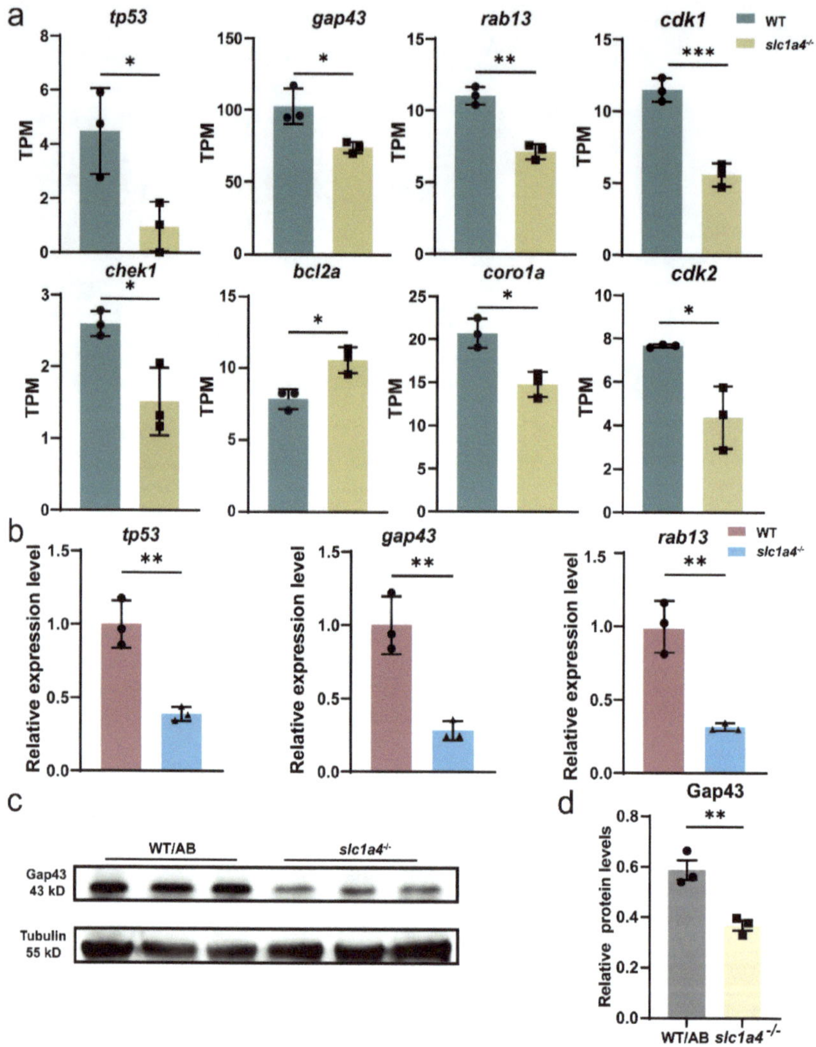

Figure 6. Slc1a4 deletion may inhibit Mauthner axons' regeneration via Gap43 suppression. (**a**) RNA-seq analysis of P53 signaling pathway-related gene expression in mutant and wild-type lines. TPMs: Transcripts Per Kilobase of exon model per Million mapped reads. Assessed by unpaired *t*-test. * $p < 0.05$. ** $p < 0.01$. *** $p < 0.001$ (**b**) qPCR detection of *tp53* and *gap43* and *rab13* expression. ** $p < 0.01$. (**c**,**d**) Gap43 was decreased in knockout group as shown via Western blot quantification. Data shown as mean ± sem. Unpaired Student's two-tailed *t*-test; ** $p < 0.01$.

3. Discussion

In this study, we investigated the role of the *slc1a4* gene in axonal regeneration using a zebrafish model with single axon injury on Mauthner cells (M-cells). Our results demonstrated that the overexpression of Slc1a4 significantly promotes axonal regeneration, while a deficiency of Slc1a4 severely impedes this process. This impairment in regeneration was associated with the downregulation of key genes in the P53 signaling pathway, including *gap43* and *rab13*. These findings suggest that Slc1a4 is a critical factor in the axonal regeneration of M-cells, potentially through its regulation of amino acid homeostasis and the P53 signaling pathway.

Both extrinsic factors and intrinsic factors are crucial for successful axonal regeneration [6,34–36]. Traditional models of axonal regeneration have predominantly utilized animal models such as rodents and invertebrates. These models have been instrumental in understanding the molecular and cellular mechanisms underlying axonal regeneration [34,37]. For instance, rodent models, particularly mice and rats, have been extensively employed to study spinal cord injuries and peripheral nerve regeneration. However, one of the significant challenges in studying axonal regeneration is the heterogeneity among neurons [38–41]. To address the issue of neuronal heterogeneity, our research focuses on the use of single M-cells. Dissimilar to what has been shown in a mouse model of axonal regeneration, after an M-cell was damaged, it showed strong regenerative ability. After two-photon laser damage to the axon, the regeneration length can reach about 500 μm at 2 days post axotomy, which is very convenient for the study of axons. In M-cells' model, single-cell overexpression technology was utilized to overexpress *slc1a4*, leading to the conclusion that slc1a4 overexpression may promote axon regeneration in single M-cells.

Many studies have shown that SLC1A4, also known as ASCT1, is a neutral amino acid transporter that has been implicated in various physiological processes, including amino acid homeostasis, neurotransmission, and cellular metabolism [12,42–44]. Previous studies have primarily focused on its role in the CNS and its involvement in neurological disorders. For instance, mutations in Slc1a4 have been linked to developmental delay, microcephaly, and hypomyelination, highlighting its critical role in brain development and function [42]. SLC1A4 is also involved in the regulation of d-serine levels in the brain, which is essential for NMDA receptor function and neurodevelopment [12]. Despite these significant findings, the role of SLC1A4 in axonal regeneration remains largely unexplored. Our research aims to fill this gap by investigating the function of Slc1a4 in neuronal axonal regeneration. Utilizing in vivo imaging techniques, we have demonstrated that Slc1a4-mediated neutral amino acid transport may influence axonal regeneration through the P53/Gap43 signaling pathway. After treatment with an inhibitor of *slc1a4*, it was found that the regeneration of M-cells in zebrafish would be significantly inhibited. The mutation of *slc1a4* inhibits axonal regeneration and motor function recovery.

Previous studies have shown that GAP43 (Growth-Associated Protein 43) is a well-established marker of axonal growth and regeneration [45–48]. This phosphoprotein plays a crucial role in axonal outgrowth, synaptic plasticity, and neural development. Predominantly expressed in growth cones of developing neurons and regenerating axons, GAP43 is a key player in neuronal repair processes [49]. The importance of GAP43 in axonal regeneration has been highlighted through various studies. For instance, the upregulation of GAP43 is associated with axonal growth in the rat spinal cord following compression injury [50–54]. Building on the established role of GAP43 in axonal regeneration, our research introduces a novel finding: the knockout of *slc1a4* significantly impacts the expression of Gap43, thereby inhibiting axonal regeneration. This discovery provides new insights into the molecular mechanisms underlying axonal regeneration. It suggests that Slc1a4 is not only involved in amino acid transport but also plays a pivotal role in the regulation of key regenerative proteins like Gap43. This finding opens up potential therapeutic avenues for enhancing axonal regeneration by targeting the Slc1a4/Gap43 pathway. In conclusion, while previous studies have firmly established the role of Gap43 in axonal regeneration, this research highlights the regulatory effect of Slc1a4 and Gap43. While transcriptomic sequencing and WB have identified Slc1a4 as a potential regulator of Gap43 expression, thereby influencing axonal regeneration, the direct experimental validation of the interaction between Slc1a4 and Gap43 remains to be established [43–46].

The use of the Mauthner-cell model has indeed provided significant advantages by addressing the issue of neuronal heterogeneity. This model allows for a more controlled and precise investigation of axonal regeneration mechanisms. However, the Mauthner-cell model is inherently limited due to its specificity and uniqueness. The findings derived from this model may not be universally applicable across different neuronal types and organisms. To gain a comprehensive understanding of the role of SLC1A4 in axonal regeneration, it is essential to extend the investigation to other models. Exploring the function of Slc1a4 in diverse neuronal systems and organisms will help to validate the findings from the Mauthner-cell model and provide broader insights into its role in axonal regeneration. Such studies would involve using various in vivo and in vitro models, including rodent models, zebrafish, and other well-established systems for studying neuronal repair. This approach will help to establish the generalizability of the Slc1a4/Gap43 interaction and its implications for axonal regeneration across different biological contexts. In conclusion, while the Mauthner-cell model has provided valuable insights into the role of Slc1a4 in axonal regeneration, further research using a variety of models is necessary to fully elucidate the molecular mechanisms involved. The direct experimental validation of the interaction between Slc1a4 and Gap43, along with studies in different neuronal systems, will enhance our understanding of the pathways regulating axonal regeneration and potentially lead to the development of targeted therapeutic strategies for neuronal repair.

Despite the significant differences in neural regeneration abilities between zebrafish and mammals, particularly humans, this study has revealed the crucial role of the Slc1a4/P53/Gap43 signaling pathway in axonal regeneration in zebrafish, offering a potential target for human spinal cord injury (SCI) treatment strategies. This discovery underscores the importance of identifying and utilizing factors, such as P53 and Gap43, that promote axonal regeneration in the research of human SCI treatment to improve neurological function recovery. Future research should focus on exploring the pathways through which Slc1a4 affects axonal regeneration in mammals and further investigating the conservation of its mechanism in regulating P53 and Gap43. This will enhance our understanding of the similarities and differences in this process across different species, thereby laying a foundation for designing more effective treatment methods for human SCI.

In summary, we discovered that Slc1a4-mediated neutral amino acid transport influences Gap43 expression through the P53 signaling pathway in zebrafish. In vivo imaging revealed that Slc1a4 plays a crucial role in axonal regeneration. The loss of Slc1a4 disrupts amino acid homeostasis, leading to the accumulation of transported neutral amino acids and subsequently reducing axonal regeneration levels. These findings provide new insights into the molecular mechanisms underlying axonal regeneration and suggest potential therapeutic targets for enhancing neuronal repair in spinal cord injuries. Despite some limitations, our research contributes to a better understanding of the intrinsic factors involved in axonal regeneration and offers a foundation for future studies aimed at improving outcomes for patients with CNS injuries.

4. Materials and Methods

4.1. Zebrafish Strains and Maintenance

In this study, adult zebrafish were maintained in an aquatic system at 28.5 °C with a 14/10 h light/dark cycle (14 h light and 10 h dark cycle). Embryos were collected from natural spawning after mating male and female zebrafish in a 3:2 ratio and raised at 28.5 °C in an incubator with 5 mM NaCl, 0.17 mM KCL, 0.33 mM $CaCl_2$, 0.33 $MgSO_4$, and 0.1% methylene blue at pH 7.0. From 2 dpf, embryos were supplemented with 0.003% N-phenylthiourea (PTU, Sigma-Aldrich, St. Louis, MO, USA) to prevent pigmentation [55–57]. The transgenic line used was Tg (Tol 056: EGFP), where M-cells express enhanced green fluorescent protein. The USTC (University of Science and Technology of China) Animal Resources Center and University Animal Care and Use Committee provided the guidelines that experiments followed. Protocols needed approval by the USTC ethics committee (license number: USTCACUC1103013).

4.2. Genome Editing

The CRISPR-mediated genome editing for generating *slc1a4*$^{-/-}$ mutants was implemented. Cas9 mRNA was synthesized in accordance with the appropriate plasmid

(108301, Addgene, Watertown, MA, USA) through the mMessage mMachine T7 Ultra Kit (AM1345 mMESSAGE mMACHINE™ T7 ULTRA, Thermo Fisher, Waltham, MA, USA). The single-guide RNA (sgRNA) that targeted the *slc1a4* sequence was fabricated by using the aforesaid plasmids with the Megashortscript T7 kit. We delicately combined the Cas9 mRNA (300 ng/μL) with the sgRNA (40 ng/μL), and thereafter, microinjected this mixture into embryos at the one-cell stage.

4.3. Single-Cell Electroporation

Prior to electroporation, 4 dpf larvae were anesthetized using ethyl 3-aminobenzoic methanesulfonate (MS222, Sigma-Aldrich) and embedded in 1% low-melting agarose (Sangon, Shanghai, China) within an electroporation chamber. A micropipette tip pulled by a micropipette puller (Sutter Instrument, Novato, CA, USA) was filled with plasmids and positioned near the M-cell soma. Electric stimulation was applied to the zebrafish larvae to deliver the plasmids, with a concentration of 120 ng/μL, into the unilateral M-cell.

4.4. Two-Photon Axotomy

Before axotomy, anesthetized zebrafish larvae at 6 dpf were fixed in 1% low-melting agarose in a chamber. A Zeiss microscope (LSM980; Carl Zeiss, Oberkochen, Germany) equipped with two photons was utilized at a wavelength of 800 nm and an intensity ranging from 15% to 35% to ablate the M-cell axons under a 25× oil immersion lens.

4.5. In Vivo Imaging

Larvae were sedated with MS222 and then embedded in 1% low-melting agarose in a chamber [58]. Larvae were photographed two days after axotomy using a confocal system (FV1000; Olympus, Tokyo, Japan) and a water immersion lens (40×, 0.85 numerical-aperture objective). Z-stack images were acquired at 3 μm intervals.

4.6. Quantitative Real-Time PCR

Total RNA was extracted from the entire larvae through the use of RNAsio (TAKARA), and approximately 1 μg of RNA was reverse-transcribed into cDNA with the assistance of HiScript II qRT SuperMix II (Vazyme, Nanjing, China). The qPCR application was carried out in a total volume of 10 μL in a contained 5 μL of ChamQ Universal SYBR Green qPCR Master Mix and 1 μL of the cDNA template on a real-time quantification system (LightCycler 96, Roche, Pleasanton, CA, USA). The mRNA expression levels were analyzed by means of the comparative Ct relative quantification method formula $2^{-\Delta\Delta CT}$, with the housekeeping gene β-actin mRNA serving as an invariant control to normalize the mRNA of the target genes [59]. This was repeated three times for each sample. All the primers used are detailed in Table 1.

Table 1. Primes used in qPCR experiments.

Primers	Sequences (5′-3′)
β-actin-qPCR-F	CATTGGCAATGAGCGTTTC
β-actin-qPCR-R	TACTCCTGCTTGCTGATCCAC
tp53-qPCR-F	TGGAGAGGAGGTCGGCAAAATCAA
tp53-qPCR-R	GACTGCGGGAACCTGAGCCTAAAT
gap43-qPCR-F	TGCTGCATCAGAAGAACTAA
gap43-qPCR-R	CCTCCGGTTTGATTCCATC
rab13-qPCT-F	GCATACTACAGAGGGGCCA
rab13-qPCT-R	CATTCGACTTACACCCGCTG
slc1a4-sgRNA-F	TAATACGACTCACTATAGGCGAAATCAACGGACACGCTGT TTTAGAGCTAGAAATAGC

Table 1. *Cont.*

Primers	Sequences (5′-3′)
slc1a4-sgRNA-R	AGCACCGACTCGGTGCCACT
slc1a4-qPCR-F	TACCATCATCCCTAGCCCGA
slc1a4-qPCR-R	ATGGAGGAGAAGAAGAGCGAAATCA

4.7. Protein Extraction and Western Blotting

Wild-type and *slc1a4*$^{-/-}$ mutant larvae at 4 dpf were collected and lysed with an RIPA (Radio Immunoprecipitation Assay Lysis) buffer supplemented with a protease inhibitor and phosphatase inhibitor (Sangon, Shanghai, China). The lysates were centrifuged and the collected supernatant was maintained on ice. In accordance with the manufacturers' instructions of the BCA Protein Assay Kit (Beyotime, Shanghai, China) the concentration of each protein sample was determined on a microplate reader. Samples were boiled for 4 min and run on a 10% SDS-PAGE gel along with a loading buffer (5×) and then transferred to a PVDF membrane. After incubation in 5% nonfat milk and TBST for 60 min at room temperature, the membranes were washed once with TBST and incubated with antibodies against Slc1a4 (1:1000; Proteintech, Wuhan, China) or Gapdh (1:2000; HuaAn, Wenzhou, Zhejiang, China) or β-tubulin (1:2000; GeneTex, Irvine, CA, USA) at 4 °C for 12 h. Subsequently, the membranes were incubated with secondary goat anti-rabbit antibodies (1:5000; Proteintech) for 1 h at room temperature. The blots were washed with TBST three times and visualized by the enhanced chemiluminescence (ECL) system (Thermo Fisher, MA, USA). The densities of the bands were quantified by ImageJ software (National Institutes of Health, Bethesda, MD, USA) and normalized to protein Gapdh or β-tubulin.

4.8. Drug Treatment

The OH-Pro (Cat.No.H54409, MCE, Shanghai, China) induces Slc1a4 downregulation. At 4 dpf, the larvae were treated with drugs by different concentrations. The concentration gradient of OH-Pro is 0.5 mM, 1 mM, and 1.5 mM. The control group is 0.1% DMSO. At 6 dpf, the axons were ablated and continued to be restored in drugs. The M-cells' axon regeneration was observed and statistically analyzed at 8 dpf.

4.9. Escape Behavior Assay

The device system was comprised of a high-speed camera (1000 fps, Revealer, Hefei, China), a computer, and a loudspeaker. The 8 dpf zebrafish larvae were placed in a Petri dish with EM and moved to a platform with suitable light. The computer was linked to the loudspeaker near the Petri dish, and the high-speed camera was adjusted appropriately. Before each test, the larvae were left for 5 min without any disturbance. Movement trajectories were induced by sound stimulation of sinusoidal waves (500 Hz, 20 ms), and video acquisition was controlled by specialized software. For the injured group, unilateral M-cells were ablated with two-photon axotomy at 6 dpf before the escape behavior assay.

4.10. Statistical Analysis

Graphs and statistical significance were analyzed using GraphPad Prism 10.2.3 software (San Diego, USA), Adobe Photoshop CC2020, and Adobe Illustrator CC 2020. Data are presented as the mean ± standard error of the mean (SEM). Experiments were analyzed using unpaired two-tailed Student's *t*-tests. Experiments with more than two groups were analyzed using one-way analyses of variance (ANOVAs), and experiments involving two independent variables were analyzed using two-way ANOVAs. Experiments were repeated at least three times. Differences were considered significant when * $p < 0.05$, ** $p < 0.01$, *** $p < 0.001$, and **** $p < 0.0001$. The figure legends provide all other pertinent information, such as sample size and precise statistical tests used.

4.11. RNA-Seq

The *slc1a4*$^{-/-}$ lines were in the experimental group and WT was in the control group. At 6 dpf, total RNA was extracted by screening for positive expression (n = 35, with each sample repeated three times). Total mRNA was enriched using Oligo (dT) beads, and then it was fragmented into short fragments with a fragmentation buffer and reversely transcribed into cDNA using the NEB Next Ultra RNA Library Prep Kit for Illumina (Cat. No. #7530, New England Biolabs, Ipswich, MA, USA). The cDNA libraries were sequenced on the Illumina sequencing platform by Sangon Biotech (Shanghai) Co., Ltd. (Shanghai, China). In order to obtain high-quality clean reads, we filtered it using fastp (version 0.18.0) [60]. Bowtie2 (version 2.2.8) was used to remove ribosome RNA (rRNA) reads by aligning to the zebrafish rRNA database. Then, HISAT2 [61] was used to make paired-end clean reads' map to the reference zebrafish genome (Ensembl_release109) [15]. RSEM was used to calculate the expression abundance, and it was normalized to Transcripts Per Kilobase of exon model per Million mapped reads (TPMs). A bioinformatic analysis was performed using RStudio version cd7011dc. The differentially expressed genes (DEGs) were identified using the DESeq2 package version 1.42.0 [62]. The Gene Ontology (GO) and Kyoto Encyclopedia of Genes and Genomes (KEGG) enrichment analyses of DEGs were conducted using the clusterProfiler package version 1.10.1 [63–65]. The enrichment significance was determined by Fisher's exact test, and the false discovery rate (FDR) was corrected. The heatmap of DEGs was generated using the pheatmap package version 1.0.12.

Supplementary Materials: The following supporting information can be downloaded at: https://www.mdpi.com/article/10.3390/ijms252010950/s1.

Author Contributions: Conception and design: K.L., D.F., Z.Z. and B.H.; Acquisition of data: K.L., D.F. and A.H.; Data analysis: K.L. and D.F.; Manuscript draft and revision: K.L., D.F., A.H., Z.S., J.Z., X.T. and B.H. All authors have read and agreed to the published version of the manuscript.

Funding: This work was funded by the National Natural Science Foundation of China, No. 82071357, the Center for Advanced Interdisciplinary Science and Biomedicine of IHM (QYZD20220002) and the Ministry of Science and Technology of China, No. 2019YFA0405600 (to B.H.).

Institutional Review Board Statement: All protocols were approved by the University of Science and Technology Animal Experimentation Ethics Committee (license number: USTCACUC1103013).

Informed Consent Statement: Not applicable.

Data Availability Statement: All data generated or analyzed during this study are included in this published article and its Supplementary Information files.

Acknowledgments: Thanks to the Core Facility Center for Life Sciences, University of Science and Technology of China. Thanks to Shin-ichi Higashijima (RIkagaku KENkyusho/Institute of Physical and Chemical Research, RIKEN), who provided the Tg (Tol-056) zebrafish strain with which M-cells were labeled by enhanced green fluorescent protein (EGFP). Thanks to the Revealer (Hefei, Anhui, China) for the help with the high-speed camera.

Conflicts of Interest: The authors declare that they have no conflicts of interest.

References

1. Middleton, J.; Tran, Y.; Craig, A. Relationship Between Quality of Life and Self-Efficacy in Persons With Spinal Cord Injuries. *Arch. Phys. Med. Rehabil.* **2007**, *88*, 1643–1648. [CrossRef] [PubMed]
2. Functional Role of a Specialized Class of Spinal Commissural Inhibitory Neurons during Fast Escapes in Zebrafish | Journal of Neuroscience. Available online: https://www.jneurosci.org/content/29/21/6780.short (accessed on 25 June 2024).
3. Fawcett, J.W. The Struggle to Make CNS Axons Regenerate: Why Has It Been so Difficult? *Neurochem. Res.* **2020**, *45*, 144–158. [CrossRef] [PubMed]
4. Hilton, B.J.; Husch, A.; Schaffran, B.; Lin, T.; Burnside, E.R.; Dupraz, S.; Schelski, M.; Kim, J.; Müller, J.A.; Schoch, S.; et al. An Active Vesicle Priming Machinery Suppresses Axon Regeneration upon Adult CNS Injury. *Neuron* **2022**, *110*, 51–69.e7. [CrossRef]
5. Curcio, M.; Bradke, F. Axon Regeneration in the Central Nervous System: Facing the Challenges from the Inside. *Annu. Rev. Cell Dev. Biol.* **2018**, *34*, 495–521. [CrossRef] [PubMed]

6. Tedeschi, A.; Bradke, F. Spatial and Temporal Arrangement of Neuronal Intrinsic and Extrinsic Mechanisms Controlling Axon Regeneration. *Curr. Opin. Neurobiol.* **2017**, *42*, 118–127. [CrossRef]
7. Chondroitinase ABC Promotes Functional Recovery after Spinal Cord Injury | Nature. Available online: https://www.nature.com/articles/416636a (accessed on 10 July 2024).
8. Silver, J.; Miller, J.H. Regeneration beyond the Glial Scar. *Nat. Rev. Neurosci.* **2004**, *5*, 146–156. [CrossRef]
9. Liu, K.; Tedeschi, A.; Park, K.K.; He, Z. Neuronal Intrinsic Mechanisms of Axon Regeneration. *Annu. Rev. Neurosci.* **2011**, *34*, 131–152. [CrossRef]
10. Wu, S.; Xu, J.; Dai, Y.; Yu, B.; Zhu, J.; Mao, S. Insight into Protein Synthesis in Axon Regeneration. *Exp. Neurol.* **2023**, *367*, 114454. [CrossRef]
11. Wang, Y.; Gao, B.; Chen, X.; Shi, X.; Li, S.; Zhang, Q.; Zhang, C.; Piao, F. Improvement of Diabetes-Induced Spinal Cord Axon Injury with Taurine via Nerve Growth Factor-Dependent Akt/mTOR Pathway. *Amino Acids* **2024**, *56*, 32. [CrossRef]
12. Kaplan, E.; Zubedat, S.; Radzishevsky, I.; Valenta, A.C.; Rechnitz, O.; Sason, H.; Sajrawi, C.; Bodner, O.; Konno, K.; Esaki, K.; et al. ASCT1 (Slc1a4) Transporter Is a Physiologic Regulator of Brain <span Class=. *Proc. Natl. Acad. Sci. USA* **2018**, *115*, 9628–9633. [CrossRef]
13. Barut, B.A.; Zon, L.I. Realizing the Potential of Zebrafish as a Model for Human Disease. *Physiol. Genom.* **2000**, *2*, 49–51. [CrossRef] [PubMed]
14. Zebrafish: An Emerging Model System for Human Disease and Drug Discovery—Kari—2007—Clinical Pharmacology & Therapeutics—Wiley Online Library. Available online: https://ascpt.onlinelibrary.wiley.com/doi/full/10.1038/sj.clpt.6100223 (accessed on 10 July 2024).
15. The Zebrafish Reference Genome Sequence and Its Relationship to the Human Genome | Nature. Available online: https://www.nature.com/articles/nature12111 (accessed on 25 June 2024).
16. Cells | Free Full-Text | Know How to Regrow—Axon Regeneration in the Zebrafish Spinal Cord. Available online: https://www.mdpi.com/2073-4409/10/6/1404 (accessed on 10 July 2024).
17. Gonzalez, D.; Allende, M.L. Current Advances in Comprehending Dynamics of Regenerating Axons and Axon–Glia Interactions after Peripheral Nerve Injury in Zebrafish. *Int. J. Mol. Sci.* **2021**, *22*, 2484. [CrossRef] [PubMed]
18. Li, J.-H.; Shi, Z.-J.; Li, Y.; Pan, B.; Yuan, S.-Y.; Shi, L.-L.; Hao, Y.; Cao, F.-J.; Feng, S.-Q. Bioinformatic Identification of Key Candidate Genes and Pathways in Axon Regeneration after Spinal Cord Injury in Zebrafish. *Neural Regen. Res.* **2020**, *15*, 103. [CrossRef] [PubMed]
19. Takahashi, M.; Narushima, M.; Oda, Y. In Vivo Imaging of Functional Inhibitory Networks on the Mauthner Cell of Larval Zebrafish. *J. Neurosci.* **2002**, *22*, 3929–3938. [CrossRef]
20. Kimmel, C.B.; Sessions, S.K.; Kimmel, R.J. Morphogenesis and Synaptogenesis of the Zebrafish Mauthner Neuron. *J. Comp. Neurol.* **1981**, *198*, 101–120. [CrossRef]
21. Wang, Z.; Wang, X.; Shi, L.; Cai, Y.; Hu, B. Wolfram Syndrome 1b Mutation Suppresses Mauthner-Cell Axon Regeneration via ER Stress Signal Pathway. *Acta Neuropathol. Commun.* **2022**, *10*, 184. [CrossRef] [PubMed]
22. Becker, T.; Becker, C.G. Axonal Regeneration in Zebrafish. *Curr. Opin. Neurobiol.* **2014**, *27*, 186–191. [CrossRef]
23. Kim, D.; Lee, Y.-R.; Choi, T.-I.; Kim, S.-H.; Kang, H.-C.; Kim, C.-H.; Lee, S. Comparative Proteome Research in a Zebrafish Model for Vanishing White Matter Disease. *Int. J. Mol. Sci.* **2021**, *22*, 2707. [CrossRef]
24. Hwang, W.Y.; Fu, Y.; Reyon, D.; Maeder, M.L.; Tsai, S.Q.; Sander, J.D.; Peterson, R.T.; Yeh, J.-R.J.; Joung, J.K. Efficient Genome Editing in Zebrafish Using a CRISPR-Cas System. *Nat. Biotechnol.* **2013**, *31*, 227–229. [CrossRef]
25. Cui, N.; Faure, G.; Singh, A.; Macrae, R.; Zhang, F. Microfluidic Enrichment and Computational Analysis of Rare Sequences from Mixed Genomic Samples for Metagenomic Mining. *CRISPR J.* **2022**, *5*, 677–684. [CrossRef]
26. Scheldeman, C.; Mills, J.D.; Siekierska, A.; Serra, I.; Copmans, D.; Iyer, A.M.; Whalley, B.J.; Maes, J.; Jansen, A.C.; Lagae, L.; et al. mTOR-Related Neuropathology in Mutant *Tsc2* Zebrafish: Phenotypic, Transcriptomic and Pharmacological Analysis. *Neurobiol. Dis.* **2017**, *108*, 225–237. [CrossRef] [PubMed]
27. Banu, S.; Gaur, N.; Nair, S.; Ravikrishnan, T.; Khan, S.; Mani, S.; Bharathi, S.; Mandal, K.; Kuram, N.A.; Vuppaladadium, S.; et al. Understanding the Complexity of Epimorphic Regeneration in Zebrafish: A Transcriptomic and Proteomic Approach. *Genomics* **2021**, *114*, 110300. [CrossRef] [PubMed]
28. How Stimulus Direction Determines the Trajectory of the Mauthner-Inthated Escape Response in a Teleost Fish | Journal of Experimental Biology | The Company of Biologists. Available online: https://journals.biologists.com/jeb/article/161/1/469/6421/How-Stimulus-Direction-Determines-the-Trajectory (accessed on 26 June 2024).
29. Liu, Y.-C.; Bailey, I.; Hale, M.E. Alternative Startle Motor Patterns and Behaviors in the Larval Zebrafish (Danio Rerio). *J. Comp. Physiol. A* **2012**, *198*, 11–24. [CrossRef] [PubMed]
30. Roberts, A.C.; Reichl, J.; Song, M.Y.; Dearinger, A.D.; Moridzadeh, N.; Lu, E.D.; Pearce, K.; Esdin, J.; Glanzman, D.L. Habituation of the C-Start Response in Larval Zebrafish Exhibits Several Distinct Phases and Sensitivity to NMDA Receptor Blockade. *PLoS ONE* **2011**, *6*, e29132. [CrossRef] [PubMed]
31. Herrero-Turrión, M.J.; Rodríguez-Martín, I.; López-Bellido, R.; Rodríguez, R.E. Whole-Genome Expression Profile in Zebrafish Embryos after Chronic Exposure to Morphine: Identification of New Genes Associated with Neuronal Function and Mu Opioid Receptor Expression. *BMC Genomics* **2014**, *15*, 874. [CrossRef]

32. Yeo, M.-K.; Pak, S.-W. Exposing Zebrafish to Silver Nanoparticles during Caudal Fin Regeneration Disrupts Caudal Fin Growth and P53 Signaling. *Mol. Cell. Toxicol.* **2008**, *4*, 311–317.
33. Hu, Z.; Holzschuh, J.; Driever, W. Loss of DDB1 Leads to Transcriptional P53 Pathway Activation in Proliferating Cells, Cell Cycle Deregulation, and Apoptosis in Zebrafish Embryos. *PLoS ONE* **2015**, *10*, e0134299. [CrossRef]
34. He, Z.; Jin, Y. Intrinsic Control of Axon Regeneration. *Neuron* **2016**, *90*, 437–451. [CrossRef]
35. Sun, F.; He, Z. Neuronal Intrinsic Barriers for Axon Regeneration in the Adult CNS. *Curr. Opin. Neurobiol.* **2010**, *20*, 510–518. [CrossRef]
36. Mahar, M.; Cavalli, V. Intrinsic Mechanisms of Neuronal Axon Regeneration. *Nat. Rev. Neurosci.* **2018**, *19*, 323–337. [CrossRef]
37. PTEN Deletion Enhances the Regenerative Ability of Adult Corticospinal Neurons—PubMed. Available online: https://pubmed.ncbi.nlm.nih.gov/20694004/ (accessed on 2 October 2024).
38. Smith, T.P.; Sahoo, P.K.; Kar, A.N.; Twiss, J.L. Intra-Axonal Mechanisms Driving Axon Regeneration. *Brain Res.* **2020**, *1740*, 146864. [CrossRef] [PubMed]
39. Kiyoshi, C.; Tedeschi, A. Axon Growth and Synaptic Function: A Balancing Act for Axonal Regeneration and Neuronal Circuit Formation in CNS Trauma and Disease. *Dev. Neurobiol.* **2020**, *80*, 277–301. [CrossRef] [PubMed]
40. Renthal, W.; Tochitsky, I.; Yang, L.; Cheng, Y.-C.; Li, E.; Kawaguchi, R.; Geschwind, D.H.; Woolf, C.J. Transcriptional Reprogramming of Distinct Peripheral Sensory Neuron Subtypes after Axonal Injury. *Neuron* **2020**, *108*, 128–144.e9. [CrossRef] [PubMed]
41. Neural Regeneration Research. Available online: https://journals.lww.com/nrronline/fulltext/2021/16010/Neuroregeneration_and_functional_recovery_after.13.aspx/1000 (accessed on 12 July 2024).
42. Mutations in SLC1A4, Encoding the Brain Serine Transporter, Are Associated with Developmental Delay, Microcephaly and Hypomyelination | Journal of Medical Genetics. Available online: https://jmg.bmj.com/content/52/8/541.short (accessed on 13 July 2024).
43. Pujol-Giménez, J.; Mirzaa, G.; Blue, E.E.; Albano, G.; Miller, D.E.; Allworth, A.; Bennett, J.T.; Byers, P.H.; Chanprasert, S.; Chen, J.; et al. Dominant-Negative Variant in SLC1A4 Causes an Autosomal Dominant Epilepsy Syndrome. *Ann. Clin. Transl. Neurol.* **2023**, *10*, 1046–1053. [CrossRef] [PubMed]
44. Odeh, M.; Sajrawi, C.; Majcher, A.; Zubedat, S.; Shaulov, L.; Radzishevsky, A.; Mizrahi, L.; Chung, W.K.; Avital, A.; Hornemann, T.; et al. A New Type of Blood-Brain Barrier Aminoacidopathy Underlies Metabolic Microcephaly Associated with SLC1A4 Mutations. *Brain J. Neurol.* **2024**, awae134. [CrossRef] [PubMed]
45. Watson, D.C.; Bayik, D.; Storevik, S.; Moreino, S.S.; Sprowls, S.A.; Han, J.; Augustsson, M.T.; Lauko, A.; Sravya, P.; Røsland, G.V.; et al. GAP43-Dependent Mitochondria Transfer from Astrocytes Enhances Glioblastoma Tumorigenicity. *Nat. Cancer* **2023**, *4*, 648–664. [CrossRef]
46. Zuo, Z.; Fan, B.; Zhang, Z.; Liang, Y.; Chi, J.; Li, G. Interleukin-4 Protects Retinal Ganglion Cells and Promotes Axon Regeneration. *Cell Commun. Signal. CCS* **2024**, *22*, 236. [CrossRef]
47. Hu, M.; Veldman, M.B. Intraocular Axon Regeneration in a Model of Penetrating Eye Injury. *J. Ocul. Pharmacol. Ther. Off. J. Assoc. Ocul. Pharmacol. Ther.* **2023**, *39*, 563–571. [CrossRef]
48. Williams, R.R.; Venkatesh, I.; Pearse, D.D.; Udvadia, A.J.; Bunge, M.B. MASH1/Ascl1a Leads to GAP43 Expression and Axon Regeneration in the Adult CNS. *PLoS ONE* **2015**, *10*, e0118918. [CrossRef]
49. Skene, J.H.P. Axonal Growth-Associated Proteins. *Annu. Rev. Neurosci.* **1989**, *12*, 127–156. [CrossRef]
50. Curtis, R.; Green, D.; Lindsay, R.M.; Wilkin, G.P. Up-Regulation of GAP-43 and Growth of Axons in Rat Spinal Cord after Compression Injury. *J. Neurocytol.* **1993**, *22*, 51–64. [CrossRef] [PubMed]
51. Schaden, H.; Stuermer, C.A.O.; Bähr, M. Gap-43 Immunoreactivity and Axon Regeneration in Retinal Ganglion Cells of the Rat. *J. Neurobiol.* **1994**, *25*, 1570–1578. [CrossRef]
52. Zhang, Y.; Bo, X.; Schoepfer, R.; Holtmaat, A.J.D.G.; Verhaagen, J.; Emson, P.C.; Lieberman, A.R.; Anderson, P.N. Growth-Associated Protein GAP-43 and L1 Act Synergistically to Promote Regenerative Growth of Purkinje Cell Axons in Vivo. *Proc. Natl. Acad. Sci. USA* **2005**, *102*, 14883–14888. [CrossRef] [PubMed]
53. Kusik, B.W.; Hammond, D.R.; Udvadia, A.J. Transcriptional Regulatory Regions of *Gap43* Needed in Developing and Regenerating Retinal Ganglion Cells. *Dev. Dyn.* **2010**, *239*, 482–495. [CrossRef]
54. Carriel, V.; Garzón, I.; Campos, A.; Cornelissen, M.; Alaminos, M. Differential Expression of GAP-43 and Neurofilament during Peripheral Nerve Regeneration through Bio-Artificial Conduits: GAP-43 and Neurofilament during Nerve Regeneration. *J. Tissue Eng. Regen. Med.* **2017**, *11*, 553–563. [CrossRef]
55. Kim, D.-C.; Kim, S.; Hwang, K.-S.; Kim, C.-H. P-Coumaric Acid Potently Down-Regulates Zebrafish Embryo Pigmentation: Comparison of in Vivo Assay and Computational Molecular Modeling with Phenylthiourea. *Biomed. Sci. Lett.* **2017**, *23*, 8–16. [CrossRef]
56. Li, Z.; Ptak, D.; Zhang, L.; Walls, E.K.; Zhong, W.; Leung, Y.F. Phenylthiourea Specifically Reduces Zebrafish Eye Size. *PLoS ONE* **2012**, *7*, e40132. [CrossRef] [PubMed]
57. Huang, W.-C.; Hsieh, Y.-S.; Chen, I.-H.; Wang, C.-H.; Chang, H.-W.; Yang, C.-C.; Ku, T.-H.; Yeh, S.-R.; Chuang, Y.-J. Combined Use of MS-222 (Tricaine) and Isoflurane Extends Anesthesia Time and Minimizes Cardiac Rhythm Side Effects in Adult Zebrafish. Available online: https://www.liebertpub.com/doi/10.1089/zeb.2010.0653 (accessed on 26 June 2024).

58. Livak, K.J.; Schmittgen, T.D. Analysis of Relative Gene Expression Data Using Real-Time Quantitative PCR and the $2^{-\Delta\Delta CT}$ Method. *Methods* **2001**, *25*, 402–408. [CrossRef]
59. Ultrafast One-pass FASTQ Data Preprocessing, Quality Control, and Deduplication Using Fastp—Chen—2023—iMeta—Wiley Online Library. Available online: https://onlinelibrary.wiley.com/doi/10.1002/imt2.107 (accessed on 25 June 2024).
60. Rapid and Accurate Alignment of Nucleotide Conversion Sequencing Reads with HISAT-3N. Available online: https://genome.cshlp.org/content/31/7/1290 (accessed on 25 June 2024).
61. Graph-Based Genome Alignment and Genotyping with HISAT2 and HISAT-Genotype | Nature Biotechnology. Available online: https://www.nature.com/articles/s41587-019-0201-4 (accessed on 25 June 2024).
62. Love, M.I.; Huber, W.; Anders, S. Moderated Estimation of Fold Change and Dispersion for RNA-Seq Data with DESeq2. *Genome Biol.* **2014**, *15*, 550. [CrossRef]
63. Yu, G.; Wang, L.-G.; Han, Y.; He, Q.-Y. clusterProfiler: An R Package for Comparing Biological Themes Among Gene Clusters. *OMICS J. Integr. Biol.* **2012**, *16*, 284–287. [CrossRef]
64. clusterProfiler 4.0: A Universal Enrichment Tool for Interpreting Omics Data—ScienceDirect. Available online: https://www.sciencedirect.com/science/article/pii/S2666675821000667?via=ihub (accessed on 26 June 2024).
65. Kanehisa, M.; Goto, S. KEGG: Kyoto Encyclopedia of Genes and Genomes. *Nucleic Acids Res.* **2000**, *28*, 27–30. [CrossRef] [PubMed]

International Journal of
Molecular Sciences

MDPI

Article

Short- and Long-Term Neurobehavioral Effects of Developmental Exposure to Valproic Acid in Zebrafish

Marina Ricarte [1,2], Niki Tagkalidou [1], Marina Bellot [2], Juliette Bedrossiantz [1], Eva Prats [3], Cristian Gomez-Canela [2], Natalia Garcia-Reyero [4] and Demetrio Raldúa [1,*]

1 Institute for Environmental Assessment and Water Research (IDAEA-CSIC), 08034 Barcelona, Spain; marina.ricarte@idaea.csic.es (M.R.); niki.tagkalidou@idaea.csic.es (N.T.); juliette.bedrossiantz@idaea.csic.es (J.B.)
2 Department of Analytical and Applied Chemistry, School of Engineering, Institut Químic de Sarrià, Universitat Ramon Llull, 08017 Barcelona, Spain; marina.bellot@iqs.url.edu (M.B.); cristian.gomez@iqs.url.edu (C.G.-C.)
3 Research and Development Center (CID-CSIC), 08034 Barcelona, Spain; eva.prats@cid.csic.es
4 Institute for Genomics, Biocomputing & Biotechnology (IGBB), Mississippi State University, Starkville, MS 39762, USA; natalia@icnanotox.org
* Correspondence: demetrio.raldua@idaea.csic.es

Abstract: Autism spectrum disorder (ASD) is a neurodevelopmental disorder characterized by impairments in social interaction and communication, anxiety, hyperactivity, and interest restricted to specific subjects. In addition to the genetic factors, multiple environmental factors have been related to the development of ASD. Animal models can serve as crucial tools for understanding the complexity of ASD. In this study, a chemical model of ASD has been developed in zebrafish by exposing embryos to valproic acid (VPA) from 4 to 48 h post-fertilization, rearing them to the adult stage in fish water. For the first time, an integrative approach combining behavioral analysis and neurotransmitters profile has been used for determining the effects of early-life exposure to VPA both in the larval and adult stages. Larvae from VPA-treated embryos showed hyperactivity and decreased visual and vibrational escape responses, as well as an altered neurotransmitters profile, with increased glutamate and decreased acetylcholine and norepinephrine levels. Adults from VPA-treated embryos exhibited impaired social behavior characterized by larger shoal sizes and a decreased interest for their conspecifics. A neurotransmitter analysis revealed a significant decrease in dopamine and GABA levels in the brain. These results support the potential predictive validity of this model for ASD research.

Keywords: valproic acid; animal model; zebrafish; autism spectrum disorder; social behavior; neurotransmitter

1. Introduction

Autism spectrum disorder (ASD) is a neurodevelopmental disorder characterized by impairments in social interaction and communication, anxiety, hyperactivity, and interest restricted to specific subjects [1–3]. It is believed that the emergence of autism involves a combination of anatomical brain abnormalities, genetic anomalies, and neurochemical imbalances [2,4,5]. Epidemiological studies emphasize the significance of genetic factors, revealing that autistic disorder ranks among the most genetically influenced neuropsychiatric conditions [6]. While candidate genes implicated in autism primarily involve proteins that regulate neuronal network patterning and the balance between excitatory and inhibitory signaling, ASD is highly polygenic, with no specific gene or locus associated with a large group of patients [6–8]. Approximately 20% of patients exhibit genetic mutations affecting known major genes, contributing to the complexity of clinical manifestations and etiology, without a specific biological hallmark identified for this disorder [2]. In addition to the genetic factors, multiple environmental factors have been related to the development of ASD [1,3,6,7].

Animal models in ASD research can serve as crucial tools for understanding the complexity of the disorder, as they allow testing specific hypotheses of its causes and identifying potential therapeutic strategies [9,10]. However, challenges persist in creating models that faithfully represent the complexity of ASD due to its unclear etiology and heterogeneous symptomatology [9]. Replicating ASD symptoms through animal models is crucial for the identification of the neurobiological basis associated with it [2].

Molecular genetic models allow for the simulation of core features of human autism, including the disruption of glutamatergic synaptic transmission, loss of inhibitory GABAergic interneurons, and impairments in synaptic plasticity [11]. However, while mammalian models expressing autism-linked genetic mutations have been developed, their utility is constrained by the associated costs. Thus, simpler models, such as *Danio rerio*, are proposed for initial mechanistic and screening studies [7]. Zebrafish, a widely used model in biomedical research, emerges as an ideal candidate for behavioral screening in ASD research, as its predominant social behavior along with other advantages make it a valuable tool for investigating the neurobiological basis of the disorder [12–14].

Environmental factors, such as drug and toxin exposure; viral infections; and immune dysfunctions during pre- and postnatal stages can influence ASD risks [6–8]. For instance, prenatal exposure to valproic acid (VPA) has been associated with an increased incidence of autism [7,15]. Because of the association observed in humans between maternal treatment with VPA and ASD, the use of prenatal VPA exposure has been used to build chemical models in different animal models, including zebrafish, to study ASD [9,16,17]. VPA dysregulates key transcription factors; signal transduction pathways; inositol metabolism; and direct modulation of epigenetic regulators, specifically histone deacetylases (HDACs), resulting in more compact DNA-histone packaging and reduced gene transcription [8,9]. Moreover, VPA exposure has been reported to lead to changes in gamma-aminobutyric acid (GABA) levels in the central nervous system [18–20].

In this study, an integrative approach combining behavioral analysis and neurotransmitters profile has been used, for the first time, to determine the effects of early-life exposure to VPA both in the larval and adult stages. With this aim, 4 h post fertilization (hpf) zebrafish embryos have been exposed to 48 mM VPA for 48 h, and then, transferred to clean fish water. The effect of early-life exposure to VPA on basal locomotor activity and the escape response evoked by visual and vibrational stimuli were determined at 8 days post-fertilization. At the adult stage (3 months), social behavior (shoaling and social preference tests) was analyzed, and metabolomic changes in the brain were determined.

2. Results

2.1. Systemic and Developmental Effects in Zebrafish Larvae Exposed to VPA during Early Development

Embryos exposed to VPA exhibited a significant delay in hatching time compared to controls. At 48 hpf, around 40% of the control embryos, but none of those exposed to VPA, had hatched (Supplementary Figure S1A; $z = -8.33$, $p < 0.0001$). At 72 hpf, 95% of the control and 80% of the VPA-exposed embryos had hatched (Supplementary Figure S1B; $z = -5.75$, $p < 0.0001$).

When the toxicity of the VPA treatment was determined in 7 dpf larvae, the VPA-exposed group exhibited a similar cumulative mortality to the control group (Supplementary Figure S1C). However, VPA exposure led to a mild but significant increase in phenotypic abnormalities, such as pericardial and yolk sac edemas (Supplementary Figure S1D; 0.83% and 4.65%, median values for control and VPA, respectively; $z = -3.52$, $p < 0.0001$).

VPA exposure during early development (4–48 hpf) also led to a significant increase in heart rate when larvae reached 8 dpf, compared to the corresponding controls (Supplementary Figure S2; $t\ (31) = 3.504$, $p = 0.0014$).

2.2. Neurobehavioral Effects in Zebrafish Larvae Exposed to VPA during Early Development

For neurobehavioral assessment, 8 dpf zebrafish larvae without any evidence of systemic toxicity were selected. Moreover, no differences in the standard length were found between the control and the VPA-treated groups [t (18) = 1.852, p = 0.08]. As shown in Figure 1, the exposure of embryos to VPA from 4 to 48 hpf led to significant changes in all three analyzed behaviors. While BLA increased in the treated larvae ($z = -2.92$, $p = 0.0034$), the escape response evoked by a visual and a vibrational stimulus decreased ($z = -2.35$, $p = 0.0185$ for VMR; $z = -3.86$, $p < 0.0001$ for VSR).

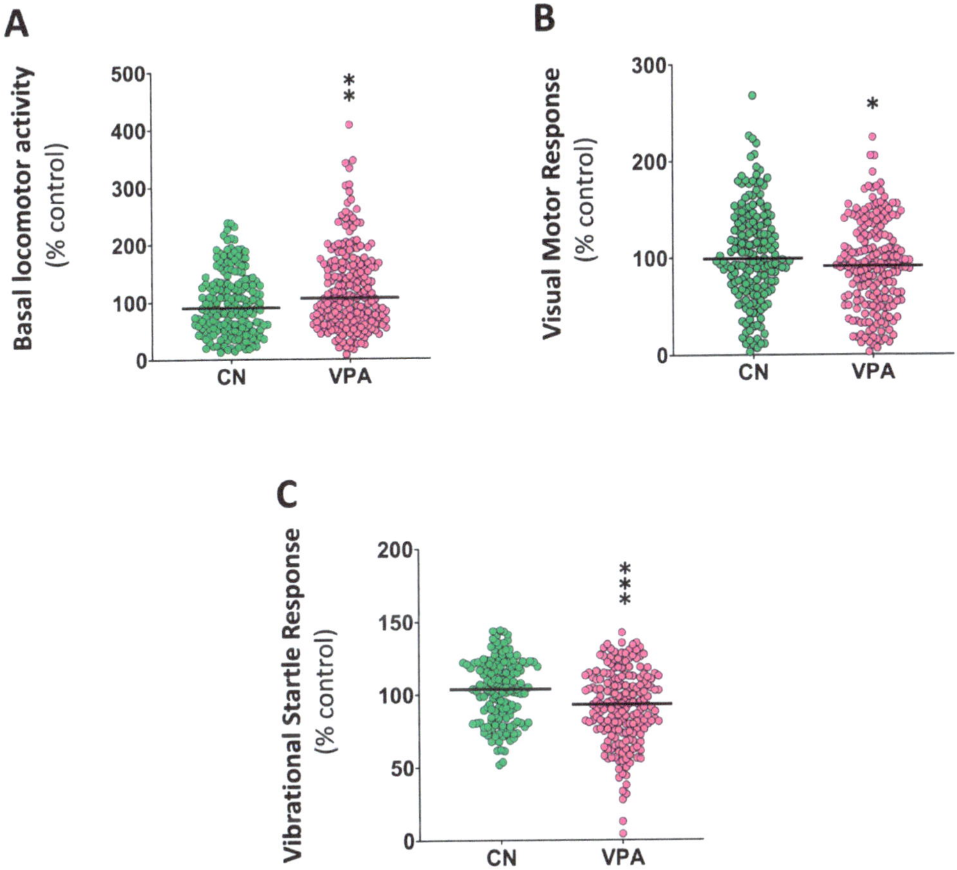

Figure 1. Effects of embryonic exposure to VPA on the behavior of zebrafish larvae. (**A**) Percentage of basal locomotor activity relative to the control; (**B**) percentage of light-off visual motor response relative to the control; (**C**) percentage of vibrational startle response relative to the control. Data reported as a scatter plot with the median (n = 160–209). * $p < 0.05$, ** $p < 0.01$, *** $p < 0.001$; Mann–Whitney U test. Data from 3 independent experiments.

The following step was used to determine if the observed changes in the different behaviors might be related to changes in the neurotransmitter profile of larvae exposed to VPA during early development. As shown in Figure 2 and Supplementary Table S1, larval heads exposed to VPA during early development presented a significant reduction in the levels of acetylcholine ($U(N_{control} = 7, N_{VPA} = 7) = 6.00$, $p = 0.017$) and norepinephrine ($U(N_{control} = 7, N_{VPA} = 7) = 4.00$, $p = 0.007$), whereas the levels of glutamate increased in this group ($U(N_{control} = 7, N_{VPA} = 7) = 48.00$, $p = 0.001$). VPA-exposed larvae also showed a significant decrease in epinephrine levels (U($N_{control}$ = 7, N_{VPA} = 7) = 2.00, p = 0.002; Supplementary Table S1).

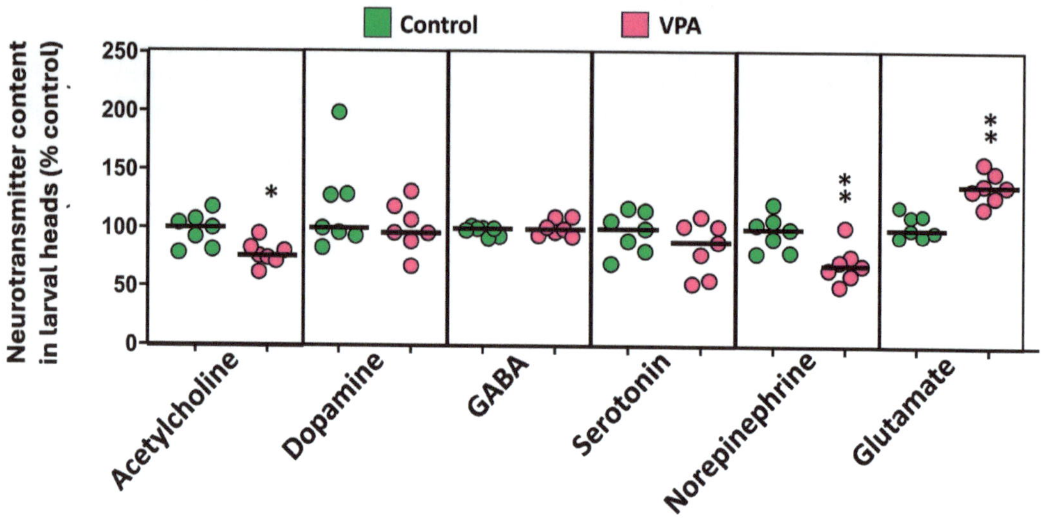

Figure 2. Effects of embryonic exposure to VPA on the neurotransmitter profile of zebrafish larvae. The neurotransmitters analyzed in the larval heads include catecholamines (dopamine, norepinephrine, and epinephrine), serotonin, acetylcholine, GABA, aspartic acid, and glutamate. Data reported as a scatter plot with the median (n = 7) * $p < 0.05$, ** $p < 0.01$; Mann–Whitney U test. Data from two independent experiments.

2.3. *Neurobehavioral Effects in Adult Zebrafish Exposed to VPA during Early Development*

The adult zebrafish selected for behavioral analysis had a similar body length (19.97 ± 0.28 mm for the control and 18.95 ± 0.34 for the VPA-treated group), and no differences in the total distance moved in 6 min were found between the VPA-exposed and control fish (Supplementary Figure S2).

The effects of early-life exposure to VPA on adult social behavior were explored using two experimental paradigms, the shoaling test and the social preference test. The shoaling test (Figure 3A) showed significantly higher average interfish distance (U($N_{control}$ = 18, N_{VPA} = 16) = 261.00, $p = 1.28 \times 10^{-5}$) and farthest interfish distance (t (32) = −8.667, $p = 6.65 \times 10^{-10}$), a behavioral phenotype consistent with social isolation, in the adults exposed to VPA during early development. The social preference test (Figure 3B) showed a significant decrease in both the time spent and distance moved by treated fish in the zone closest to conspecifics (time: t (31) = 2.350, p = 0.025; distance: t (31) = 2.348, p = 0.025), and a concomitant increase in the time spent and the distance moved by these fish in the empty virtual zone (time: t (31) = −2.112, p = 0.043; distance: t (31) = −2.125, p = 0.042). These results are consistent with the social isolation phenotype suggested by the shoaling test results.

As shown in Figure 4, when the levels of different neurotransmitters in the brain were analyzed, a significant decrease in the levels of GABA (U($N_{control}$ = 6, N_{VPA} = 6) = 3.00, p = 0.015) and dopamine (U($N_{control}$ = 6, N_{VPA} = 6) = 5.00, p = 0.041) was found in VPA-exposed fish compared to controls. A trend to a reduction in glutamate levels was also found in the brain of the exposed fish (U($N_{control}$ = 6, N_{VPA} = 6) = 6.00, p = 0.065).

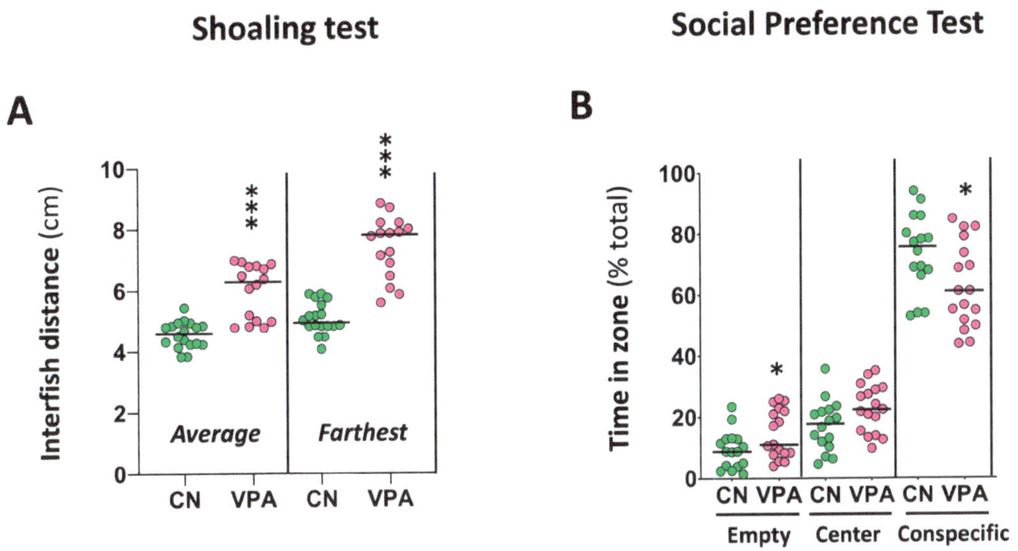

Figure 3. Effect of early-life VPA exposure on social behavior. (**A**) VPA exposure during the early embryonic development increases the average interfish distance and the farthest interfish distance in the shoaling test. (**B**) VPA exposure during early embryonic development decreases the time that fish spend in the zone closest to conspecifics, increasing, however, the time spent in the empty zone in the social preference test (SPT). In the SPT, the time spent in each virtual zone for each fish was normalized to the total time of the assay. Data reported as a scatter plot with the median (n = 16–18) * $p < 0.05$, *** $p < 0.001$; t-test or Mann–Whitney U test. Data from two independent experiments.

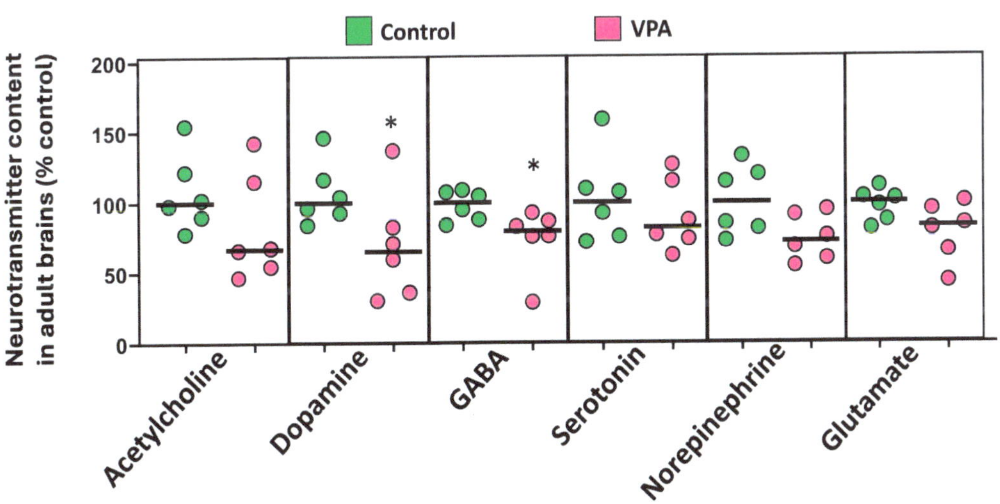

Figure 4. Effects of embryonic exposure to VPA on a neurotransmitter profile in the brain of an adult zebrafish. The neurotransmitters analyzed include catecholamines (dopamine and norepinephrine), serotonin, acetylcholine, GABA, and glutamate. Data reported as a scatter plot with the median (n = 6) * $p < 0.05$; Mann–Whitney U test. Data from two independent experiments.

3. Discussion

VPA is a medication widely used to treat epilepsy, bipolar disorder, and migraines [21]. Despite its therapeutic effect, exposure to VPA during the first three months of pregnancy can result in a rare condition known as fetal valproate syndrome (FVS) [21]. Children with FVS also have an increased risk of ASD [22,23]. In fact, VPA is commonly used to build chemical models of ASD in different animal species, including zebrafish [12,24]. In this study, an integrative approach combining behavioral analysis and neurotransmitter profiles has been used, for the first time, to determine the effects of early-life exposure to VPA both in the larval and adult stages. First of all, we have found that the exposure of 4 hpf zebrafish embryos to 48 μM VPA for 48 h did not result in mortality when this endpoint was evaluated at 7 dpf. Consistently, no mortality in zebrafish embryos has been reported by other authors using the same [12] or similar exposure conditions [25] to VPA. However, the effect of VPA exposure on zebrafish embryo mortality is, anyway, controversial. Whereas in one report, embryos exposed to 40 μM VPA from 8 to 120 hpf showed a 96% mortality at 7 dpf [26], in another report, the exposure of 8 hpf zebrafish embryos for 100 h to 1500 μM VPA did not result in mortality [16].

In our study, we have found that exposure of zebrafish embryos during the first 48 h of development to VPA led to a mild (4.65%) but significant increase in phenotypic abnormalities [21]. In contrast, no phenotypic effects have been reported in exposed embryos by other authors using similar exposure conditions [12,25]. While morphological alterations were also not reported after exposure of embryos to 75 μM VPA from 4 to 120 hpf [17] or to 50 μM VPA from 8 to 108 hpf [16], exposure to 30 μM VPA from 8 to 120 hpf showed 30.5% malformations.

The observed variability in mortality and malformation rate among these studies based on waterborne exposure to VPA could be related to the differences in the final pH of the experimental solutions. VPA is a weak acid with a pK_a of 4.86, and depending on the pH of the experimental solution, the chemical will be ionized (low bioavailability) or non-ionized (high bioavailability). The percentage of non-ionized VPA molecules is higher at low pH (39% of VPA is in its non-ionized form at pH 5.0) than at the physiological pH 6.3 (3.1% non-ionized) [27]. Therefore, VPA uptake should be significantly higher in studies using experimental solutions with low pH than in those using experimental solutions with a final pH above 7. In fact, when we performed the first experiments with VPA, using our standard fish water to prepare the experimental solution of VPA (pH 6.5), we found a huge mortality of exposed embryos. At pH 6.5, about 1.96% of the VPA is expected to be in the form of valproate (non-ionized form) and, therefore, bioavailable. The pH of the fish water was then raised to 7.5 in both fish water and experimental solutions by adding bicarbonate, and at this pH, the non-ionized fraction should be only 0.23%. When this water was used to prepare the experimental solutions, we found no mortality and a very limited percentage of malformations. In addition to the potential differences in VPA bioavailability, the sensitivity of the zebrafish strain and batch could have also played a role, as some malformations were observed on the controls too.

The hyperactivity found in VPA-exposed larvae in this study aligns with the results obtained by most of the previous studies using this chemical [12,16,26,28] and may be related to the dysregulation of dopaminergic and glutamatergic neurotransmission in certain brain regions, as suggested for ADHD (attention-deficit hyperactivity disorder), a disorder related to and co-occurring with ASD [29–32]. However, no effects on BLA have been reported in *shank3b*$^{-/-}$ zebrafish larvae, a result suggesting that hyperactivity is a VPA effect unrelated to ASD [14].

In this study, we have also found a decrease in the light-off VMR in larvae exposed to VPA during early development, a result consistent with that reported by Bailey et al. [33] in 6 dpf larvae exposed to 30–50 μM VPA from 4 to 120 hpf, but not with the significant increase reported by Joseph et al. [26] in 5 and 7 dpf larvae exposed from 8 to 108 hpf to 5–10 μM VPA and by Baronio et al. [34] in 5 dpf larvae exposed to 25 μM VPA from 10 to 24 hpf. A similar decrease in light-off VMR was also reported in *shank3b*$^{-/-}$ zebrafish larvae, suggesting that this effect of VPA could be related to ASD. The decrease in the VMR could be related to the excitatory–inhibitory imbalance in the visual cortex suggested to happen in ASD patients [34]. We also observed a significant decrease in the vibrational-evoked startle response in the larvae exposed to VPA during early embryonic development. Interestingly, Gupta et al. [25], in a seminal article on neuroanatomical mapping of the zebrafish brain, found that 6 dpf zebrafish larvae exposed to 50 μM VPA during early development presented a loss of glutamatergic signal in part of the statoacoustic ganglion, leading to a decreased acoustic startle responsiveness. Glutamate dysregulation is thought to be implicated in the changes in startle response through the NMDA and mGlu5 receptors [34–37].

It is difficult to explain the behavioral changes found in the VPA-exposed larvae based on the observed changes in the neurotransmitter profiles found in their heads. For instance, hyperactivity has been linked to the increase in the dopamine, acetylcholine [38,39], and/or norepinephrine [40,41] levels in the zebrafish larvae. However, the levels of acetylcholine and norepinephrine in the heads of the exposed larvae decreased, and no changes were found in dopamine levels. The observed hyperactivity is also not explained by the increased glutamate levels in the head of VPA-exposed larvae, as glutamate has been reported to reduce the motor activity of zebrafish larvae [42]. In our opinion, the difficulty of correlating behavioral changes with changes in the neurotransmitter profile observed in the larval heads has two components. First, the UHPLC-MS/MS analysis used in this study provides information on the entire set of neurotransmitters present in the larval head, and not only those released in the synaptic spaces, which are the ones directly involved in behavior. This is a clear limitation of the zebrafish model, in which, due to its small size, it is not possible to use microdialysis [43] to selectively determine neurotransmitters in the synaptic clefts of specific CNS nuclei, as for example is conducted in rats. The second limitation of this methodology is that the analysis is performed on the entire head, not discrete nuclei. It is often reported that a chemical increases the levels of a neurotransmitter in some nuclei and decreases them in others. Therefore, even if the total pool of these neurotransmitters does not change in the context of the whole brain, small differences in the neurotransmitter released at specific nuclei of the brain could explain the observed behavioral changes.

The decrease in the social behavior found in this study in adult zebrafish (90 dpf) exposed to VPA during early embryonic development is consistent with previous studies reporting increased size of the shoals [33] and decreased preference for being near conspecifics [12] in VPA-treated zebrafish. Similar disorders of social behavior have been described in genetic models of ASD constructed in zebrafish [13,14]. Interestingly, the decrease found in the dopamine levels in the brain of the VPA-treated fish has also been reported in different mammalian models [44,45]. Since the dopaminergic system is involved in social behavior, the observed decrease in dopamine levels could be behind the impaired social behavior found in ASD [44,46,47]. Moreover, the significant decrease in GABA levels found in the brain of adult fish exposed to VPA during early development is consistent with the decrease in this neurotransmitter reported in ASD patients [48,49].

In summary, the neurobehavioral short- and long-term effects of developmental exposure to valproic acid in zebrafish have been analyzed, with special emphasis on determining the relationship between behavioral and neurochemical changes. Our results suggest the importance of the pH of the experimental solution on the bioavailability of the VPA and, therefore, on the severity of the observed effects. Moreover, our results show that the developed ASD model exhibited an impaired social behavior, characterized by increases in the average inter-fish distance and in the shoaling test and by decreases in the time spent and distance moved in the closest zone to conspecifics and in the social preference test. The impairment of social behavior combined with the significant decrease found in the brain levels of dopamine and GABA in adult zebrafish treated with VPA during early development strongly suggests that this model could be of great interest for both the study of FVS and ASD. However, additional efforts are needed to determine the predictive validity of this model in ASD research.

4. Materials and Methods

4.1. Zebrafish Housing and Husbandry

Adult zebrafish were supplied by Pisciber BSF (Terrassa, Barcelona, Spain) and maintained at the Research and Development Center of the Spanish Research Council (CID-CSIC) facilities in fish water [FW: reverse-osmosis purified water containing 90 μg/mL of Instant Ocean® (Aquarium Systems, Sarrebourg, France), 0.58 mM $CaSO_4 \cdot 2H_2O$, pH 6.5] at 28 ± 1 °C. A 12:12 light–dark photoperiod was used. The fish were fed twice a day with flake food (TetraMin, Tetra, Germany).

For breeding, a series of polycarbonate crossing tanks were prepared with a total of five adult zebrafish in each one, with the ratio of female:male being 3:2. After a treatment with 0.1% methylene blue, the embryos were maintained in crystallizing dishes with fish water inside the incubator (at 28 °C, 12:12 light–dark photoperiod). At 4 hpf, the embryos at the sphere and dome stages [50] were selected and transferred to 6-well cell culture plates (20 embryos/well) with the exposure medium.

All procedures were approved by the Institutional Animal Care and Use Committees at the CID-CSIC and conducted in accordance with the institutional guidelines under a license from the local government (agreement number 11336).

4.2. Experimental Procedure

Sodium valproate (VPA; CAS: 1069-66-5) was purchased from Sigma-Aldrich (St Louis, MO, USA). For treatment, an exposure protocol similar to that described by Zimmermann et al. (2015) was used [12]. Basically, the day of the experiment, a fresh 48 μM VPA solution was prepared in FW, and then, pH was adjusted to 7.5 with a 1 M $NaHCO_3$ solution. At 4 hpf, embryos at sphere-dome stages were exposed to the VPA solution (treated group) or to FW (also adjusted to pH 7.5, control group) for 48 h without medium renewal. Zimmermann et al. (2015) reported that this concentration of VPA is stable in fish water for at least 48 h [12]. At 52 hpf, the embryos were removed from the experimental solutions, washed, and transferred to pH 7.5 FW (28 °C and 12L:12D photoperiod) until 8 dpf, when the behavioral effects of early exposure to VPA were initially tested. Afterwards, the larvae were placed in a 2 L tank with fish water and fed twice a day. After three months, when the zebrafish reached the adult state, social behavior was determined. Finally, the fish were euthanized by inducing a hypothermic shock in ice-chilled water (2–4 °C), and the brains were immediately dissected and individually stored at −80 °C for neurotransmitter and transcriptional analysis.

4.3. Phenotypic Analysis

Live embryos and larvae were examined with a Nikon SMZ 1500 stereomicroscope (Nikon, Champigny sur Marne, France) every 24 h to observe the phenotype, recording the lethality, malformations, and hatching time in both the control and VPA-exposed groups. The standard length of the animals from both experimental groups was determined at 8, 30, 60, and 90 dpf from pictures taken with a GigE camera mounted on the stereomicroscope using the GIMP software (version 2.10.32).

4.4. Neurobehavioral Assay in Larvae

Behavioral assays including basal locomotor activity (BLA), light-off visual motor response (VMR), and vibrational startle response (VSR) assay were conducted and analyzed in a DanioVision platform driven by EthoVision XT 13 software (Noldus, Wageningen, The Netherlands). The behavioral assays in the larvae were conducted as previously described [51,52].

At day 7, larvae with no signs of systemic toxicity were placed in 48-well microplates, with 1 larva per well in 1 mL FW. For the behavioral analysis, the larvae were first acclimated to the new environment inside the observation chamber for 10 min in darkness, after which the plate received one vibrational stimulus by means of a solenoid (tapping stimulus). The distance (cm) moved in response to the tapping routine corresponds to the vibrational startle response (VSR). After delivering the tapping, the larvae were maintained in the dark without any stimuli for 15 min. The last 10 min in the dark were used to calculate their basal locomotor activity (BLA), corresponding to the total distance moved during that time. The routine continued by switching a light on for 10 min and then turning it off for 15 more minutes. The light-off visual motor response (VMR) was calculated as the distance moved in the first two minutes of darkness minus the distance moved in the last two minutes of light. After the behavioral battery, the larvae were euthanized by inducing a hypothermic shock in ice-chilled water (2–4 °C) and then their heads were cut. Eight pools of fifteen heads were prepared for each experimental condition and stored at −80 °C.

4.5. Neurobehavioral Assays in Adults

All tests were conducted in an isolated behavioral room at 27–28 °C. To assess the effects on social behavior, a shoaling and social preference test (SPT) was performed following the protocol described by Bedrossiantz et al. [53]. Ethovision XT 13.0 (Noldus, Wageningen, The Netherlands) was used for the video-tracking analysis. After these tests, the fish were euthanized by inducing hypothermic shock in ice-chilled water (2–4 °C), and the brain of each individual was dissected and transferred into an Eppendorf tube. Eight brains of each concentration were used for the analysis. During all of the processes, the samples were maintained on ice. When all the samples were prepared, they were kept in a freezer at −80 °C until the day of the extraction.

4.6. Neurochemical Analysis by UHPLC-MS/MS

For the neurochemical extraction of larval heads and adult brains and UHPLC-MS/MS, a protocol similar to that described by Ricarte et al. [51] was used. Basically, samples were homogenized by means of a bead mill (TissueLyser LT, Quiagen, Hilden, Germany) and centrifuged. The supernatant was filtered, using 0.22 µm nylon filter, into chromatographic vials that were kept at −20 °C until the analysis. To extract and conduct the analysis of neurotransmitters, acetonitrile (ACN), HPLC-MS grade, was supplied from VWR chemicals Prolabo (Leuven, Belgium); formic acid (FA) from Fisher Scientific (Loughborough, UK); and ammonium formate from Sigma-Aldrich (St. Louis, MO, USA). Ultra-pure water was obtained through the Millipore Milli-Q purification system (Millipore, Bedford, MA, USA).

The neurochemical content of the extract was determined by UHPLC–MS/MS, using conditions described elsewhere [54,55]. A BEH Amide column was used for separation and elution, and the detection was performed in MRM mode with ESI+, ensuring specificity during detection and quantification. For the calibration curve, pure reference standards were used: serotonin hydrochloride (5-HT), dopamine hydrochloride (DA), γ-aminobutyric acid (GABA), epinephrine (Epi), and acetylcholine (ACh) were supplied by Sigma-Aldrich (St. Louis, MO, USA). Glutamic acid (Glu) was supplied by BLD Pharmatech (Shanghai, China), and norepinephrine (NE) was obtained from Tocris Bioscience (Ellisville, MO, USA). A mixture of isotopically labelled standards was used as an internal standard. 5-HIAA-d5, 5-HTP-d4, 5HT-d4, NE-d6, DA-1,1,2,2-d4, and 3-MT-d4 were purchased from Toronto Research Chemicals (TRC, Toronto, ON, Canada).

4.7. Statistical Analysis

The data were analyzed with IBM SPSS v29 (Statistical Package 2010, Chicago, IL, USA) and GraphPad Prism 9 for Windows (GraphPad software Inc., La Jolla, CA, USA) and plotted with GraphPad Prism 9 for Windows (GraphPad software Inc., La Jolla, CA, USA).

In order to determine if the samples followed a normal distribution, a Shapiro–Wilk test was used. For normally distributed groups, an Unpaired t-test was used for determining statistical significance and one-way ANOVA followed by Dunnett's as a multiple comparison test. When parametric assumptions could not be made, statistical significance was determined by a Mann–Whitney U test and a Kruskal–Wallis test followed by Dunn–Bonferroni's test to see if there were any differences between more than two groups. Significance was set at $p < 0.05$.

Supplementary Materials: The following supporting information can be downloaded at https://www.mdpi.com/article/10.3390/ijms25147688/s1.

Author Contributions: Conceptualization: D.R., C.G.-C., and N.G.-R.; Methodology: M.B., J.B., and N.G.-R.; Formal Analysis: D.R., M.R., and M.B.; Investigation: M.R., N.T., and E.P.; Resources, D.R. and C.G.-C.; Writing—Original Draft Preparation: D.R., M.R., N.T., M.B., J.B., E.P., C.G.-C., and N.G.-R.; Writing—Review and Editing: D.R., M.R., N.T., M.B., J.B., E.P., C.G.-C., and N.G.-R.; Visualization, D.R., M.R., and J.B.; Supervision, D.R. and C.G.-C.; Project Administration, D.R.; Funding Acquisition, D.R. and C.G.-C. All authors have read and agreed to the published version of the manuscript.

Funding: This work was supported by "Agencia Estatal de Investigación" from the Spanish Ministry of Science and Innovation (projects PID2020-113371RB-C21 and TED2021-130845B-C3) and IDAEA-CSIC, Severo Ochoa Centre of Excellence (CEX2018-000794-S). Juliette Bedrossiantz was supported by a PhD grant (PRE2018-083513) co-financed by the Spanish Government and the European Social Fund (ESF).

Institutional Review Board Statement: All procedures were approved by the Institutional Animal Care and Use Committees at the CID-CSIC and conducted in accordance with the institutional guidelines under a license from the local government (agreement number 11336).

Informed Consent Statement: Not applicable.

Data Availability Statement: The data supporting the findings of this study are available within the manuscript and its Supplementary Material file or will be made available from the corresponding author upon request.

Conflicts of Interest: The authors declare no conflicts of interest.

References

1. Bölte, S.; Girdler, S.; Marschik, P.B. The contribution of environmental exposure to the etiology of autism spectrum disorder. *Cell Mol. Life Sci.* **2019**, *76*, 1275–1297. [CrossRef] [PubMed]
2. Fernández, M.; Mollinedo-Gajate, I.; Peñagarikano, O. Neural Circuits for Social Cognition: Implications for Autism. *Neuroscience* **2018**, *370*, 148–162. [CrossRef] [PubMed]
3. Polšek, D.; Jagatic, T.; Cepanec, M.; Hof, P.R.; Šimić, G. Recent developments in neuropathology of autism spectrum disorders. *Transl. Neurosci.* **2011**, *2*, 256–264. [CrossRef] [PubMed]
4. Hettinger, J.A.; Liu, X.; Schwartz, C.E.; Michaelis, R.C.; Holden, J.J.A. A DRD1 haplotype is associated with risk for autism spectrum disorders in male-only affected sib-pair families. *Am. J. Med. Genet. B Neuropsychiatr. Genet.* **2008**, *147B*, 628–636. [CrossRef]
5. Marotta, R.; Risoleo, M.C.; Messina, G.; Parisi, L.; Carotenuto, M.; Vetri, L.; Roccella, M. The Neurochemistry of Autism. *Brain Sci.* **2020**, *10*, 163. [CrossRef] [PubMed]
6. Andres, C. Molecular genetics and animal models in autistic disorder. *Brain Res. Bull.* **2002**, *57*, 109–119. [CrossRef]
7. Keil-Stietz, K.; Lein, P.J. Gene × environment interactions in autism spectrum disorders. *Curr. Top. Dev. Biol.* **2023**, *152*, 221–284. [CrossRef] [PubMed]
8. Meshalkina, D.A.; Kizlyk, M.N.; Kysil, E.V.; Collier, A.D.; Echevarria, D.J.; Abreu, M.S.; Barcellos, L.J.G.; Song, C.; Warnick, J.E.; Kyzar, E.J.; et al. Zebrafish models of autism spectrum disorder. *Exp. Neurol.* **2018**, *299*, 207–216. [CrossRef]
9. Ranger, P.; Ellenbroek, B.A. Perinatal Influences of Valproate on Brain and Behaviour: An Animal Model for Autism. *Curr. Top. Behav. Neurosci.* **2016**, *29*, 363–386. [CrossRef]
10. Roullet, F.I.; Crawley, J.N. Mouse models of autism: Testing hypotheses about molecular mechanisms. *Curr. Top. Behav. Neurosci.* **2011**, *7*, 187–212. [CrossRef]
11. Topál, J.; Román, V.; Turcsán, B. The dog (*Canis familiaris*) as a translational model of autism: It is high time we move from promise to reality. *Wiley Interdiscip. Rev. Cogn. Sci.* **2019**, *10*, e1495. [CrossRef] [PubMed]
12. Zimmermann, F.F.; Gaspary, K.V.; Leite, C.E.; De Paula Cognato, G.; Bonan, C.D. Embryological exposure to valproic acid induces social interaction deficits in zebrafish (*Danio rerio*): A developmental behavior analysis. *Neurotoxicol. Teratol.* **2015**, *52*, 36–41. [CrossRef] [PubMed]
13. Kim, O.H.; Cho, H.J.; Han, E.; Hong, T.I.; Ariyasiri, K.; Choi, J.H.; Hwang, K.S.; Jeong, Y.M.; Yang, S.Y.; Yu, K.; et al. Zebrafish knockout of Down syndrome gene, DYRK1A, shows social impairments relevant to autism. *Mol. Autism* **2017**, *8*, 50. [CrossRef] [PubMed]
14. Liu, C.X.; Li, C.Y.; Hu, C.C.; Wang, Y.; Lin, J.; Jiang, Y.H.; Li, Q.; Xu, X. CRISPR/Cas9-induced shank3b mutant zebrafish display autism-like behaviors. *Mol. Autism.* **2018**, *9*, 23. [CrossRef] [PubMed]
15. Dufour-Rainfray, D.; Vourc'h, P.; Tourlet, S.; Guilloteau, D.; Chalon, S.; Andres, C.R. Fetal exposure to teratogens: Evidence of genes involved in autism. *Neurosci. Biobehav. Rev.* **2011**, *35*, 1254–1265. [CrossRef] [PubMed]

16. Chen, J.; Lei, L.; Tian, L.; Hou, F.; Roper, C.; Ge, X.; Zhao, Y.; Chen, Y.; Dong, Q.; Tanguay, R.L.; et al. Developmental and behavioral alterations in zebrafish embryonically exposed to valproic acid (VPA): An aquatic model for autism. *Neurotoxicol. Teratol.* **2018**, *66*, 8–16. [CrossRef] [PubMed]
17. Dwivedi, S.; Medishetti, R.; Rani, R.; Sevilimedu, A.; Kulkarni, P.; Yogeeswari, P. Larval zebrafish model for studying the effects of valproic acid on neurodevelopment: An approach towards modeling autism. *J. Pharmacol. Toxicol. Methods* **2019**, *95*, 56–65. [CrossRef] [PubMed]
18. Chomiak, T.; Turner, N.; Hu, B. What We Have Learned about Autism Spectrum Disorder from Valproic Acid. *Patholog. Res. Int.* **2013**, *2013*, 712758. [CrossRef] [PubMed]
19. Ghodke-Puranik, Y.; Thorn, C.F.; Lamba, J.K.; Leeder, J.S.; Song, W.; Birnbaum, A.K.; Altman, R.B.; Klein, T.E. Valproic acid pathway: Pharmacokinetics and pharmacodynamics. *Pharmacogenet. Genom.* **2013**, *23*, 236–241. [CrossRef]
20. Terbach, N.; Williams, R.S.B. Structure-function studies for the panacea, valproic acid. *Biochem. Soc. Trans.* **2009**, *37*, 1126–1132. [CrossRef]
21. Bialer, M.; Yagen, B. Valproic acid: Second generation. *Neurotherapeutics* **2007**, *4*, 130. [CrossRef] [PubMed]
22. Williams, G.; King, J.; Cunningham, M.; Stephan, M.; Kerr, B.; Hersh, J.H. Fetal valproate syndrome and autism: Additional evidence of an association. *Dev. Med. Child Neurol.* **2001**, *43*, 202–206. [CrossRef] [PubMed]
23. Tayanloo-Beik, A.; Hamidpour, S.K.; Abedi, M.; Shojaei, H.; Tavirani, M.R.; Namazi, N.; Larijani, B.; Arjmand, B. Zebrafish Modeling of Autism Spectrum Disorders, Current Status and Future Prospective. *Front. Psychiatry* **2022**, *13*, 911770. [CrossRef] [PubMed]
24. Mabunga, D.F.N.; Gonzales, E.L.T.; Kim, J.; Kim, K.C.; Shin, C.Y. Exploring the Validity of Valproic Acid Animal Model of Autism. *Exp. Neurobiol.* **2015**, *24*, 285–300. [CrossRef] [PubMed]
25. Gupta, T.; Marquart, G.D.; Horstick, E.J.; Tabor, K.M.; Pajevic, S.; Burgess, H.A. Morphometric analysis and neuroanatomical mapping of the zebrafish brain. *Methods* **2018**, *150*, 49–62. [CrossRef]
26. Joseph, T.P.; Zhou, F.; Sai, L.Y.; Chen, H.; Lin, S.L.; Schachner, M. Duloxetine ameliorates valproic acid-induced hyperactivity, anxiety-like behavior, and social interaction deficits in zebrafish. *Autism Res.* **2022**, *15*, 27–41. [CrossRef] [PubMed]
27. Terbach, N.; Shah, R.; Kelemen, R.; Klein, P.S.; Gordienko, D.; Brown, N.A.; Wilkinson, C.J.; Williams, R.S.B. Identifying an uptake mechanism for the antiepileptic and bipolar disorder treatment valproic acid using the simple biomedical model Dictyostelium. *J. Cell Sci.* **2011**, *124*, 2267–2276. [CrossRef]
28. Zellner, D.; Padnos, B.; Hunter, D.L.; MacPhail, R.C.; Padilla, S. Rearing conditions differentially affect the locomotor behavior of larval zebrafish, but not their response to valproate-induced developmental neurotoxicity. *Neurotoxicol. Teratol.* **2011**, *33*, 674–679. [CrossRef] [PubMed]
29. Bai, W.J.; Luo, X.G.; Jin, B.H.; Zhu, K.S.; Guo, W.Y.; Zhu, X.Q.; Qin, X.; Yang, Z.X.; Zhao, J.J.; Chen, S.R.; et al. Deficiency of transmembrane AMPA receptor regulatory protein γ-8 leads to attention-deficit hyperactivity disorder-like behavior in mice. *Zool. Res.* **2022**, *43*, 851. [CrossRef]
30. Maltezos, S.; Horder, J.; Coghlan, S.; Skirrow, C.; O'Gorman, R.; Lavender, T.J.; Mendez, M.A.; Mehta, M.; Daly, E.; Xenitidis, K.; et al. Glutamate/glutamine and neuronal integrity in adults with ADHD: A proton MRS study. *Transl. Psychiatry* **2014**, *4*, e373. [CrossRef]
31. Naaijen, J.; Bralten, J.; Poelmans, G.; Glennon, J.C.; Franke, B.; Buitelaar, J.K.; Faraone, S.; Asherson, P.; Banaschewski, T.; P Ebstein, R.; et al. Glutamatergic and GABAergic gene sets in attention-deficit/hyperactivity disorder: Association to overlapping traits in ADHD and autism. *Transl. Psychiatry* **2017**, *7*, e999. [CrossRef] [PubMed]
32. Cheng, J.; Liu, A.; Shi, M.Y.; Yan, Z. Disrupted Glutamatergic Transmission in Prefrontal Cortex Contributes to Behavioral Abnormality in an Animal Model of ADHD. *Neuropsychopharmacology* **2017**, *42*, 2096–2104. [CrossRef] [PubMed]
33. Bailey, J.M.; Oliveri, A.N.; Karbhari, N.; Brooks, R.A.J.; De La Rocha, A.J.; Janardhan, S.; Levin, E.D. Persistent behavioral effects following early life exposure to retinoic acid or valproic acid in zebrafish. *Neurotoxicology* **2016**, *52*, 23–33. [CrossRef] [PubMed]
34. Montanari, M.; Martella, G.; Bonsi, P.; Meringolo, M. Autism Spectrum Disorder: Focus on Glutamatergic Neurotransmission. *Int. J. Mol. Sci.* **2022**, *23*, 3861. [CrossRef] [PubMed]
35. Rojas, D.C. The role of glutamate and its receptors in autism and the use of glutamate receptor antagonists in treatment. *J. Neural Transm.* **2014**, *121*, 891–905. [CrossRef] [PubMed]
36. Wright, J.W.; Harding, J.W. Contributions of Matrix Metalloproteinases to Neural Plasticity, Habituation, Associative Learning and Drug Addiction. *Neural Plast.* **2009**, *2009*, 579382. [CrossRef] [PubMed]
37. Faria, M.; Bedrossiantz, J.; Prats, E.; Garcia, X.R.; Gómez-Canela, C.; Piña, B.; Raldúa, D. Deciphering the mode of action of pollutants impairing the fish larvae escape response with the vibrational startle response assay. *Sci. Total Environ.* **2019**, *672*, 121–128. [CrossRef] [PubMed]
38. Mora-Zamorano, F.X.; Svoboda, K.R.; Carvan, M.J. The nicotine-evoked locomotor response: A behavioral paradigm for toxicity screening in zebrafish (*Danio rerio*) embryos and eleutheroembryos exposed to methylmercury. *PLoS ONE* **2016**, *11*, e0154570. [CrossRef] [PubMed]
39. Kalueff, A.V. *The Rights and Wrongs of Zebrafish: Behavioral Phenotyping of Zebrafish*; Springer: Berlin/Heidelberg, Germany, 2017; ISBN 9783319337746.

40. Abbas, K.; Saputra, F.; Suryanto, M.E.; Lai, Y.H.; Huang, J.C.; Yu, W.H.; Chen, K.H.C.; Lin, Y.T.; Hsiao, C. Der Evaluation of effects of ractopamine on cardiovascular, respiratory, and locomotory physiology in animal model zebrafish larvae. *Cells* **2021**, *10*, 2449. [CrossRef]
41. Basnet, R.M.; Zizioli, D.; Taweedet, S.; Finazzi, D.; Memo, M. Zebrafish larvae as a behavioral model in neuropharmacology. *Biomedicines* **2019**, *7*, 23. [CrossRef]
42. McCutcheon, V.; Park, E.; Liu, E.; Wang, Y.; Wen, X.Y.; Baker, A.J. A Model of Excitotoxic Brain Injury in Larval Zebrafish: Potential Application for High-Throughput Drug Evaluation to Treat Traumatic Brain Injury. *Zebrafish* **2016**, *13*, 161–169. [CrossRef]
43. Chefer, V.I.; Thompson, A.C.; Zapata, A.; Shippenberg, T.S. Overview of Brain Microdialysis. *Curr. Protoc. Neurosci.* **2009**, *47*, 7.1.1–7.1.28. [CrossRef] [PubMed]
44. Kuo, H.Y.; Liu, F.C. Pathophysiological Studies of Monoaminergic Neurotransmission Systems in Valproic Acid-Induced Model of Autism Spectrum Disorder. *Biomedicines* **2022**, *10*, 560. [CrossRef]
45. Ali, E.H.A.; Elgoly, A.H.M. Combined prenatal and postnatal butyl paraben exposure produces autism-like symptoms in offspring: Comparison with valproic acid autistic model. *Pharmacol. Biochem. Behav.* **2013**, *111*, 102–110. [CrossRef]
46. Pavăl, D. A Dopamine Hypothesis of Autism Spectrum Disorder. *Dev. Neurosci.* **2017**, *39*, 355–360. [CrossRef] [PubMed]
47. Nguyen, M.; Roth, A.; Kyzar, E.J.; Poudel, M.K.; Wong, K.; Stewart, A.M.; Kalueff, A.V. Decoding the contribution of dopaminergic genes and pathways to autism spectrum disorder (ASD). *Neurochem. Int.* **2014**, *66*, 15–26. [CrossRef] [PubMed]
48. Schür, R.R.; Draisma, L.W.R.; Wijnen, J.P.; Boks, M.P.; Koevoets, M.G.J.C.; Joëls, M.; Klomp, D.W.; Kahn, R.S.; Vinkers, C.H. Brain GABA levels across psychiatric disorders: A systematic literature review and meta-analysis of 1H-MRS studies. *Hum. Brain Mapp.* **2016**, *37*, 3337–3352. [CrossRef]
49. Purkayastha, P.; Malapati, A.; Yogeeswari, P.; Sriram, D. A Review on GABA/Glutamate Pathway for Therapeutic Intervention of ASD and ADHD. *Curr. Med. Chem.* **2015**, *22*, 1850–1859. [CrossRef]
50. Kimmel, C.B.; Ballard, W.W.; Kimmel, S.R.; Ullmann, B.; Schilling, T.F. Stages of embryonic development of the zebrafish. *Dev. Dyn.* **1995**, *203*, 253–310. [CrossRef]
51. Ricarte, M.; Prats, E.; Montemurro, N.; Bedrossiantz, J.; Bellot, M.; Gómez-Canela, C.; Raldúa, D. Environmental concentrations of tire rubber-derived 6PPD-quinone alter CNS function in zebrafish larvae. *Sci. Total Environ.* **2023**, *896*, 165240. [CrossRef]
52. Bellot, M.; Manen, L.; Prats, E.; Bedrossiantz, J.; Barata, C.; Gómez-Canela, C.; Antolin, A.A.; Raldúa, D. Short-term exposure to environmental levels of nicotine and cotinine impairs visual motor response in zebrafish larvae through a similar mode of action: Exploring the potential role of zebrafish α7 nAChR. *Sci. Total Environ.* **2024**, *912*, 169301. [CrossRef] [PubMed]
53. Bedrossiantz, J.; Prats, E.; Raldúa, D. Neurotoxicity Assessment in Adult *Danio rerio* using a Battery of Behavioral Tests in a Single Tank. *J. Vis. Exp.* **2023**, *2023*, 65869. [CrossRef] [PubMed]
54. Gómez-Canela, C.; Tornero-Cañadas, D.; Prats, E.; Piña, B.; Tauler, R.; Raldúa, D. Comprehensive characterization of neurochemicals in three zebrafish chemical models of human acute organophosphorus poisoning using liquid chromatography-tandem mass spectrometry. *Anal. Bioanal. Chem.* **2018**, *410*, 1735–1748. [CrossRef]
55. Prats, E.; Gómez-Canela, C.; Ben-Lulu, S.; Ziv, T.; Padrós, F.; Tornero, D.; Garcia-Reyero, N.; Tauler, R.; Admon, A.; Raldúa, D. Modelling acrylamide acute neurotoxicity in zebrafish larvae. *Sci. Rep.* **2017**, *7*, 13952. [CrossRef] [PubMed]

International Journal of *Molecular Sciences*

MDPI

Article

Characterisation of an Adult Zebrafish Model for *SDHB*-Associated Phaeochromocytomas and Paragangliomas

Jasmijn B. Miltenburg [1], Marnix Gorissen [2], Inge van Outersterp [1], Iris Versteeg [2], Alex Nowak [2], Richard J. Rodenburg [3], Antonius E. van Herwaarden [4], Andre J. Olthaar [4], Benno Kusters [5], Catleen Conrad [6], Henri J. L. M. Timmers [1] and Margo Dona [1,*]

1 Department of Internal Medicine, Radboud University Medical Center, 6525AG Nijmegen, The Netherlands; jasmijn.miltenburg@radboudumc.com (J.B.M.)
2 Department of Plant and Animal Biology, Radboud Institute for Biological and Environmental Sciences, Radboud University, 6525AJ Nijmegen, The Netherlands
3 Departments of Pediatrics and Genetics, Radboud Center for Mitochondrial Medicine, Translational Metabolic Laboratory, Radboud University Medical Center, 6525AG Nijmegen, The Netherlands
4 Department of Laboratory Medicine, Radboud University Medical Center, 6525AG Nijmegen, The Netherlands
5 Department of Pathology, Radboud University Medical Center, 6525AG Nijmegen, The Netherlands
6 Institute of Clinical Chemistry and Laboratory Medicine, Medical Faculty and University Hospital Carl Gustav Carus, Technische Universität Dresden, 01307 Dresden, Germany
* Correspondence: margo.dona@radboudumc.com

Abstract: Phaeochromocytomas and paragangliomas (PPGLs) are rare neuroendocrine tumours arising from chromaffin cells. Pathogenic variants in the gene *succinate dehydrogenase subunit B (SDHB)* are associated with malignancy and poor prognosis. When metastases arise, limited treatment options are available. The pathomechanism of *SDHB*-associated PPGL remains largely unknown, and the lack of suitable models hinders therapy development. Germline heterozygous *SDHB* pathogenic variants predispose to developing PPGLs with a life-long penetrance of around 50%. To mimic the human disease phenotype, we characterised adult heterozygous *sdhb* mutant zebrafish as a potential model to study *SDHB*-related PPGLs. Adult *sdhb* mutant zebrafish did not develop an obvious tumour phenotype and were anatomically and histologically like their wild-type siblings. However, *sdhb* mutants showed significantly increased succinate levels, a major hallmark of *SDHB*-related PPGLs. While basal activity was increased during day periods in mutants, mitochondrial complex activity and catecholamine metabolite levels were not significantly different. In conclusion, we characterised an adult in vivo zebrafish model, genetically resembling human carriers. Adult heterozygous *sdhb* mutants mimicked their human counterparts, showing systemic elevation of succinate levels despite the absence of a tumour phenotype. This model forms a promising basis for developing a full tumour phenotype and gaining knowledge of the pathomechanism behind *SDHB*-related PPGLs.

Keywords: zebrafish; phaeochromocytoma; paraganglioma; *succinate dehydrogenase subunit B* (*SDHB*)

1. Introduction

Phaeochromocytomas and paragangliomas (PPGLs) are rare types of neuroendocrine cancers arising from chromaffin cells in the adrenal medulla or extra-adrenal paraganglia, respectively. In human patients with PPGLs of sympathetic origin, most signs and symptoms are due to the tumour's catecholamine-secreting properties, resulting in paroxysmal palpitations, headache, hyperhidrosis, nausea, hypertension, and hyperglycaemia [1]. Approximately 40% of patients affected by PPGL are predisposed to this condition as a result of a heterozygous germline pathogenic variant in one of more than 20 identified susceptibility genes [2]. It is hypothesised that the acquisition of a second (somatic) mutation in addition to the existing germline mutation is one of the involved pathomechanisms of tumorigenesis [3]. Mutations in the various susceptibility genes are associated with distinct molecular and clinical phenotypes, and tumours can therefore be subdivided into three molecular clusters, with mutations causing (1) pseudohypoxic signalling, (2) aberrant kinase signalling, or (3) aberrant Wnt signalling [3]. Of all identified PPGL cases, 20% show mutations in one of the subunits of the *succinate dehydrogenase* (*SDH*) complex, leading to SDH deficiency and a pseudohypoxic phenotype. Mutations can occur in genes encoding one of four *SDH* subunits, *SDHA-D*, collectively referred to as *SDHx* mutations. Mutations in the gene *succinate dehydrogenase subunit B* (*SDHB*) are the most common and are associated with the highest malignancy rate. Other tumours associated with mutations in *SDHB* are renal cell carcinoma (RCC), gastrointestinal stromal tumour (GIST), and pituitary adenoma [4]. On average, the age of onset for *SDHB*-associated PPGLs is between thirty and fifty years old, but they may arise at any age [2]. When tumours arise, disease progression is monitored closely. In addition to universal monitoring techniques, such as imaging and symptom monitoring, disease (progression) in endocrine active tumours can be followed by the metabolisation of the tumour's secreted catecholamines into metanephrines, measured in blood plasma or urine [5]. In the absence of metastases, management usually consists of surgical removal of the tumour. When metastases arise, in up to 50% of *SDHB*-related PPGL, treatment is challenging and is usually in a palliative setting [6].

SDHB encodes one of four subunits of the SDH complex, which is involved in the citric acid (TCA) cycle (converting succinate to fumarate) and the electron transport chain. Defects in SDH therefore lead to an accumulation of succinate and defective energy metabolism [7,8]. Succinate subsequently acts as an oncometabolite, altering the gene expression of genes involved in multiple pathways, including cell migration, cell invasion, and blood vessel formation [9]. Additionally, succinate inhibits demethylation, leading to epigenetic changes through genome hypermethylation [10].

The gaps in knowledge of the precise pathomechanism make it difficult to treat patients and develop new targeted therapies. Therefore, there is an urgent need to unravel the pathophysiological mechanisms behind *SDHB*-associated PPGL tumorigenesis.

To study the pathophysiology behind *SDHB*-associated PPGLs, a suitable model system is essential. Several model systems have been employed, from cell culture to canines, with varying levels of success [11]. In murine models, a homozygous mutation in *Sdhb* is embryonically lethal, whereas heterozygosity does not lead to a disease phenotype. Allografted mice models using *SDHB*-silenced cell lines showed tumour growth and developed metastases [12]. However, because of the genetic make-up of the allografted cells, their suitability as a disease model is limited. To date, the most accurate model system is a xenograft model, where PPGL cells derived from primary tumours in irradiated germline *Sdhb*-mutated rats were xenografted into immunocompromised mice. Resulting tumours recapitulated several human *SDHB*-associated PPGL characteristics, such as increased succinate levels, metabolic reprogramming, and transcriptional alterations related to defective electron transport and hypoxia [13]. These animals still lack natural tumour development following an *SDHB* mutation and therefore fail to recapitulate the human disease situation faithfully.

Recently, we proposed zebrafish as a model organism for *SDHB*-associated PPGLs. Generally, zebrafish have been established as an excellent cancer model system [14]. Apart from the obvious advantages of zebrafish as a model organism, such as transparency during early development, rapid development, and ease of inducing genetic alterations, there is a high degree of conservation of genes involved in cancer development, such as tumour suppressor genes, oncogenes, and cell-cycle genes. Moreover, their tumours are histologically similar to their human counterparts [14]. In the zebrafish inter-renal gland—the fish equivalent of the adrenal gland—chromaffin cells line the major blood vessels. Functionally, these chromaffin cells are homologous to mammalian adrenal medullary cells, synthesising and secreting catecholamines [15]. In addition, *sdhb* has been conserved in zebrafish and is a functional orthologue of human *SDHB* [16].

Previously, we developed a zebrafish model by CRISPR-Cas9 genome editing. In these zebrafish, homozygous *sdhb* mutant larvae showed human PPGL characteristics such as increased succinate levels and impaired energy metabolism, while their heterozygous and wild-type (WT) siblings showed a normal phenotype. However, homozygous larvae showed a severely decreased lifespan (up to 14 days), hindering possible observations of tumour formation in later life stages [16]. Therefore, the aim of the current study was to characterise the adult heterozygous *sdhb* mutant zebrafish. Adult mutants were characterised with respect to their behavioural phenotype, as well as anatomically, histologically, endocrinologically, and metabolically, to evaluate their suitability as a model for mimicking natural PPGL development in human carriers of pathogenic variants in *SDHB*.

2. Results

2.1. Lifespan and In Vivo Phenotypical Monitoring

A total of 288 heterozygous *sdhb* mutant zebrafish and 385 WT zebrafish were monitored from the start of adulthood until death or a maximum of 2 years of age. No significant differences were observed in the 2-year survival rate (%) between heterozygous *sdhb* mutants (93%) and WT zebrafish (88%) (Figure 1). Deceased fish were either found dead, or a humane endpoint was applied. Of the observed heterozygous *sdhb* mutant zebrafish, 29 showed abnormalities during their lifespan, as opposed to two of their WT siblings (Figure S1). Five of those abnormalities were lumps, defined as abnormal growth, which were all identified in heterozygous *sdhb* mutant zebrafish of 1 year of age and older, without histological evidence of PPGL formation.

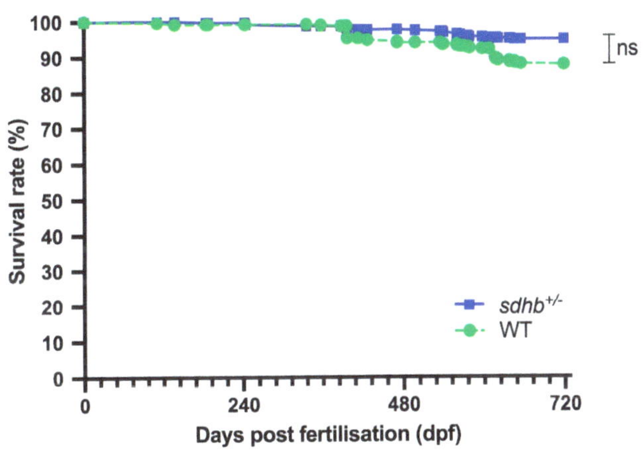

Figure 1. Monitoring of adult heterozygous *sdhb* mutant and WT zebrafish. Survival in heterozygous *sdhb* mutants and WT zebrafish. No significant difference ($p > 0.05$) was observed in the survival rate (%) between heterozygous *sdhb* mutants (n = 288) and WT zebrafish (n = 385). Statistical significance was tested with a log-rank (Mantel–Cox) test.

2.2. Behavioural Analyses

During general observations, no obvious alterations in swimming ability or behaviour were observed in heterozygous *sdhb* mutant zebrafish compared to their WT siblings. To investigate their swimming behaviour in more detail, 1.5-year-old fish were subjected to two locomotor tests for 5 days each. Basal activity, defined as seconds moved per hour, was significantly higher in heterozygous *sdhb* mutant zebrafish compared with their WT siblings during the day ($p < 0.05$) but not during the night. There were no differences in swimming velocity between WT and mutants during the day ($p = 0.08$) or night periods (Figure 2A,B).

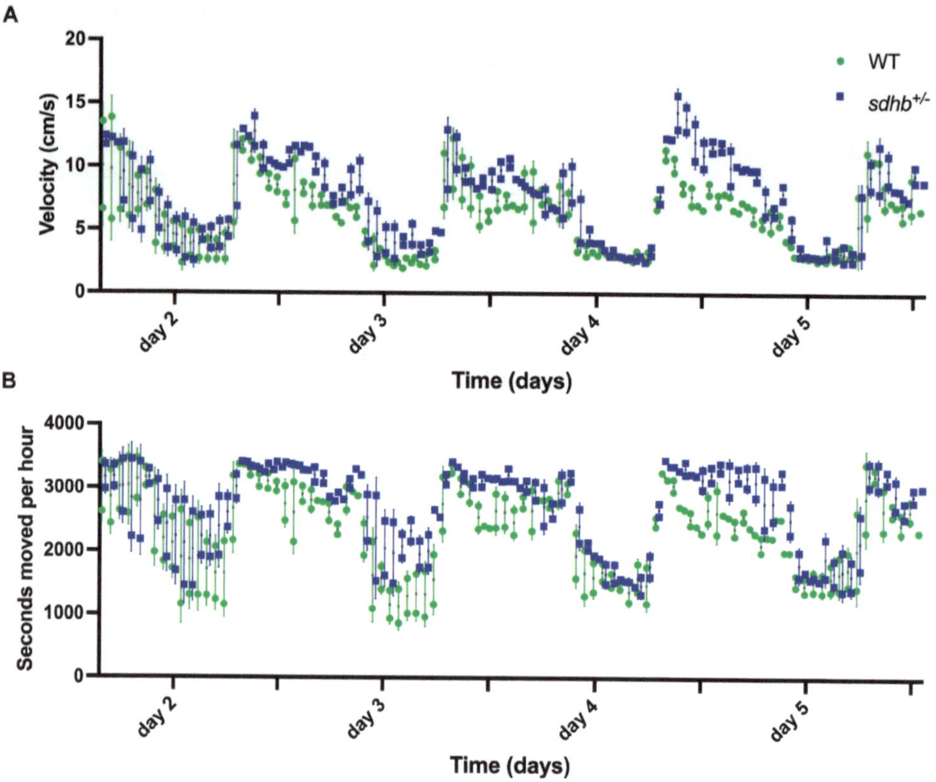

Figure 2. Behavioural analysis of adult heterozygous *sdhb* mutant (n = 5) and WT (n = 5) zebrafish. Swimming behaviour was tracked in two locomotor tests of 5 days. Ticks indicate 12-hour intervals. (**A**) Swimming velocity in centimetres per second. (**B**) Basal activity is displayed as cumulative seconds of movement. Basal activity was increased in heterozygous *sdhb* mutants during day periods ($p < 0.05$). There was no difference in swimming velocity between mutants and WT siblings during the day ($p = 0.08$) or night periods. The area under the curve (AUC) was used to evaluate statistical significance.

2.3. Anatomy, Histology, and Chromaffin Cell Morphology

General anatomy, kidney morphology, and chromaffin cell morphology were analysed in heterozygous *sdhb* mutant and WT zebrafish (representative pictures shown in Figure 3A,B,D,E). H&E stainings of 34 heterozygous *sdhb* mutant zebrafish and 35 WT siblings were studied. No major overall anatomical differences were found between the heterozygous *sdhb* mutant and WT zebrafish. In addition, no differences in kidney morphology were observed between the two groups concerning tubular necrosis, tubular regeneration, haemorrhage of the interstitium, or tubular lumen enlargement. Ten heterozygous sdhb mutants and nine WT siblings were analysed in higher detail for chromaffin morphology (Figure 3C,F). No significant differences in the fluorescent area (region of interest; ROI), number of nuclei within that area, or fluorescence intensity were observed between the two groups (Figure 3G–I; Figure S2). Fluorescent area per nucleus was used as a proxy of cell size (Figure 3J, cell area per nucleus). There was a trend towards an increase in approximate cell size in heterozygous *sdhb* mutants ($p = 0.0933$).

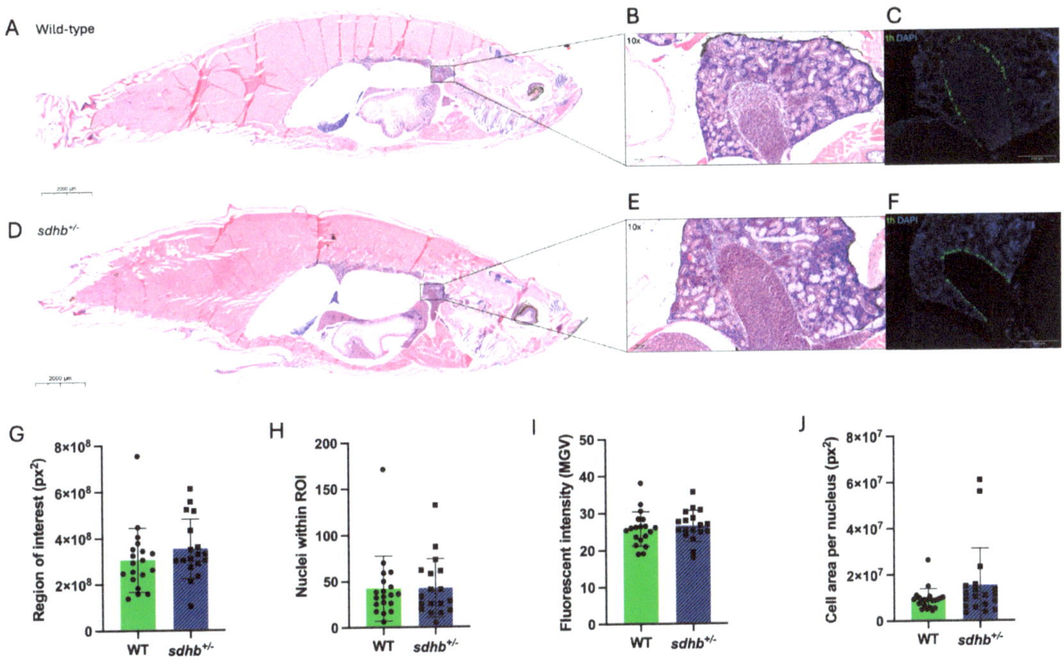

Figure 3. General histology, kidney morphology, and chromaffin cell morphology of adult heterozygous *sdhb* mutant and WT zebrafish. (**A**,**D**) General histology of adult heterozygous *sdhb* mutant (n = 34) and WT (n = 35) zebrafish (Representative images; H&E staining, 0.8× magnification). (**B**,**E**) Head kidney morphology (H&E staining, 10× magnification). (**C**,**F**) Immunohistochemical staining of tyrosine hydroxylase in the head kidney of heterozygous *sdhb* mutant (n = 10) and WT (n = 9) zebrafish (10× magnification). (**G–J**) Quantification of tyrosine hydroxylase staining (ImageJ2). Student's *t* test was used to evaluate statistical significance. No significant differences were found in any of the analyses.

2.4. Mitochondrial, Metabolomic, and Endocrine Phenotype

Muscle tissue metabolite profiling showed significantly elevated succinate levels and succinate/fumarate ratio in heterozygous *sdhb* mutant zebrafish compared with WTs ($p < 0.05$, Figure 4A). No differences were observed for the other metabolites (fumarate, iso-citrate, lactate, malate, aspartate, and asparagine) measured. Comparing mitochondrial complex activities, no significant differences in complex II activity were observed in heterozygous *sdhb* mutant zebrafish (Figure 4B; $p = 0.13$). Lastly, analyses of catecholamine metabolites 3-methoxytyramine (3-MT) and normetanephrine (NMN) revealed no differences in 3-MT and NMN levels in fish water of heterozygous *sdhb* mutant and WT zebrafish (Figure 4C). Metanephrine levels were under the limit of detection and were therefore excluded from analysis.

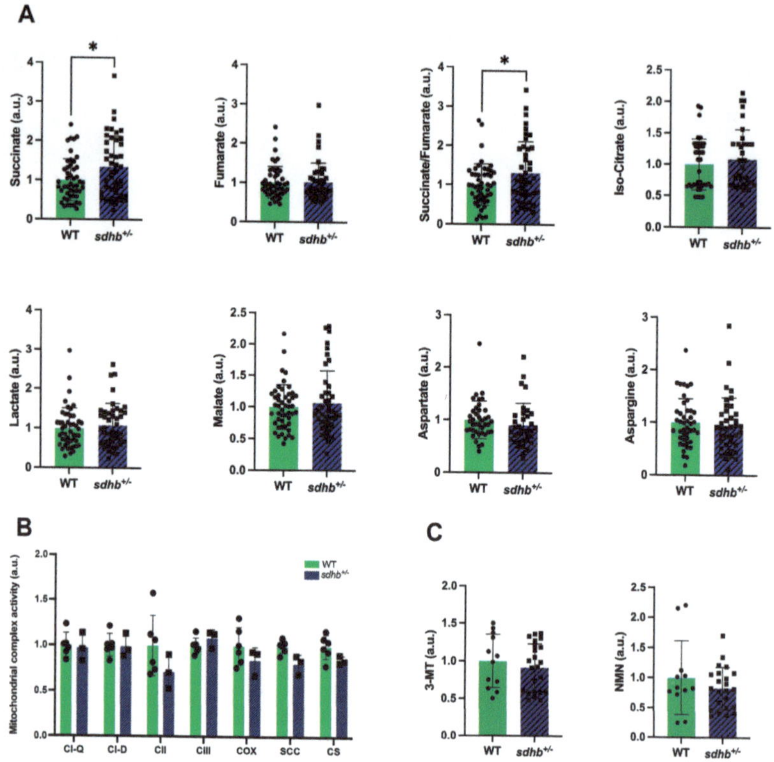

Figure 4. Metabolomics, mitochondrial complex activity, and metanephrine levels of adult heterozygous *sdhb* mutant and WT zebrafish. All values are standardised relative to the associated WT mean and expressed in arbitrary units (a.u.). (**A**) Abundance of metabolites in muscle tissue of heterozygous *sdhb* mutant (n = 44) and WT (n = 49) zebrafish (n is the total of three replicates). Succinate levels and succinate/fumarate ratios were significantly increased in heterozygous *sdhb* mutants compared with their WT siblings. (**B**) Mitochondrial complex activity in heterozygous *sdhb* mutant (n = 3) and WT (n = 6) zebrafish. Activity levels for all mitochondrial complexes were not significantly different. CI-D: complex I; CI-Q: quinone; CII: complex II; CIII: complex III; COX: complex V/cytochrome c oxidase; SCC: cytochrome P-450; CS: citrate synthase. (**C**) Catecholamine metabolite levels in the swimming water of heterozygous *sdhb* mutant (n = 23) and WT (n = 12) zebrafish. 3-MT: 3-Methoxytyramine; NMN: Normetanephrine. Student's *t* test for metabolomics analysis and catecholamine metabolite analysis, Mann–Whitney U test for mitochondrial complex activity analysis. * $p < 0.05$.

3. Discussion

In this study, we characterised adult heterozygous *sdhb* mutant zebrafish to evaluate their potential as a model to study *SDHB*-related PPGLs. We showed that heterozygous mutants have a similar lifespan as their WT siblings. Mutant zebrafish were anatomically and histologically similar to their WT siblings. Behaviourally, *sdhb* mutant zebrafish showed higher basal activity compared with their WT siblings during day periods, with no differences in swimming velocity during day or night. The number of chromaffin cells in WT and heterozygous fish was similar, while there was a trend towards an increase in approximate cell size in heterozygous *sdhb* mutants. Moreover, there was no difference between heterozygous *sdhb* mutant and WT zebrafish in mitochondrial complex II activity. Notably, succinate levels were increased in the muscle tissue of heterozygous *sdhb* mutants. The remaining TCA cycle metabolites and mitochondrial complexes showed no differences in level or activity, nor did metanephrine(s) concentrations in swim water.

Genetically, our model represents the human *SDHB* carrier situation, possessing one mutated and one WT allele. In humans, mutations of *SDHB* are inherited in an autosomal dominant manner and portray incomplete penetrance, resulting in not all individuals with a pathogenic variant developing disease-related characteristics. Recent studies report 21% of *SDHB* carriers having developed at least one tumour at age 50, compared with 42% at age 70 [17]. Thus, of the currently published human *SDHB* mutation carriers, around half are expected to develop tumours during their lifetime. There are no known environmental risk factors for the development of PPGLs that explain why some carriers develop tumours and others do not.

During the two-year follow-up in this study, there were no differences in survival between heterozygous *sdhb* mutants and WT zebrafish. Of the limited number of abnormalities observed, there were five observations of abnormal growth, all observed in heterozygous *sdhb* mutants. We could provide no histological evidence of PPGL formation. While you could speculate that monitoring of the entire life span could provide us with more insight into possible natural PPGL development, it is questionable if tumours would arise at a later stage and if they would then be specifically PPGLs or solely age-related, let alone impractical to wait more than two years for natural tumour development. We expected a two-year time frame of study, which translates to 50 to 75% of the zebrafish lifespan, to be sufficient for tumour development, if this process were to occur naturally. Other germline-mutated zebrafish cancer models have been shown to develop tumours within twelve months of age, but these contain additional mutations to the mutation of interest, such as a knockout of p53 [18]. In addition to previously mentioned incomplete penetrance observed in humans, there are of course interspecies differences between zebrafish and humans that could explain the absence of tumour development. Moreover, laboratory animals are kept in sterile environments compared with regular human life, where we are exposed to all sorts of harmful environmental factors. It might thus be necessary to aggravate the genetic phenotype with treatments that accelerate the tumorigenesis process, such as SDH inhibitors.

A major hallmark of *SDHB*-related PPGLs is SDH deficiency and the ensuing accumulation of succinate. Succinate accumulation leads to the inhibition of α-ketoglutarate-dependent dioxygenases such as prolyl hydroxylase, which, under normoxic conditions, targets the transcription factor hypoxia-inducible factor for degradation. Prolyl hydroxylase (PHD) inhibition results in hypoxia inducible factor (HIF) stabilisation, inducing a pseudohypoxic response under normoxic conditions and altering the expression of genes involved in angiogenesis, cell migration, and cell invasion [9]. To investigate (early) signs of chromaffin cell hyperplasia, proliferation, or invasiveness, we analysed heterozygous *sdhb* mutant zebrafish and their WT siblings histologically. We observed no differences in chromaffin cell numbers or morphology between heterozygous *sdhb* mutant and WT zebrafish, providing no evidence of chromaffin cell hyperplasia or proliferation.

An important characteristic of PPGLs is the increased production and secretion of catecholamines. Zebrafish and mammalian endocrine systems are highly similar. While there are differences in endocrine gland structure, cellular processes are highly conserved. The endocrine function of chromaffin cells was histologically approximated by quantification of the staining intensity of tyrosine hydroxylase (th), a rate-limiting enzyme in the catecholamine synthesis pathway in chromaffin cells. However, we did not observe an increase in staining intensity in the chromaffin cells of heterozygous *sdhb* mutant zebrafish. In humans, TH is used solely as a marker for chromaffin cells. While, in our zebrafish model, we too use th as a marker for chromaffin cells, and staining intensity might also provide a proxy for transcript abundance and thus endocrine activity. This is reflected by studies in a rat-derived cell line for pheochromocytoma, where *Sdhb* silencing and consequent loss of complex II activity was associated with increased *Th* expression [19]. In addition, previous studies have shown that, in rats and a rat-derived PPGL cell line, (pseudo)hypoxia is associated with changes in *Th* mRNA levels and *Th* activity through altered phosphorylation [20,21]. The absence of differences in staining intensity in our model could possibly reflect that the systemic increase in succinate levels, measured in muscle tissue of heterozygous *sdhb* mutants in the presence of sustained complex II activity, is insufficient to drive altered chromaffin cell morphology or endocrine function, and additional triggers are necessary.

A limitation of this quantification is the high variability between histological sections because of variations in the sectioning plane. This causes high variation in the visualisation of chromaffin cells, complicating the quantification of chromaffin cells. However, histological analysis does allow us to observe tumour development when any arise. A transgenic line, fluorescently labelling chromaffin cells, could provide a solution for this, allowing for their isolation via fluorescence-activated cell sorting (FACS). Next to the exact quantification of chromaffin cell numbers, this would enable additional analyses such as gene expression analyses by quantitative PCR or methylation assays. The high variability in the histological sections now forces us to look at group averages, while interindividual differences are large, also because of the aforementioned incomplete genetic penetrance. Moreover, looking at the individual would allow for the correlation of multiple measurements, such as relating chromaffin cell functionality to succinate levels.

To reliably assess the endocrine function of chromaffin cells, catecholamine and metanephrine levels were assessed, the latter being the metabolised product of the former. Of note is that while the methodology to measure catecholamines and their metabolites in blood plasma, urine, and tissue extracts is well established [22], we are, to our knowledge, the first to measure excreted catecholamines and their metabolic by-products in fish water. In patients, most signs and symptoms are due to the tumour's catecholamine-secreting properties. In agreement with our histological analyses, we observed no differences in metanephrine levels, probably reflecting the lack of difference in chromaffin cell mass between heterozygous *sdhb* mutant and WT zebrafish.

SDH deficiency and succinate accumulation in *SDHB*-related PPGLs is accompanied by a decrease in other TCA cycle metabolites, including fumarate, cis-aconitate, and isocitrate [23]. Consequently, an increased succinate/fumarate ratio is used as a metabolic marker to diagnose *SDHB*-related PPGLs [24]. Interestingly, we found a significant increase in succinate and the succinate/fumarate ratio in heterozygous *sdhb* mutants compared with WT zebrafish, with no differences in other TCA cycle metabolites, indicating at least impaired SDH enzymatic activity despite the presence of a WT allele. In this aspect, again, this model is a faithful and unique representation of the human *SDHB* carrier situation, as *SDHx* mutation carriers often have elevated succinate levels in the absence of tumours or other symptoms [25,26]. This is as opposed to complete homozygosity leading to complete SDH deficiency and a clear metabolic phenotype, such as in Leigh syndrome [27].

The presumed residual SDH activity is reflected by the mitochondrial complex activity. We observed no significant differences between heterozygous mutants and WT zebrafish measured in muscle tissue. This can be most likely attributed to the functional copy present in heterozygous mutants but might be partly attributed to the analysed tissue type or small sample size. This agrees with earlier observations in the larval stages of this model, where there was almost complete loss of complex II activity in homozygous *sdhb* mutant larvae but no decrease in heterozygous mutants [16]. Again, our model reflects the human *SDHB* carrier situation by portraying elevated succinate levels in the absence of a decrease in complex II activity.

Impaired energy metabolism is another characteristic of *SDHB*-related PPGLs because of the role of SDH in the electron transport chain. *SDHx*-mutated tumours have been shown to have impaired ATP production, increased activity of complexes I, III, and IV, and an increased number of morphologically and functionally abnormal mitochondria [7]. Indeed, this was reflected in the larval stages of this model, where energy metabolism, illustrated by swimming behaviour, was shown to be impaired in homozygous *sdhb* mutant larvae with no differences in swimming behaviour between heterozygous *sdhb* mutant and WT larvae [16]. In addition, succinate has been reported to be involved in energy homeostasis through succinate receptor 1 (SUCNR1) signalling [28,29]. We showed that, upon ageing, succinate accumulates in heterozygous mutants. These mutants also show increased basal activity during day periods, with no differences in swimming velocity between mutant and WT zebrafish, neither during the day nor the night periods. This difference in behaviour cannot be solely attributed to energy metabolism but could be the cumulative result of many additional factors, such as anxiety or stress levels.

We were unable to discern evidence of PPGL tumour development in adult heterozygous *sdhb* mutant zebrafish. In humans, a second hit is often necessary to develop PPGLs [3]. This is reflected in mice models, where systemic *Sdhb* heterozygosity is insufficient to induce tumour formation [10,11], *Sdhb* knock-out in medullary adrenal cells solely led to increased succinate levels in the absence of other PPGL hallmarks, but additional *Nf1* mutations induced *SDHx*-like PPGLs [30]. The introduction of additional mutations could be beneficial to induce tumour formation in our model. However, the molecular genetics underlying PPGL development are complex, resulting in different molecular tumour subtypes for different pathogenic variants [3]. Therefore, to faithfully recapitulate the human disease situation, it is crucial to reflect the patient situation genetically. Compared with the above-mentioned animal models for *SDHB*-related PPGLs, our model is the first adult model to recapitulate the relevant genetic background faithfully while partially recapitulating a disease phenotype. Genetically and metabolically, our model represents the human *SDHB* carrier situation, where heterozygosity is associated with an increase in succinate levels in the absence of any other symptoms [23]. Although no natural tumour development has been confirmed as of yet, this model forms a solid base for future research, where we aim to aggravate the current phenotype by altering the involved signalling pathways or inducing additional mutations in a tissue-specific manner. The current absence of natural tumour development allows for the investigation of the trigger necessary for tumour initiation when a heterozygous *SDHB* pathogenic variant is already present.

In conclusion, we characterised an adult in vivo zebrafish model for *SDHB*-related PPGLs, showing a partial disease phenotype through elevated succinate levels. However, increased succinate could not be correlated to other phenotypic abnormalities, indicating that *sdhb* heterozygosity in itself is insufficient to cause a full PPGL phenotype. While the current model does not allow for investigating tumour initiation and development, it is a unique model representing the human carrier situation genetically and metabolically, enabling research into risk factors associated with *SDHB* heterozygosity. In future studies, additional mutations to stimulate a second hit might be necessary to induce a full disease phenotype. In addition, we intend to aggravate the current partial phenotype by altering involved pathways, thereby hopefully identifying the trigger necessary for tumour formation. Altogether, we present an adult zebrafish model with promising prospects to gain knowledge of the pathomechanism behind *SDHB*-related PPGLs.

4. Materials and Methods

4.1. Zebrafish Husbandry

Experimental procedures were conducted in accordance with institutional guidelines and National and European legislation. Ethical approval of the experiments was granted by Radboud University's Institutional Animal Care and Use Committee (IACUC, application number RU-DEC 2020-0030). A germline heterozygous *sdhb* mutation was introduced in zebrafish larvae (*Danio rerio*), in an Oregon AB background, through CRISPR-Cas9 genome editing by our group previously ($sdhb^{rmc200}$; ZFIN: ZDB-FISH-220110-2; [16]). This resulted in zebrafish harbouring a 13-base pair deletion at the exon1-intron1 boundary, leading to a frameshift. Heterozygosity of the mutant allele or WT *sdhb* was confirmed through PCR amplification. Eggs were obtained from natural spawnings. Larvae were maintained and raised by standard methods [31].

4.2. Genotyping

Adult zebrafish were briefly anaesthetised in 2-phenoxyethanol (0.1%, v/v). To obtain genomic DNA, a small section of the tail fin was cut using microscissors. The sample was placed in 25 µL lysis buffer (40 mM NaOH, 0.2 mM EDTA), followed by incubation at 95 °C for 30 min. After incubation, samples were neutralised by adding 75 µL 0.33 mM Tris-HCl. The extracted genomic DNA was used as a PCR template. Primer sequences used to amplify *sdhb* exon 1 were 5′-CATGGCGGCTGTGTGTTTCT-3′ (fw) and 5′-ACGGGTTATTGATGAGCGGTG-3′ (rv). As input, 15.1 µL water, 0.2 µL Dreamtaq (5 U/µL, Thermo Scientific™, Waltham, MA, USA), 0.5 µL dNTPs (10 µM), 1 µL fw primer (10 µM), 1µL rv primer (10 µM), 2 µL PCR buffer, and 1 µL of genomic DNA were used. The cycling conditions were as follows: 94 °C 2 min, 35 cycles of 94 °C 15 s, 58 °C 30 s, 72 °C 30 s, and 72 °C 2 min. The obtained amplicons (WT: 123 base pairs (bps); heterozygous *sdhb* mutant: 123 bp and 110 bp) were visualised on a 2% Tris-borate-EDTA (TBE) agarose gel.

4.3. Behavioural Assessment

Using EthoVision XT7 (Noldus Information Technologies, Wageningen, The Netherlands), the locomotion of adult zebrafish aged 1 to 1.5 years old was tracked. This study was conducted using five heterozygous *sdhb* mutant zebrafish and five WT siblings, placed in two separate aquaria (15 × 30 cm) on an infrared light source, equipped with an Area scan camera (Basler, Ahrensburg, Germany, 106588). The zebrafish were tracked for 5 days. Tracking was performed twice with the same batch of fish. Cumulative time moved (s) and velocity (cm/s) were used as read-outs for behaviour. The data were analysed using Graphpad Prism software (Version 10.1.1 for macOS).

4.4. Histology and Immunohistochemistry

Adult zebrafish were euthanized using an overdose of 2-phenoxyethanol (0.1% *v*/*v*). After euthanasia, a ventral incision was made to ensure complete fixation. Fish were fixated in 4% paraformaldehyde (Sigma, Saint Louis, MO, USA) overnight and subsequently rinsed in tap water (1 h) before a 3-day decalcification in Mol-DECALCIFIER (Menarini Diagnostics, Florence, Italy). After another tap water rinse (1 h), the tissue was kept in 4% paraformaldehyde for sample transfer to tissue processing. Tissue processing from paraformaldehyde to paraffin was performed using the overnight MAGNUS protocol as follows: after a 60-min fixation at 37 °C and a 70 min fixation at 50 °C, samples were subsequently incubated in ethanol (60 min, 55 °C), isopropanol (110 min, 55 °C), and finally, isopropanol (245 min, 65 °C). Vaporisation was performed at 500 mBar (1.5 min), after which the samples were impregnated with wax in a 7-step impregnation series as follows: (1) 25 min, 500 mBar, 70 °C, (2) 15 min, 400 mBar, 70 °C, (3) 10 min, 300 mBar, 70 °C, (4) 10 min, 200 mBar, 70 °C, (5) 10 min, 150 mBar, 70 °C, (6) 150 min, 100 mBar, 65 °C, and (7) 20 min, 800 mBar, 65 °C.

To assess morphology, paraffin sections (5 µm) were mounted on SUPERFROST® PLUS Menzel-Gläser (Thermo Scientific™, Waltham, MA, USA) slides and dried overnight at 37 °C. The slides were stained with haematoxylin and eosin as described previously [32] and imaged with a Zeiss Axioskop (Zeiss, Oberkochen, Germany) light microscope.

To visualise chromaffin cells for the detection of PPGLs and/or hyperplasia, selected slides based on correct morphology were stained for tyrosine hydroxylase (TH), the rate-limiting enzyme in the catecholamine synthesis pathway. Selected sections were deparaffinized with xylene (2 × 10 min), followed by rehydration in decreasing ethanol dilutions (96, 90, 80, 70 and 50%, 3 min each), demineralised water (3 min), and phosphate-buffered saline (PBS). After antigen retrieval (10 mM Citrate buffer, pH 6.0), the slides were blocked with 2% bovine serum albumin (BSA) in PBS for 1 h at room temperature (RT). The slides were washed with PBS (3 × 5 min) and incubated with Anti-Tyrosine Hydroxylase Antibody (Merck, Rahway, NJ, USA, AB1542; 1:1000 in 2% BSA in PBS) overnight at 4 °C. The slides were then washed again with PBS (3 × 5 min) before incubation with the secondary antibody (Alexa Fluor 488 Donkey Anti-Sheep; Thermo Fisher Scientific, Waltham, MA, USA; 1:1000 in 2% BSA in PBS) for 1 h at RT in the dark. From there on, the slides were kept in a dark environment. After washing with PBS (3 × 5 min), the slides were counter-stained with Sudan Black (0.1% in 70% ethanol) for 10 min at RT. After washing with PBS

(3 × 5 min), the slides were coverslipped using DAPI-Fluoromount-G (ITK Diagnostics, Uithoorn, The Netherlands) and dried overnight at RT before imaging head kidney regions, containing chromaffin cells, and other regions of interest. Imaging was performed using a Zeiss Axio Imager (M1) (Zeiss, Oberkochen, Germany) equipped with an AxioCam MRm.

Image analysis was performed using FIJI software (ImageJ2, version 2.12.0/1.54f), excluding images of an incorrect anatomical location or containing major staining artefacts affecting downstream image analysis. An automated script was generated to perform objective quantitative image analysis (10.6084/m9.figshare.25522921). The fluorescent area (region of interest; ROI), number of nuclei in the fluorescent area (maxima), and staining intensity of the fluorescent area (mean grey value; MGV) were used as read-outs for chromaffin cell morphology.

4.5. Mitochondrial Complex Activity Measurements

Mitochondrial complex activity levels were measured in the muscle tissue of WT and heterozygous *sdhb* mutant adult zebrafish. After euthanasia, approximately 10 mg of muscle tissue was taken from the tail of the zebrafish. The samples were weighed, snap-frozen in liquid N2, and stored at −80 °C. Defrosted material was homogenised in 1450 µL 10 mM Tris-HCl, pH 7.6. Then, 250 µL 1.5 M sucrose was added, followed by 10 min 14,000× *g* centrifugation at 2 °C. The supernatant was discarded, and the pellet was resuspended in 200 µL 10 mM Tris-HCl, pH 7.6. The enzymatic activities of OXPHOS complexes I to V, citrate synthase, and the sample protein content were assayed spectrophotometrically as previously described [33]. Assays were performed in duplicate using a Konelab 20XT auto-analyser (Thermo Fisher Scientific, Waltham, MA, USA).

4.6. Mass Spectrometry of TCA Cycle Metabolites and Amino Acids

Using a mass spectrometry-based assay, metabolites involved in the TCA cycle were measured in the muscle tissue of WT and heterozygous *sdhb* mutant adult zebrafish. After euthanasia, approximately 10 mg of muscle tissue was taken from the tail of the zebrafish. The samples were weighed, homogenised in 100 µL methanol together with an internal standard mixture, centrifuged to precipitate insolubles, and dried in a speed vac concentrator. Metabolite extracts were resuspended in the mobile phase and analysed by liquid chromatography with tandem mass spectrometry (LC-MS/MS) as described elsewhere [34]. Data were normalised based on weight.

4.7. Metanephrine Measurements

To detect PPGL-related catecholamine excess, catecholamine metabolite levels were measured in the fish water of WT and heterozygous *sdhb* mutant adult zebrafish. Zebrafish were individually housed in 500 mL water for three days before sample collection. The samples were snap-frozen in liquid nitrogen and stored at −80 °C until analysis. Catecholamine metabolite levels were measured in fish water after solid-phase extraction (SPE) of 1 mL sample volume using a liquid chromatography–mass spectrometry (LCMS) based method, as described previously [35].

4.8. Statistical Analysis

Graphpad Prism software (Version 10.1.1 for macOS) was used to generate scatter plots, calculate mean values, and perform statistical analyses. The data were tested for Gaussian distribution. The log-rank (Mantel–Cox) test was used for the survival curve analysis, student's *t* test for metabolomics analyses, catecholamine metabolite analysis, and immunohistochemical staining quantification, and the Mann-Whitney U test for mitochondrial complex activity analyses. Behavioural data were statistically tested using area under the curve (AUC) analysis. R software (R 4.3.2 GUI 1.80 Big Sur ARM build (8281)) was used to generate density plots.

Supplementary Materials: The following supporting information can be downloaded at https://www.mdpi.com/article/10.3390/ijms25137262/s1.

Author Contributions: Conceptualisation, M.D., H.J.L.M.T. and M.G.; formal analysis, M.D., J.B.M., I.v.O., I.V., A.N., R.J.R., A.E.v.H., A.J.O., B.K. and C.C.; writing—original draft preparation, J.B.M.; writing—review and editing, M.D., M.G. and H.J.L.M.T.; supervision, H.J.L.M.T., M.D. and M.G.; funding acquisition, M.D. and H.J.L.M.T. All authors have read and agreed to the published version of the manuscript.

Funding: This work was supported by the Paradifference Foundation and ZonMwVeni (09150162010197).

Institutional Review Board Statement: Ethical approval of the experiments was granted by Radboud University's Institutional Animal Care and Use Committee (IACUC, application number RU-DEC 2020-0030, approval date: 4-5-2021).

Data Availability Statement: The raw data supporting the conclusions of this article will be made available by the authors upon request.

Acknowledgments: We thank the Experimental Mass Spectrometry unit at the Institute of Clinical Chemistry and Laboratory Medicine, University Hospital Carl Gustav Carus at TU Dresden, Dresden, Germany, and, in particular, Mirko Peitzsch for the analysis of TCA cycle metabolites.

Conflicts of Interest: The authors declare no conflicts of interest. The funders had no role in the design of this study; in the collection, analyses, or interpretation of data; in the writing of this manuscript; or in the decision to publish the results.

References

1. Reisch, N.; Peczkowska, M.; Januszewicz, A.; Neumann, H.P. Pheochromocytoma: Presentation, diagnosis and treatment. *J. Hypertens.* **2006**, *24*, 2331–2339. [CrossRef] [PubMed]
2. Jhawar, S.; Arakawa, Y.; Kumar, S.; Varghese, D.; Kim, Y.S.; Roper, N.; Elloumi, F.; Pommier, Y.; Pacak, K.; Del Rivero, J. New Insights on the Genetics of Pheochromocytoma and Paraganglioma and Its Clinical Implications. *Cancers* **2022**, *14*, 594. [CrossRef] [PubMed]
3. Fishbein, L.; Leshchiner, I.; Walter, V.; Danilova, L.; Robertson, A.G.; Johnson, A.R.; Lichtenberg, T.M.; Murray, B.A.; Ghayee, H.K.; Else, T.; et al. Comprehensive Molecular Characterization of Pheochromocytoma and Paraganglioma. *Cancer Cell* **2017**, *31*, 181–193. [CrossRef] [PubMed]
4. Andrews, K.A.; Ascher, D.B.; Pires, D.E.V.; Barnes, D.R.; Vialard, L.; Casey, R.T.; Bradshaw, N.; Adlard, J.; Aylwin, S.; Brennan, P.; et al. Tumour risks and genotype-phenotype correlations associated with germline variants in succinate dehydrogenase subunit genes SDHB, SDHC and SDHD. *J. Med. Genet.* **2018**, *55*, 384–394. [CrossRef]
5. Schreiner, F.; Beuschlein, F. Disease monitoring of patients with pheochromocytoma or paraganglioma by biomarkers and imaging studies. *Best Pract. Res. Clin. Endocrinol. Metab.* **2020**, *34*, 101347. [CrossRef] [PubMed]
6. Nölting, S.; Bechmann, N.; Taieb, D.; Beuschlein, F.; Fassnacht, M.; Kroiss, M.; Eisenhofer, G.; Grossman, A.; Pacak, K. Personalized management of pheochromocytoma and paraganglioma. *Endocr. Rev.* **2022**, *43*, 199–239. [CrossRef]
7. Dona, M.; Neijman, K.; Timmers, H. MITOCHONDRIA: Succinate dehydrogenase subunit B-associated phaeochromocytoma and paraganglioma. *Int. J. Biochem. Cell Biol.* **2021**, *134*, 105949. [CrossRef] [PubMed]
8. Pollard, P.J.; Brière, J.J.; Alam, N.A.; Barwell, J.; Barclay, E.; Wortham, N.C.; Hunt, T.; Mitchell, M.; Olpin, S.; Moat, S.J.; et al. Accumulation of Krebs cycle intermediates and over-expression of HIF1alpha in tumours which result from germline FH and SDH mutations. *Hum. Mol. Genet.* **2005**, *14*, 2231–2239. [CrossRef]
9. Brahimi-Horn, C.; Pouysségur, J. The role of the hypoxia-inducible factor in tumor metabolism growth and invasion. *Bull. Cancer* **2006**, *93*, E73–E80.
10. Letouzé, E.; Martinelli, C.; Loriot, C.; Burnichon, N.; Abermil, N.; Ottolenghi, C.; Janin, M.; Menara, M.; Nguyen, A.T.; Benit, P. SDH mutations establish a hypermethylator phenotype in paraganglioma. *Cancer Cell* **2013**, *23*, 739–752. [CrossRef]
11. Lepoutre-Lussey, C.; Thibault, C.; Buffet, A.; Morin, A.; Badoual, C.; Bénit, P.; Rustin, P.; Ottolenghi, C.; Janin, M.; Castro-Vega, L.J.; et al. From Nf1 to Sdhb knockout: Successes and failures in the quest for animal models of pheochromocytoma. *Mol. Cell Endocrinol.* **2016**, *421*, 40–48. [CrossRef] [PubMed]
12. Liu, Y.; Pang, Y.; Zhu, B.; Uher, O.; Caisova, V.; Huynh, T.T.; Taieb, D.; Hadrava Vanova, K.; Ghayee, H.K.; Neuzil, J.; et al. Therapeutic Targeting of SDHB-Mutated Pheochromocytoma/Paraganglioma with Pharmacologic Ascorbic Acid. *Clin. Cancer Res.* **2020**, *26*, 3868–3880. [CrossRef]
13. Powers, J.F.; Cochran, B.; Baleja, J.D.; Sikes, H.D.; Pattison, A.D.; Zhang, X.; Lomakin, I.; Shepard-Barry, A.; Pacak, K.; Moon, S.J.; et al. A xenograft and cell line model of SDH-deficient pheochromocytoma derived from Sdhb+/− rats. *Endocr. Relat. Cancer* **2020**, *27*, 337–354. [CrossRef] [PubMed]
14. Amatruda, J.F.; Shepard, J.L.; Stern, H.M.; Zon, L.I. Zebrafish as a cancer model system. *Cancer Cell* **2002**, *1*, 229–231. [CrossRef] [PubMed]

15. Menke, A.L.; Spitsbergen, J.M.; Wolterbeek, A.P.; Woutersen, R.A. Normal anatomy and histology of the adult zebrafish. *Toxicol. Pathol.* **2011**, *39*, 759–775. [CrossRef] [PubMed]
16. Dona, M.; Waaijers, S.; Richter, S.; Eisenhofer, G.; Korving, J.; Kamel, S.M.; Bakkers, J.; Rapizzi, E.; Rodenburg, R.J.; Zethof, J.; et al. Loss of sdhb in zebrafish larvae recapitulates human paraganglioma characteristics. *Endocr. Relat. Cancer* **2021**, *28*, 65–77. [CrossRef]
17. Rijken, J.A.; Niemeijer, N.D.; Jonker, M.A.; Eijkelenkamp, K.; Jansen, J.C.; van Berkel, A.; Timmers, H.; Kunst, H.P.M.; Bisschop, P.; Kerstens, M.N.; et al. The penetrance of paraganglioma and pheochromocytoma in SDHB germline mutation carriers. *Clin. Genet.* **2018**, *93*, 60–66. [CrossRef]
18. Patton, E.E.; Widlund, H.R.; Kutok, J.L.; Kopani, K.R.; Amatruda, J.F.; Murphey, R.D.; Berghmans, S.; Mayhall, E.A.; Traver, D.; Fletcher, C.D.; et al. BRAF mutations are sufficient to promote nevi formation and cooperate with p53 in the genesis of melanoma. *Curr. Biol.* **2005**, *15*, 249–254. [CrossRef]
19. Saito, Y.; Ishii, K.A.; Aita, Y.; Ikeda, T.; Kawakami, Y.; Shimano, H.; Hara, H.; Takekoshi, K. Loss of SDHB Elevates Catecholamine Synthesis and Secretion Depending on ROS Production and HIF Stabilization. *Neurochem. Res.* **2016**, *41*, 696–706. [CrossRef]
20. Gozal, E.; Shah, Z.A.; Pequignot, J.-M.; Pequignot, J.; Sachleben, L.R.; Czyzyk-Krzeska, M.F.; Li, R.C.; Guo, S.-Z.; Gozal, D. Tyrosine hydroxylase expression and activity in the rat brain: Differential regulation after long-term intermittent or sustained hypoxia. *J. Appl. Physiol.* **2005**, *99*, 642–649. [CrossRef]
21. Bechmann, N.; Poser, I.; Seifert, V.; Greunke, C.; Ullrich, M.; Qin, N.; Walch, A.; Peitzsch, M.; Robledo, M.; Pacak, K. Impact of extrinsic and intrinsic hypoxia on catecholamine biosynthesis in absence or presence of HIF2α in pheochromocytoma cells. *Cancers* **2019**, *11*, 594. [CrossRef] [PubMed]
22. Grouzmann, E.; Lamine, F. Determination of catecholamines in plasma and urine. *Best Pract. Res. Clin. Endocrinol. Metab.* **2013**, *27*, 713–723. [CrossRef] [PubMed]
23. Richter, S.; Peitzsch, M.; Rapizzi, E.; Lenders, J.W.; Qin, N.; de Cubas, A.A.; Schiavi, F.; Rao, J.U.; Beuschlein, F.; Quinkler, M.; et al. Krebs cycle metabolite profiling for identification and stratification of pheochromocytomas/paragangliomas due to succinate dehydrogenase deficiency. *J. Clin. Endocrinol. Metab.* **2014**, *99*, 3903–3911. [CrossRef] [PubMed]
24. Rao, J.U.; Engelke, U.F.; Sweep, F.C.; Pacak, K.; Kusters, B.; Goudswaard, A.G.; Hermus, A.R.; Mensenkamp, A.R.; Eisenhofer, G.; Qin, N.; et al. Genotype-specific differences in the tumor metabolite profile of pheochromocytoma and paraganglioma using untargeted and targeted metabolomics. *J. Clin. Endocrinol. Metab.* **2015**, *100*, E214–E222. [CrossRef] [PubMed]
25. Hobert, J.A.; Mester, J.L.; Moline, J.; Eng, C. Elevated plasma succinate in PTEN, SDHB, and SDHD mutation-positive individuals. *Genet. Med.* **2012**, *14*, 616–619. [CrossRef] [PubMed]
26. Lamy, C.; Tissot, H.; Faron, M.; Baudin, E.; Lamartina, L.; Pradon, C.; Al Ghuzlan, A.; Leboulleux, S.; Perfettini, J.L.; Paci, A.; et al. Succinate: A Serum Biomarker of SDHB-Mutated Paragangliomas and Pheochromocytomas. *J. Clin. Endocrinol. Metab.* **2022**, *107*, 2801–2810. [CrossRef] [PubMed]
27. Hoekstra, A.S.; Bayley, J.P. The role of complex II in disease. *Biochim. Biophys. Acta* **2013**, *1827*, 543–551. [CrossRef] [PubMed]
28. Astiarraga, B.; Martínez, L.; Ceperuelo-Mallafré, V.; Llauradó, G.; Terrón-Puig, M.; Rodríguez, M.M.; Casajoana, A.; Pellitero, S.; Megía, A.; Vilarrasa, N.; et al. Impaired Succinate Response to a Mixed Meal in Obesity and Type 2 Diabetes Is Normalized After Metabolic Surgery. *Diabetes Care* **2020**, *43*, 2581–2587. [CrossRef] [PubMed]
29. Hochachka, P.W.; Dressendorfer, R.H. Succinate accumulation in man during exercise. *Eur. J. Appl. Physiol. Occup. Physiol.* **1976**, *35*, 235–242. [CrossRef]
30. Armstrong, N.; Storey, C.M.; Noll, S.E.; Margulis, K.; Soe, M.H.; Xu, H.; Yeh, B.; Fishbein, L.; Kebebew, E.; Howitt, B.E.; et al. SDHB knockout and succinate accumulation are insufficient for tumorigenesis but dual SDHB/NF1 loss yields SDHx-like pheochromocytomas. *Cell Rep.* **2022**, *38*, 110453. [CrossRef]
31. Kimmel, C.B.; Ballard, W.W.; Kimmel, S.R.; Ullmann, B.; Schilling, T.F. Stages of embryonic development of the zebrafish. *Dev. Dyn.* **1995**, *203*, 253–310. [CrossRef] [PubMed]
32. Fischer, A.H.; Jacobson, K.A.; Rose, J.; Zeller, R. Hematoxylin and eosin staining of tissue and cell sections. *CSH Protoc.* **2008**, *2008*, pdb.prot4986. [CrossRef] [PubMed]
33. Rodenburg, R.J. Biochemical diagnosis of mitochondrial disorders. *J. Inherit. Metab. Dis.* **2011**, *34*, 283–292. [CrossRef] [PubMed]
34. Richter, S.; Gieldon, L.; Pang, Y.; Peitzsch, M.; Huynh, T.; Leton, R.; Viana, B.; Ercolino, T.; Mangelis, A.; Rapizzi, E.; et al. Metabolome-guided genomics to identify pathogenic variants in isocitrate dehydrogenase, fumarate hydratase, and succinate dehydrogenase genes in pheochromocytoma and paraganglioma. *Genet. Med.* **2019**, *21*, 705–717. [CrossRef]
35. Peitzsch, M.; Novos, T.; Kaden, D.; Kurlbaum, M.; van Herwaarden, A.E.; Müller, D.; Adaway, J.; Grouzmann, E.; McWhinney, B.; Hoad, K. Harmonization of LC-MS/MS measurements of plasma free normetanephrine, metanephrine, and 3-methoxytyramine. *Clin. Chem.* **2021**, *67*, 1098–1112. [CrossRef]

International Journal of
Molecular Sciences

MDPI

Review

Zebrafish Congenital Heart Disease Models: Opportunities and Challenges

Dixuan Yang †, Zhenjie Jian †, Changfa Tang, Zhanglin Chen, Zuoqiong Zhou ‡, Lan Zheng *,‡ and Xiyang Peng *,‡

State Key Laboratory of Developmental Biology of Freshwater Fish, Key Laboratory of Physical Fitness and Exercise Rehabilitation of Hunan Province, College of Physical Education, Hunan Normal University, Changsha 410000, China; 202220152984@hunnu.edu.cn (D.Y.); 202030172020@hunnu.edu.cn (Z.J.); changfatang@hunnu.edu.cn (C.T.); zhanglinchen@hunnu.edu.cn (Z.C.); zhouzuoqiong@hunnu.edu.cn (Z.Z.)
* Correspondence: lanzheng@hunnu.edu.cn (L.Z.); xiyangpeng@hunnu.edu.cn (X.P.)
† These authors contributed equally to this work.
‡ These authors also contributed equally to this work.

Abstract: Congenital heart defects (CHDs) are common human birth defects. Genetic mutations potentially cause the exhibition of various pathological phenotypes associated with CHDs, occurring alone or as part of certain syndromes. Zebrafish, a model organism with a strong molecular conservation similar to humans, is commonly used in studies on cardiovascular diseases owing to its advantageous features, such as a similarity to human electrophysiology, transparent embryos and larvae for observation, and suitability for forward and reverse genetics technology, to create various economical and easily controlled zebrafish CHD models. In this review, we outline the pros and cons of zebrafish CHD models created by genetic mutations associated with single defects and syndromes and the underlying pathogenic mechanism of CHDs discovered in these models. The challenges of zebrafish CHD models generated through gene editing are also discussed, since the cardiac phenotypes resulting from a single-candidate pathological gene mutation in zebrafish might not mirror the corresponding human phenotypes. The comprehensive review of these zebrafish CHD models will facilitate the understanding of the pathogenic mechanisms of CHDs and offer new opportunities for their treatments and intervention strategies.

Keywords: zebrafish model; congenital heart defects; single-defect heart disease; heart disease syndrome

1. Introduction

Heart disease is a significant global cause of mortality and reduced life expectancy, with the current global prevalence of congenital heart defects (CHDs) being 9.410%, which is lower than that in China (4.905%). CHDs are structural abnormalities occurring during embryonic and neonatal heart development, accounting for approximately 30% of cardiac cases in infancy, and mortality rates are associated with the severity of these defects [1]. These defects can manifest in various heart parts, such as the heart lining, the septum between the atria and ventricles, the valves, and major arteries and veins [2]. CHDs exhibit variability in their prevalence, severity, and affected cardiac tissues and can occur alone or as part of syndromic diseases impacting multiple organs [3]. Animal models, particularly the zebrafish, have become significant in investigating the pathogenesis of CHDs, owing to their small size, high reproduction rate, rapid growth cycle, and genetic similarity to humans, with 82% of related genes having orthologous genes in humans [4]. Over the last two decades, zebrafish models have been extensively used in studies involving heart development and cardiovascular diseases [5]. Researchers have effectively used gene editing techniques to establish numerous zebrafish CHD models, enabling the verification of genome-wide association study data in patients with cardiovascular disease and facilitating clinical treatment investigations [6]. In this context, this review evaluated the application of zebrafish in CHD studies, offering insights for further leveraging this model to explore CHD pathogenesis, clinical diagnosis, and treatment.

2. Relevant Sections

In this section, we discussed the advantages and disadvantages of using zebrafish as a model for CHDs in vertebrates, as well as the early development of the zebrafish heart. We also summarized the former point into a table (Table 1).

Table 1. Summary of advantages and disadvantages of zebrafish as a model for CHDs in vertebrates.

Advantages	Disadvantages
Similar to the electrophysiology of the human heart.	Absence of pulmonary circulation in the double-chambered heart.
Embryos can survive for a certain period even with impaired cardiovascular function.	Central venous pressure differs from that in humans.
Heart can regenerate after being injured.	The ventricles are primarily filled through atrial contraction.
Embryonic heart develops rapidly.	There are still differences in the ionic flow.
The formation pattern of the aortic arch is similar to that of mammals.	Lack of sarcolemmal T-tubules in cardiomyocytes.
Embryos and young fish are optically transparent.	Force–frequency relationship is distinct from that in humans.
The usability and accessibility of gene editing.	Lack a conduction system with specialized Purkinje fibers.
High homology with human genes.	Relatively limited antibodies and reagents.
Embryo development outside the body.	Fewer fibroblasts in adult fish hearts.
Ease, speed, and affordability of maintenance and breeding.	Genome duplication.

2.1. *Advantages of Zebrafish as a Model of Vertebrate CHDs*

2.1.1. Cardiovascular System Has Physiologically Advantageous Characteristics

Zebrafish is significant among model organisms owing to its similarity in heart rate to that of humans. An adult zebrafish exhibits clear P, QRS, and T waves on electrocardiogram analysis [7], closely resembling human cardiac electrophysiology [8–10]. In contrast, traditional animal models, such as mice, excessively rely on the cardiovascular system for obtaining oxygen during early embryonic development. When CHDs occur in mouse embryos, they can rapidly cause poor overall health or even death, with limited regenerative capacity post-birth, occurring only within the first 7 days [11–13]. Conversely, zebrafish embryos can survive for a period even without a fully functional cardiovascular system, obtaining oxygen through passive diffusion. Notably, the zebrafish heart can fully regenerate after injury [14–17]. Moreover, the zebrafish embryo's heart develops considerably faster than that of other model animals, beating for just one day as opposed to a week for mice [5,18]. The zebrafish has six pairs of aortic arch arteries, unlike mammals that possess only five pairs. Nonetheless, the formation and pattern of aortic arch arteries in zebrafish closely resemble that in mammals [19–21]. These unique characteristics make zebrafish an excellent model for studying various aspects of cardiovascular development, CHDs, and cardiac regeneration mechanisms.

2.1.2. Imaging Technology Is Mature

Optogenetics involves measuring and manipulating cellular activity using genetically encoded light-sensitive proteins [22,23]. In terms of cardiology, optogenetics has wide applications [24]. With their optical transparency, zebrafish embryos and larvae enable the visualization of morphological changes in the cardiovascular system during development through transmitted light or fluorescence imaging techniques [20]. Additionally, two-photon microscopy facilitates the imaging and analysis of cardiac functions in zebrafish [25]. Leveraging transgenic zebrafish technology, researchers can simultaneously observe instantaneous cardiac calcium changes and contractions in juvenile fish [26,27]. Zebrafish have become significant in cardiac optogenetic studies with technological advancements. For example, researchers use optical tools and transgenic zebrafish possessing light-gated ion channels (rhodopsin) to locate pacemaker points in the heart and apply specific light-stimulation methods. This approach enables the optical control of the heart rate, the simulation of various heart disease states (such as tachycardia and bradycardia), and the elucidation of organ function emergence during development [28]. For example, researchers utilized optogenetics to study the impact of the zebrafish Popeye domain containing a gene 2 knockout. This led to irregular atrial and ventricular activities, atrioventricular node conduction failure, and varying degrees of atrioventricular blockage [29]. With the continuous improvement of optogenetic techniques [24], research into the pathological or physiological structure and function of the heart using zebrafish has become feasible.

2.1.3. Gene Editing Tools Are Mature

Comparing the human reference genome with the zebrafish genome revealed that approximately 70% of human genes have at least one clear zebrafish ortholog [4]. Recent studies have further revealed a more comprehensive map of zebrafish genetic data, building upon the known genome sequence of zebrafish [30]. Forward and reverse genetics methods have mature application technologies in zebrafish embryos. The forward genetics method involves inducing random mutations through radiation, chemical treatment, or the insertion of exogenous DNA to obtain a strain with a stable pathological phenotype, enabling the exploration of the genetic basis of diseases. For example, forward genetics methods have been pivotal in screening and identifying mutations that cause valve defects and aortic coarctation [31,32]. Conversely, reverse genetics involves using gene editing technology to directionally alter previously identified genes and analyze the consequential effects of these genetic changes on organisms [33]. An example of this approach includes using the clustered regularly interspaced short palindromic repeats (CRISPR)/Cas9 technology to develop a *heg1* knockout zebrafish strain, revealing the regulatory role of *heg1* in heart failure and thrombosis and its potential application in cardiovascular drug screening [34]. The high degree of genetic conservation, ease of genetic manipulation, external development, and availability of large numbers of embryos daily accelerate research on CHDs, making zebrafish an excellent model for studying CHDs. In the Section 3 below, we focus on reviewing the use of gene editing to model CHDs in zebrafish, which will contribute to our understanding of the mechanism of CHDs.

2.2. *Disadvantages of Zebrafish as a Model of Vertebrate CHDs*

The heart is the first organ to develop and function in the embryo of zebrafish. In contrast to those in mammals, the coronary vessels in zebrafish develop within the first few weeks post-fertilization, and the myocardial trabeculae do not undergo compaction, remaining as permanent structures [5,35]. In addition, some limitations stem from differences between zebrafish and mammalian anatomy and physiology; zebrafish have only two wall chambers, lack a pulmonary circulation, and lack a conduction system with specialized Purkinje fibers, which limits the generation of models of septal defects or conduction system diseases. Furthermore, in fish, diastolic ventricular filling is primarily determined by atrial contraction, unlike the central venous pressure in humans [36].

Most of the calcium responsible for ventricular cardiomyocyte contraction in humans originates from intracellular sarcoplasmic reticulum stores; however, in zebrafish sarcoplasmic reticulum, calcium release is limited [37]. The lack of sarcoplasmic T-tubules is less dependent on sarcoplasmic reticulum calcium cycling and more dependent on sarcoplasmic T-type calcium currents [38]. The majority of calcium sources are extracellular and enter primarily through sarcoplasmic T-type calcium channels [38,39]. In contrast, in humans, the calcium-induced generation of sarcolemmal calcium currents occurs through L-type calcium channels, while T-type calcium channels are significantly absent [40]. In the healthy mammalian heart, the force–frequency relationship is positive; that is, the force of myocardial contraction increases with increasing heart rate [41]. These differ from observations in zebrafish [42]. These aspects need to be considered when using zebrafish to model human myocardial contractility-related diseases. Despite the similarity of the electrophysiological basis of the heart between zebrafish and humans, as previously recounted, there are still differences in the flow of ions, that is, the depolarizing flow of sodium ions, calcium ions, and the repolarizing flow of potassium ions [39]. Therefore, a better understanding of zebrafish cardiac electrophysiology and its limitations is needed to model diseases related to human cardiac electrophysiology. Notably, there has been successful modeling of human genetic repolarization disorders using zebrafish [7,43,44]. Notably, this may be a model of choice for cardiac electrophysiology associated with abnormal repolarization but may be less suitable for the study of depolarization disorders or calcium-regulated arrhythmias [45]. Moreover, the adult zebrafish heart contains a small number of fibroblasts [46]. While various stimuli trigger fibrosis in mammals, they do not produce similar results in zebrafish, limiting our capacity to mimic specific fibrotic pathologies [47,48].

During teleost evolution, the zebrafish genome underwent a duplication event, leading to the existence of two copies of several genes that typically perform redundant functions [49,50]. This genomic duplication poses a challenge in generating mutants, as it necessitates the simultaneous knockdown of both copies of the genes. Compared to mammalian models, the availability of antibodies and other reagents for zebrafish is relatively limited. In addition, some zebrafish genes have no homologs of known human genes but exercise functions similar to other genes. This should be remembered when researching a specific gene related to CHDs in zebrafish.

2.3. Early Heart Development in Zebrafish

During zebrafish embryonic development, the early heart originates from the ventrolateral cardiogenic zone of embryos at 3–4 h post-fertilization (hpf). By 5 hpf, atrial, ventricular, and endocardial progenitor cells cluster in the ventrolateral region before migrating toward the midline and dorsal side. In this migration process, the cells reach both sides of the midline by 16 hpf. Subsequently, at 20 hpf, the precursor cells on both sides gradually coalesce to form the cardiac cone, transitioning into the cardiac tube. By 24 hpf, the atrial cell population is distinctly positioned to the left and in front of the ventricular cell population. Visible atrial and ventricular cavities begin to form at 30 hpf. At 36 hpf, the embryonic heart adopts a circular morphology, signaling the progressive differentiation and specialization of the heart tube regions. Simultaneously, the head-end develops a primitive outflow tract, and endocardial cells start to form at the atrioventricular intersection, with matrix materials moving closer to the cardiac cushion tissue. By 48 hpf, the endocardial cushion tissue matures into a functional valve. Figure 1 shows the visual representation of these developmental stages from a to i.

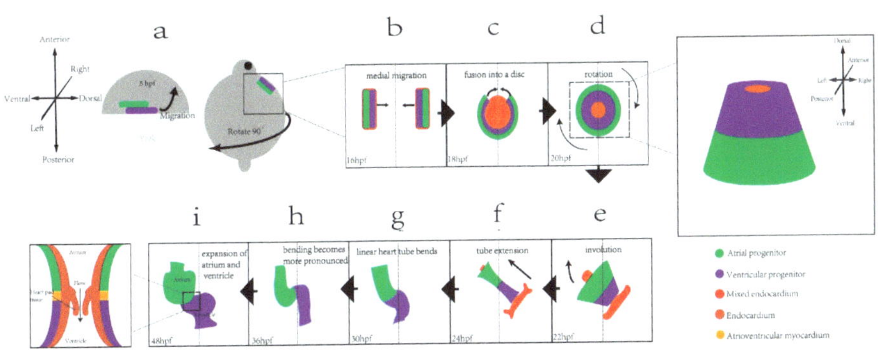

Figure 1. Zebrafish heart development process. (**a**) Cardiac precursor cells are initially on the ventrolateral side, while ventricular precursor cells are on the dorsal side and edge of atrial precursor

cells. Endocardial precursor cells are on the medial side of both precursor cells; (**b**) cardiac precursor cells migrate medially; (**c**) intracardiac mesangial cell groups form the cardiac disc at the midline, and the ventricular and atrial precursor cell groups migrate toward the midline; (**d**) the myocardium begins to fuse from the posterior portion of the cardiac disc formed by the endocardium and continues to surround the endocardium to form the cardiac cone. For the ventricular anterior, the somatic cell group is at the tip, and the atrial precursor cell group is at the bottom; (**e**) the cardiac cone rotates and tilts for the atrial precursor cells to be located to the left of the ventricular precursor cells, and the endocardium and cardiomyocytes fuse; (**f**) the cardiac cone extends for the atrial precursor cell group to be located to the left and in front of the ventricular precursor cell group; (**g**) the heart tube undergoes circularization to form a slightly "S"-shaped tube, and the atrial and ventricular cavities are formed; (**h**) the heart undergoes circularization to form a morphology; and (**i**) the chamber expands, and endocardial cells and the matrix form heart cushion tissue, which form a valve and perform valve function.

3. Discussion

Heart defects result from an intricate interplay of genetic and environmental factors. During heart development, any slight mistake may lead to deformity or dysfunction. Mutations can affect cardiac development alone or produce a series of defects, including developmental syndromes. We discuss the genes related to single-defects and syndromes (Table 2) and highlight that the zebrafish embryo serves as a valuable model for studying the impact of teratogens on heart development.

Table 2. Summary of zebrafish CHD model.

Simulated Human Disease	Human Gene Orthologs	Knock Zebrafish Gene	Zebrafish Main Phenotype	References
Heart chamber size defects	*NR2F1*	*nr2f1a*	The atria and atrioventricular canals are smaller.	[51]
	SEMA3F	*sema3fb*	The cardiac chambers become smaller; the cardiac muscle cells in the chamber become smaller.	[52]
Looping defects	*NODAL*	*spaw*	Abnormal cardiac looping.	[53,54]
	DAND5	*dand5*	Abnormal cardiac looping and displacement.	[53]
	PKD2	*pkd2*	Asymmetrical defects in the internal organs (including the heart) and brain.	[55]
	WNT3a	*wnt3a*	Cardiac asymmetrical defects.	[56]
	HAS2	*has2*	Displacement of the heart tube to the midline.	[57]
Atrioventricular valve defects	*KLF2*	*klf2a*	Valvular hypoplasia; absence of leaflets.	[58]
	PKD1	*pkd1a*	Increased incidence of retrograde blood flow.	[59]
	TNNI1	*tnni1b*	Phenotypes such as slow heart rate, lack of valve leaflets, and cardiac tube malformations.	[60]
	NFATC1	*nfatc1*	Atrial enlargement; retrograde blood flow in the atrioventricular canals.	[61,62]
Outflow tract stenosis	*TRAF7*	*traf7*	Embryonic heart displacement; pericardial edema and facial and cranial defects.	[63]
	PAK1	*pak1*	Outflow tract obstruction.	[64]
	ACVRL1	*acvrl1*	Outflow tract stenosis.	[65]

Table 2. *Cont.*

Simulated Human Disease	Human Gene Orthologs	Knock Zebrafish Gene	Zebrafish Main Phenotype	References
Tetralogy of Fallot	*BVES*	*bves*	Abnormal circularization; ventricular outflow tract stenosis.	[66]
	WDR62	*wdr62*	Abnormal circularization; narrowed heart chambers and thin walls.	[67]
CHARGE syndrome	*CHD7*	*chd7*	Abnormal branch of first branch arch.	[21]
Noonan syndrome	*PTPN11*	*ptpn11a*	Cardiac edema; craniofacial defects.	[68,69]
Alagille syndrome	*JAG1;AG2*	*jag1a;jag2b*	Cardiac edema.	[70,71]
	MIB1	*mib1*	Periaortic hemorrhage; cranial hemorrhage; and pericardial sac dilatation.	[72,73]
Axenfeld–Rieger syndrome	*FOXC1*	*foxc1a*	Cardiac edema; hypoplastic ventricles; short outflow tract; and poor ventricular contractility.	[74,75]
	FOXC1	*foxc1b*	Cardiac looping defect.	[75–77]
	PITX2	*pitx2*	Abnormal cardiac morphology; cardiac arrhythmia; and fibrosing cardiomyopathy.	[78]

3.1. *Zebrafish Single-Defect CHD Model*

3.1.1. Heart Chamber Size Defects

During the process of chamber formation, an abnormal expression or mutations of certain chamber development-related genes, such as nuclear receptor subfamily 2 group F member 2 (*NR2F2*), may cause human CHDs. Proper cardiac chamber size and proportions are essential for effective cardiac function in vertebrates, as blood circulation relies on the continuous contraction of the atria and ventricles [79–81]. The two-chamber heart of zebrafish provides a simpler model for studying mechanisms related to chamber size regulation compared with the four-chamber hearts of vertebrates [82–85]. Using zebrafish models, previous studies have investigated genes and transcription factors involved in regulating chamber size.

NR2F proteins, such as NR2F1 and NR2F2, are highly conserved solitary nuclear receptors. NR2F1 is crucial in neurodevelopment, and NR2F2 is implicated in cardiac genesis [86]. The expression levels of both proteins overlap in the atrial cardiomyocytes of the heart [87–90]. NR2F2 has been particularly highlighted as being associated with CHDs [79]. Notably, it has been shown that, in zebrafish, Nr2f1a promotes atrial development and differentiation at a level equivalent to NR2F2 in mammals [51]. The atrium of *nr2f1a* mutants was observed to be smaller than that of the wild-type zebrafish, indicating a potential regulatory role of NR2F1a in atrial size and development through bone morphogenetic protein (Bmp) signaling, as revealed by a recent study [51]. This study introduces novel avenues for investigating the etiology of CHDs associated with atrial size.

Semaphorins are a large family of secreted or membrane-associated glycoproteins [91]. Among them, class 3 semaphorins (Sema3s) are used as guidance signals for cardiovascular development [92,93]. Sema3 signaling is mediated through the plexin receptor and its coreceptors, neuropilins [91]. In previous studies, a *sema3fb* knockout zebrafish model was established to investigate the molecular mechanism of chamber-specific cardiomyocyte differentiation [52]. The findings revealed that at 3 hpf, both cardiac chambers of the *sema3fb* knockout fish were smaller than those of the wild-type, with the ventricular differentiation process remaining unaffected by Sema3fb deletion; however, the cardiomyocytes in the atria were smaller than their wild-type counterparts. This discrepancy may be caused by a disruption in the expression of specific differentiation genes [52]. The secretion of Sema3 signals by cardiomyocytes potentially facilitates the establishment of a boundary between the ventricle and the atrium via spatially specific regulatory signals. This activity is essential for the normal chamber development to its appropriate size [52]. Studies involving zebrafish models have revealed an association between CHDs and aberrations in Sema3 signaling, offering valuable insights for the clinical understanding, diagnosis, and treatment of CHD-related conditions [94,95].

3.1.2. Left and Right Asymmetry (Looping Defects)

Heart development is significantly influenced by left–right pattern disorders, with deviations in this pattern potentially causing heart defects [96]. An illustration of this is the development of abnormal asymmetry in the heart, causing defects in cardiac circulation and leading to CHDs during embryonic development, such as left or right atrial isomerism [97]. Zebrafish, similar to vertebrates, such as humans, mice, and chickens, undergo heart development guided by a left–right asymmetric pattern with a comparable anatomy and molecular mechanisms [98,99]. Notably, the various stages of cardiac asymmetry, commencing from the formation of the heart field to the heart tube and looping, are observable through microscopic examination [100]. Researchers have meticulously investigated the distinctive stages of asymmetric cardiac development through the zebrafish model, scrutinizing the associated events, processes, and molecular mechanisms.

The asymmetric development of the heart in zebrafish starts from the Kupffer's vesicle (KV) [101]. This is a transient structure that marks the first break in the left–right symmetry within the embryo [102]. Within the KV, cilia act as mechanical sensors of shear force, generating a directional liquid flow known as the nodal flow; the perception of mechanical forces by the cilia is pivotal in establishing left–right patterning, and *wnt3a* and *8a* modulate cilia generation [56,101,103]. Polycystin-2 (PKD2) is crucial for intraciliary calcium oscillations within the left–right organizer. These mechanical stimuli, transmitted through ciliary structures, influence calcium transients, regulating the expression of vital asymmetric molecules [55,103,104]. However, *curly up* (*cup*) affects the gene homologous to zebrafish *pkd2*, which encodes a Ca^{2+}-activated non-specific cation channel Pkd2 [55], and *southpaw* (*spaw*) is an early asymmetrically expressed gene around the KV [54]. During the 4–6 somite stage, Spaw is symmetrically expressed around the KV and gradually expands toward the left lateral plate mesoderm (LPM) from the 10–12 somite stage; this asymmetric expression depends on the leftward nodal flow within the KV [105]. *spaw* expression is regulated by *left–right determination factor 1* (*lefty1*), Bmp signaling, and *dand5*. Bmp signaling can modulate *spaw* expression through *lefty1*, influencing *spaw* expression in the LPM, while inducing *lefty1* expression in the midline [106], and *lefty1* should maintain a specific ratio with *spaw* expression [107–109]. Furthermore, a localized Bmp protein source induces heart tube localization, whereas the asymmetric expression of hyaluronan synthase 2 affects the correct guidance of locally expressed Bmp proteins to the heart tube [57].

The characteristic feature of lateral defects is the left–right patterning defect during embryonic development. A study used exome sequencing to screen for candidate genes in 70 patients with CHDs and lateral defects, identifying candidate genes *TRIP11*, *DNHD1*, *CFAP74*, and *EGR4* [110]. Owing to the rapid development, ease of operation, and highly conserved left–right patterning process across vertebrate species, the zebrafish was selected for analyzing candidate genes. The knockdown of zebrafish *trip11*, *dnhd1*, and *cfap74* results in significant cardiac looping abnormalities, a disturbed expression of spaw and related genes (*lefty2* and *pitx2*), and the altered formation and function of KV cilia during embryonic development, respectively. The left–right asymmetric defects of these mutant hearts can be effectively rescued by introducing the corresponding candidate mRNAs [110]. Cilia are crucial for regulating left and right patterning, and defects in genes related to their function can cause abnormal cardiac development. Therefore, zebrafish models are vital in investigating the specific roles of these genes in left–right patterning to understand CHDs resulting from human gene variations and are essential for CHD diagnosis and treatment.

3.1.3. Atrioventricular Valve Defects

Heart valves are crucial in maintaining the one-way blood flow within the heart. Defects, malformations, or aberrant expression of related genes during valve development are common factors contributing to CHDs, such as bicuspid aortic valve and mitral valve prolapse [111–113]. Therefore, understanding the developmental processes and molecular mechanisms underlying heart valve formation is crucial for developing effective treatment methods and preventive measures. The zebrafish is a valuable model for investigating early valve development because its heart can be visualized in vivo during embryonic stages with single-cell resolution [25,114]. The formation of heart valves involves collaboration between endocardial cells and the extracellular matrix [115–117]. Endocardial cells are responsive to stimuli from blood flow, activating Krüppel-like factor 2a (Klf2a)-mediated mechanosensitive channels. The disruption of these mechanically sensitive channels can impact the valve morphology [118]. *klf2a* knockdown in zebrafish embryos results in underdeveloped valves, usually lacking leaflets altogether [58]. Moreover, Klf2a expression is influenced by the PKD family, which is crucial in valve formation through its mediation by Camk2g [59,119]. Mutant larvae deficient in *pkd1a* exhibit an increased incidence of retrograde blood flow at 78 hpf, with only 66% of surviving larvae displaying elongated upper leaflets, a lower rate than that of a normal development [59].

In human patients with valvular heart disease, the expression of troponin I type 1 (*TNNI1*) is inhibited or inactivated [120]. The loss of *tnni1* inhibits the myocardial Wnt signaling pathway, causing defects in atrioventricular valve development. *tnni1b* overexpression in homozygous zebrafish embryos can partially resolve valve development defects. When *tnni1b* (a homolog of human *TNNI1*, a structural and regulatory protein involved in cardiac contraction) is knocked out in zebrafish, phenotypes, such as a slowed heart rate, lack of valve leaflets, and cardiac tube malformations, occur. Additionally, all homozygous embryos died within 1 week [84]. The expression levels of other sarcomeric genes, such as *ctnnt*, *ctnnc*, *myl7*, *myh7*, and *myh6*, remain unaffected, further confirming that abnormalities in TNNI1b contribute to the observed cardiac phenotype in the mutants [60]. However, further studies are required to determine the precise mechanism by which TNNI1b influences the myocardial Wnt signaling pathway.

NFATC1, a nuclear factor-κB-related transcription factor, has been associated with congenital heart valve septal defects in humans [121–123]. Zebrafish studies have revealed that *nfatc1* deletion results in heart valve defects, and *nfatc1* knockout mutants display enlarged atria compared with wild-type fish [61]. This atrial enlargement resembles the condition observed in patients with mitral regurgitation [62], and nfatc1 knockout zebrafish exhibit retrograde blood flow in the atrioventricular canal [61]. *nfatc1* deficiency may cause a diminished recruitment and proliferation of valve interstitial cell (VIC) precursors, causing VIC defects and disturbances in valve extracellular matrix organization [61]. *nfatc1*-mediated regulation is crucial for early VIC establishment, which is essential in valve development during the embryonic stage and subsequent normal valve function in adulthood.

3.1.4. Outflow Tract Stenosis

Neural crest cells are principal cell sources in the development of the cardiac outflow tract (OFT) [124], a critical component for proper cardiac function. Their migration and differentiation within the OFT region are crucial in cardiac development. However, abnormalities arising in the OFT during development can cause CHDs, which are prevalent among approximately 30% of cases [125,126]. Consequently, understanding the role of neural crest cells in OFT development is essential for comprehending CHD pathogenesis and exploring potential therapeutic interventions.

Cardiac neural crest cells are derived from a cell population originating from the cranial part of the neural tube [127]. Tumor necrosis factor receptor-associated factor 7 (*TRAF7*), a member of the multifunctional TRAF family, exhibits high expression levels in the neural crest and its derivatives, particularly in the cardiac OFT [63]. In zebrafish models, *traf7* knockout causes an increased probability of abnormal embryonic heart tube displacement, accompanied by noticeable pericardial edemas, irregular cardiac looping, and craniofacial anomalies [63]. While no OFT-specific abnormalities have been documented, the migration ability of neural crest cells to form OFT pads underscores their potential significance in this process [127]. Additionally, the significant reduction in the expression of the neural crest marker (*sox10*) in zebrafish embryos after *traf7* knockout indicates a possible association with the observed OFT defective phenotype [63]. Therefore, further investigations into the OFT defective phenotype in *traf7* knockout zebrafish are required.

During embryonic development, the neural crest is regulated by p21-activated protein kinase 1 (Pak1). This study demonstrated that *pak1* knockout in zebrafish embryos causes defects in neural crest development and, subsequently, blockage in the OFT. This finding highlights the significance of the Pak/Mek/Erk/Gata6 signaling pathway in normal neural crest development. Gata6 signaling activation via extracellular regulated protein kinases (Erk) is reportedly a central mechanism in this process [64]. Notably, *GATA6* mutations in humans have been associated with various CHDs, including OFT malformations [128]. This association underscores the significance of understanding the molecular pathways involved in neural crest development for preventing and treating CHDs.

Studies revealed that activin A receptor-like type 1 (Alk1), a transmembrane serine/threonine receptor kinase belonging to the TGF-β family and encoded by *ACVRL1*, is related to the occurrence of CHDs [129,130]. *acvrl1* mutations have been shown to affect zebrafish OFT morphology. While applying the zebrafish model, a study suggested a novel perspective on OFT development: biological signals generated by fluid forces can stimulate endocardial proliferation and an *acvrl1*-dependent increase in endothelial cells, contributing to the formation of the OFT cavity with a specific endocardial thickness [65]. In zebrafish embryos where Acvrl1 is knocked down, OFT endocardial cells fail to accumulate normally at 51 hpf, causing the narrowing of the OFT cavity [65]. This investigation introduces new insights for evaluating the etiology of CHDs associated with abnormal OFT morphology.

3.1.5. Tetralogy of Fallot (TOF)

The most common type of cyanotic CHD is TOF, with an incidence rate of 0.34 per 1000 live births [131]. TOF can occur in the setting of additional non-cardiac anomalies or in isolation [132]. TOF is characterized anatomically by four structural defects, namely ventricular septal defects, right OFT/pulmonary artery stenosis, an overriding aorta, and right ventricular hypertrophy [133]. A vital clinical indicator for diagnosing TOF is the degree of right ventricular outflow tract (RVOT) stenosis [134].

The first genetic cause of TOF identified was *NKX2.5* mutation, a crucial transcription factor that controls heart development [135,136]. Similarly, *GATA4* mutation can hinder its interaction with *NKX2.5*, contributing to TOF [137]. Blood vessel epicardial substance (BVES), also known as POPDC1, is a highly evolutionarily conserved membrane protein that is highly expressed in the adult vertebrate heart [138,139]. Mutations or reduced *BVES* gene functioning are found in patients with non-syndromic TOF and heart failure [140–142]. In mice, *Bves* is expressed in the developing cardiomyocytes and coronary endothelial cells and in the conduction system in adult mice [143–145]. A study has found that the rs2275289 (p.R129W) single nucleotide polymorphism in BVES reduces its own expression and likely decreases the expression of *NKX2.5* and *GATA4* [141], and in studies involving *bves* knockdown zebrafish, abnormal annularization and ventricular OFT stenosis were observed, in addition to reduced *nkx2.5* and *gata4* expression levels, causing the manifestation of TOF, a congenital heart disease [66]. Therefore, further exploration of the connection and detailed regulatory mechanisms between the downregulation of *BVES* and the occurrence of RVOT stenosis in patients with TOF is required. Another recent study highlighted WD repeat domain 62 (*WDR62*) as a novel susceptibility gene associated with CHDs through sequencing analysis in a large number of human patients with CHDs, with a relatively high mutation frequency associated with TOF [67]. Subsequently, it was demonstrated that *wdr62* was abundantly expressed in the zebrafish heart through whole-mount in situ hybridization. The knockdown of *wdr62* by morpholino in zebrafish resulted in heart defects in 80% of zebrafish, including abnormal cardiac looping, narrowed chambers, thin chamber walls, impaired OFT rotation, and TOF-related defects, as evidenced in histological sections. Insights from this study revealed the correlation between *WDR62* and CHDs, indicating that *WDR62* could influence cardiac development by affecting spindle assembly and the cell cycle of cardiomyocytes [67]. While zebrafish hearts have a relatively simple two-chamber structure and generate lower pressures than those of humans, the aberrant annularization of fish embryos observed in *bves* and *wdr62* knockdown zebrafishes, with the former also exhibiting OFT stenosis, implies potential limitations in fully recapitulating all four classic TOF phenotypes [66,67]. Malformations in the OFT during embryonic development, similar to those observed in human patients, can trigger the emergence of the other three structural defects, even if all phenotypes are not present [146–148]. Given the critical role of OFT development in the pathogenesis of TOF, emphasizing abnormalities in this process through the zebrafish model can provide valuable insights into the pathophysiology of human TOF.

3.2. Zebrafish CHD-Related Syndrome Model

3.2.1. CHARGE Syndrome

CHARGE syndrome (Coloboma, heart defects, Atresia of the choanae, retarded growth and mental development, genital anomalies, and Ear malformations), an autosomal dominant genetic disease, occurs at a frequency of 1 in 8500 to 1 in 15,000 live births worldwide [149,150]. Among individuals affected by CHARGE syndrome, approximately 80% will develop CHD [151]. These cardiac abnormalities can manifest as OFT defects, patent ductus arteriosus, atrioventricular septal defects, and aortic arch abnormalities [152]. Notably, *CHD7* mutations are identified in approximately two-thirds of patients with CHARGE syndrome, and such mutations have been detected in some individuals with isolated CHDs [152–154]. This suggests that *CHD7* and its downstream genes are possibly crucial in the development of cardiac defects in CHARGE syndrome [155]. *Chd7* heterozygous mice show various signs of CHARGE syndrome, including growth retardation and severe head bobbing, destruction of the lateral semicircular canals, septal defects, and genital abnormalities [156,157]. In contrast, *Chd7* heterozygous mice only live until 10.5 days after birth, making it impossible to study the overall function of *CHD7* and drug screening [154,156]. The recognized causes of death include growth retardation, axial rotation failure, peri-cardial swelling, and tail structure formation failure [158]. Zebrafish exhibit a lower tolerance to cardiovascular defects than other models, and *chd7* mutants display craniofacial anomalies, cardiac defects, gastrointestinal narrowing, and cranial nerve defects [21,159–161]. This is more advantageous for studying the overall function of CHD7 and cardiovascular-related diseases. Therefore, the zebrafish model is considered more suitable for studying the overall function of *chd7* and cardiovascular-related diseases than other models with a lower tolerance to cardiovascular defects [21]. The *chd7* mutant zebrafish model effectively mimics the cardiovascular phenotype observed in CHARGE and non-syndromic patients with CHD, with the mutant zebrafish displaying abnormal first-branch arch branches similar to the clinical symptoms of patients with CHARGE syndrome [21]. Despite similarities in the initial formation and pattern of arch arteries between zebrafish and mammals, zebrafish do not undergo the same complex remodeling of the aorta and pulmonary arteries [20]. As many patients with CHARGE syndrome require surgical interventions for conditions, such as cardiovascular malformations and tracheoesophageal fistula, complications, including decreased oxygen saturation, respiratory rate, and abnormal heart rate, usually arise post-operatively [162]. Furthermore, these surgical interventions expose patients to increased doses of anesthesia and frequent experiences of adverse effects during anesthesia, such as an abnormal heart rate, reduced oxygen saturation, and decreased respiratory rate [163]. Observations show that *chd7* mutant zebrafish exhibit decreased heart rates, prolonged anesthesia requirements, and elevated respiratory rates upon awakening from anesthesia [164], and a study investigated the gene expression of phox2ba and phox2bb in a zebrafish model of CHARGE syndrome. As *PHOX2B* is involved in neural crest development, similar to *CHD7*, the loss of *PHOX2B* function can lead to respiratory issues resembling those in patients with CHARGE syndrome [149,164–166]. Leveraging the *chd7* mutant zebrafish model to simulate the anesthesia process in patients with CHARGE syndrome can deepen the understanding of the molecular underpinnings of these adverse events, enhancing the safety of anesthetic procedures in this population.

3.2.2. Noonan Syndrome (NS)

Cardiac damage is a prominent NS feature, with CHDs occurring in 80% of patients [167]. The most prevalent manifestation of CHDs in NS is pulmonary valve stenosis (PVS), at approximately 40%, followed by atrial or ventricular septal defects (ASDs) and atrioventricular canal defects (AVCDs), at 8% and 15%, respectively [168]. *PTPN11* pathogenic variants are responsible for approximately 80% of cases involving PVS or ASDs [169]. Similarly, patients with AVCD and NS predominantly exhibit pathogenic variants in *PTPN11* [170]. Consequently, *PTPN11* emerges as a gene closely associated with NS-related CHDs. The (SHP2) protein encoded by *PTPN11* is distributed in the heart and regulates crucial processes, such as cell proliferation, migration, and differentiation during development [171]. In mouse *Ptpn11* knockout embryos, the embryos are damaged during implantation [172]. In zebrafish embryos, the complete absence of functional Shp2 allows for survival until 5–6 dpf; notably, the zebrafish genome encodes two *ptpn11* variants, *ptpn11a* and *ptpn11b*, and their respective encoded proteins Shp2a and Shp2b share 91% and 64% homology, respectively, with human SHP2 [173,174]. Further studies on zebrafish have revealed that double-knockout embryos for *ptpn11a* and *ptpn11b* exhibit severe cardiac edemas and craniofacial defects [68]. Creating the Shp2-D61G NS zebrafish model provides a specific disease target for different *shp2* mutants. Symptom variations in this mutant zebrafish model mirror those seen in individual human patients, with defects ranging in severity from mild to severe, including reduced body axis extension, heart edemas, craniofacial deformities, cardiac and mandibular edemas, significant developmental delays, and mortality before adulthood. Children with NS who have the D61G mutation are more likely to develop juvenile myelomonocytic leukemia-like myeloproliferative neoplasm (JMML-like MPN). Zebrafish embryos with the Shp2-D61G mutation also exhibited JMML-like MPN characteristics, including myeloid lineage expansion, mild anemia, and thrombocytopenia. Given the importance of fetal hematopoiesis in the development of JMML-like MPNs, the zebrafish Shp2-D61G mutant also serves as a dependable and distinctive model resembling JMML for studying the hematopoietic abnormalities resulting from SHP2 mutations in NS [69].

Shp2, a positive effector of Erk/mitogen-activated protein kinase (Mapk) signal transduction located downstream of most receptor tyrosine kinase (Rtks), is crucial in cellular processes [171,175]. The aberrant activation of the Mapk signaling pathway reportedly impairs ciliary function [176], particularly in NS, where pathogenic mutations in NS-related genes enhance this signaling cascade [177,178]. This emphasizes the significance of modulating Mapk signaling to prevent or ameliorate the onset of NS. One approach is the targeted inhibition of the rat sarcoma (Ras)/Mapk pathway, as demonstrated in a study that involved using trametinib, a MEK inhibitor, in infants with severe hypertrophic cardiomyopathy (HCM) and PVS associated with NS [179]. Notably, treatment with trametinib caused improvements in HCM and PVS, suggesting that the inhibition of Ras/Mapk overactivation could facilitate the remodeling of abnormal heart valves. Given the promising clinical outcomes observed with trametinib in NS, further systematic investigations are required to evaluate the impact of CHDs and their underlying biological mechanisms in NS. The zebrafish model is an attractive avenue for such studies because it can be rapidly used to assess the bioactivity and toxicity of various compounds using its absorbent gills and skin [180–182]. Zebrafish offer practical advantages over rodent models, including cost-effectiveness, ease of maintenance, and accelerated developmental processes, which can expedite study timelines and minimize animal stress induced by invasive procedures [183]. Trametinib has been less explored in zebrafish NS models, and existing studies primarily focus on its efficacy in zebrafish tumor models [184–187]. However, considering the benefits of zebrafish models and the clinical efficacy of trametinib, the zebrafish is a promising model system for investigating the intricate relationship between this drug and CHDs in NS.

3.2.3. Alagille Syndrome (AGS)

AGS is an autosomal dominant, complex multisystem disorder with a clinical incidence rate of 1/100,000 live births [188,189]. TOF is the most common complex structural abnormality in AGS, occurring in up to 16% of cases [189,190]. Other malformations include ventricular septal defects, atrial septal defects, aortic stenosis, and aortic constriction, which significantly contribute to the high morbidity and mortality in patients with AGS [189]. AGS is commonly associated with the Notch signaling pathway [191], with most patients exhibiting pathogenic mutations in JAGGED1 (*JAG1*)-encoding Notch pathway ligands or in the receptor *NOTCH2* [73]. The existing AGS zebrafish model, generated by knocking down *jagged1b* and *jagged2b*, has primarily focused on studying liver pathology [70,192]. However, recent studies have shown that *jag1b-* and *jag2b-* knockout double mutants and jag2b knockout zebrafish develop cardiac edemas and liver abnormalities [70,71]. Although the cardiac defects in the AGS zebrafish model do not sufficiently replicate CHDs in human patients with AGS, the underlying molecular mechanisms causing heart defects are evolutionarily conserved. This finding offers valuable insights into the clinical treatment and pathogenesis of human-related diseases associated with AGS.

The ubiquitination of the Notch ligand intracellular tail and its subsequent activation, a process requiring Mind bomb 1 (Mib1) and Mib2 [193,194], is crucial for cellular signaling. Similarly, MIB1 is significant in the Wnt/β-catenin signaling pathway, influencing its regulation [195]. Knockout zebrafish lacking *mib1* exhibit cardiovascular abnormalities, including periaortic and intracranial hemorrhages and pericardial sac dilatation [72]. The significant differences in phenotypic characteristics between *mib1* knockout zebrafish and individuals with AGS question the fidelity of the zebrafish model in simulating AGS [73]. Despite this limitation, the study reveals the potential use of the zebrafish mutagenesis model for efficiently evaluating novel genetic variations, particularly in cardiovascular contexts. This model can facilitate the early identification of disease-causing factors, expediting CHD diagnosis and treatment [72].

3.2.4. Axenfeld–Rieger Syndrome (ARS)

ARS is an autosomal dominant disorder characterized by ocular abnormalities. It also involves defects in cardiac development, such as mild left ventricular hypoplasia and abnormal valve and OFT formation [196–198]. The disease has an estimated prevalence of approximately 1 in 50,000 to 100,000 live births [199]. Approximately 40% of patients with ARS carry mutations in *FOXC1* or *PITX2* [200,201].

In zebrafish, there are two genes homologous to human FOXC1, namely *foxc1a* and *foxc1b*, with the proteins encoded by each sharing 66% and 55% homology with human FOXC1, respectively [202]. The genetic redundancy of *foxc1* in zebrafish has been previously reported [202]. When *foxc1a* is knocked out, juvenile fish exhibit various cardiovascular defects, including cardiac edemas, hypoplastic ventricles, shorter OFT, defective valve leaflets, and poor ventricular contractility, shortening their survival time [74,75]. These observed defects in zebrafish are similar to the clinical cardiac abnormalities in patients with ARS and *Foxc1* knockout mice [197]. Conversely, *foxc1b* knockout in zebrafish causes only a few juvenile fish to exhibit circulation defects, with no evident phenotype in adults [75–77]. Concurrent disruptions of both genes have been conducted subsequently, revealing that most double homozygous (*foxc1a* $-/-$; *foxc1b* $-/-$) zebrafish embryos develop heart defects, accompanied by more severe craniofacial malformations than those in *foxc1a* knockout homozygotes [75,203]. Furthermore, zebrafish with three mutant alleles (*foxc1a* $+/-$;

foxc1b −/−) exhibit mild pericardial edemas and low survival rates, attributed to the thicker and denser myocardial zone in the fish, which impacts cardiac function [75,203]. Notably, other allele combinations do not exhibit visible phenotypes in embryos or adults. These findings suggest that *foxc1b* may compensate for *foxc1a* deficiency later in zebrafish heart development. When *foxc1b* is entirely lost, the normal phenotype cannot be maintained, underscoring the concept of genetic buffering by two paralogs. The evaluation of mutated genes in zebrafish embryos provides valuable insights into their potential roles in patients carrying the mutation, underscoring the significance of understanding the functional implications of genetic mutations in disease pathogenesis.

Recent studies have demonstrated that zebrafish *pitx2c* mutations can cause cardiac defects similar to those in individuals with ARS [204,205]. An investigation indicated that *pitx2c* mutant zebrafish larvae did not exhibit the full spectrum of cardiac abnormalities associated with ARS; however, it confirmed that the asymmetrical development of the heart tube and morphogenesis were unaffected by the mutation [206]. In contrast, another study focusing on adult zebrafish models with *pitx2c* loss-of-function mutations revealed the development of cardiac morphological irregularities, such as atrial dysplasia, increased fibrosis, arrhythmias, and fibrotic cardiomyopathy [78]. These findings reveal the cardiac phenotype observed in a subset of clinical patients with ARS, suggesting a potential association. However, these cardiac complications in adult fish may be a secondary effect of altered cell metabolism in the heart, an event that causes cardiac dysfunction [78]. Consequently, the collective findings underscore the significance of further investigating the role of metabolic pathways in ARS pathogenesis, leveraging the zebrafish mutation model as a valuable tool for future studies.

3.3. Teratogen-Induced Models of Embryonic Heart Development Defects

Approximately 80,000 synthetic chemicals are manufactured annually, with some of them being teratogens [207]. Compared to mammalian models (such as mice), zebrafish are an excellent model for studying the effects of teratogens on heart development, owing to their high fecundity, external embryo development, and independence from the mother during embryogenesis. In the following section, we discuss the impact of teratogens on zebrafish embryo heart development.

In pregnant women, diabetes increases the risk of congenital heart defects in the fetus by three to five times [208]. Hyperglycemia is a primary teratogenic condition that can affect the expression of genes essential for heart development [209]. In animal experiments, exposure to hyperglycemia can cause fetal malformations, similar to those observed in diabetic embryos, even when the mother does not have diabetes [210–212]. These studies support the view that hyperglycemia itself is teratogenic to embryos. Exposing zebrafish embryos to hyperglycemia results in a range of cardiac abnormalities, such as an increased incidence of ventricular stagnation and valve regurgitation, decreased peripheral blood flow, pericardial edemas, increased diameter and thinning of the heart wall, and the abnormal expression of tbx5a [210]. In another study, embryos treated with high glucose concentrations exhibited an abnormal tbx5 expression, pericardiac edemas, patent hearts, and valve regurgitation [213]. TBX5 mutation in humans can cause CHDs [214], and the deletion of tbx5 results in heart defects in zebrafish embryos, causing the swelling and relaxation of heart mitochondria [215]. TBX5 is possibly the main factor involved in fetal heart malformation induced by hyperglycemia.

Fish are highly susceptible to dioxin exposure, showing toxic effects, such as the cardiovascular dysfunction observed in other vertebrates [216]. Zebrafish eggs exposed to dioxins shortly after fertilization exhibited severe cardiac abnormalities in embryos, such as reduced cardiac output, diminished peripheral blood flow, and pericardial edemas [217–219]. The damage detected decreases as time after fertilization increases [217]. Adult zebrafish injected with a lethally high dose of dioxin did not exhibit any morphological changes in their hearts [220]. Research demonstrated that dioxins significantly influence embryonic development. Exposure to dioxins reduces the number of myocardial cells, as indicated by the expression of the marker gene *cmlc2* at 48 hpf [218]. Activating aryl hydrocarbon receptor 2 in zebrafish cardiomyocytes leads to the same defects, indicating that cardiomyocytes are the targets of dioxins [221].

Alcohol, the most widely used and consumed drug in daily human life, has a wide range of irreversible side effects on human fetuses. Maternal alcohol consumption is considered to have a high teratogenic effect on fetuses, affecting their development, including the heart. Research shows that in zebrafish embryos, exposure to ethanol can result in significant increases in cardiac edemas, decreases in cardiac volume, and reductions in ventricular thickness [222]. Alteration in the Bmp and Notch signaling pathways results in abnormal valve morphology [223]. During embryogenesis, the Bmp signal transduction pathway is essential for various stages of heart development, including cardiac specification, differentiation, endocardium differentiation, cardiac ring formation, chamber morphogenesis, and outflow tract separation [17,224–227]. The Notch signaling pathway is essential for endocardial differentiation and the formation of the OFT [228–230]. Folic acid has the potential to mitigate the damage caused by ethanol to embryonic heart morphology, although it does not fully remove it [231,232]. Folic acid's antioxidant properties may weaken the oxidative stress caused by ethanol [233,234]. Thus, studying the impact of teratogens on the heart structure, morphology, and gene expression of zebrafish embryos, along with genetic techniques or drugs, can enhance our comprehension of disease origins and preventive strategies.

4. Future Directions

The zebrafish, a classic vertebrate with a high degree of genetic homology to humans, offers a valuable platform for examining the genes and mechanisms underlying CHDs. Using mature genetic and molecular technologies, researchers can efficiently screen and understand the genetic basis of human CHDs. Recognizing that certain pleiotropic gene mutations can cause numerous subclinical phenotypic effects is essential, as these events may collectively impact the fitness of affected individuals. Therefore, when specific disease candidate genes identified in patients with CHD are individually disrupted in zebrafish, corresponding phenotypes may manifest. In contrast, in humans, mutations in a single gene may not always cause observable effects and can involve multiple genes, highlighting the cumulative impact of pleiotropic genes. While the knockout of a single gene in zebrafish may not fully replicate the phenotypes in patients with heart disease-related syndromes, it may unveil evolutionarily conserved regulatory mechanisms governing vital developmental events in cardiac syndrome phenotypes. At the same time, human pluripotent stem cell models are used to corroborate findings obtained from the zebrafish model in investigating events, as the human genetic background can be preserved [235]. A further understanding of the defective phenotypes of gene deletions can offer insights into the origins of other phenotypes in humans, revealing potential treatment strategies for CHDs. Furthermore, by tracking cell lineage in regions affected by gene mutations associated with specific diseases, researchers can infer how genes influence other conditions during embryonic development. In recent years, several single nucleotide variants (SNVs) linked to CHDs have been discovered in humans, although the functions of many remain uncertain [236]. Preserving a substantial number of coding SNVs in zebrafish allows for quickly prioritizing disease variants in F0 and confirming their initial functional relevance in F1 [237]. The new base editing framework developed in this study is applicable to a wide range of SNV-susceptible traits in zebrafish, contributing to direct candidate validation and the prioritization of detailed mechanistic downstream studies [237]. The potential of zebrafish as a model for identifying the effects of SNV on cellular function during morphogenesis would be valuable. Leveraging the strengths of zebrafish in cardiac studies to establish a model for human CHDs can help elucidate the mechanisms underpinning these conditions and generate innovative approaches for their treatment and management.

Author Contributions: D.Y. and Z.J.: Conceptualization, Roles/Writing—Original draft; Z.C.: Visualization; X.P.: Writing—review and editing, Funding acquisition; Z.Z.: Supervision, Writing—review and editing, Funding acquisition, C.T. and L.Z.: Resources, Funding acquisition. All authors have read and agreed to the published version of the manuscript.

Funding: This research was funded by the National Natural Science Foundation of China, grant numbers 32100919, 81801392, and 32371182. Key Research and Development Program of Hunan Province, grant number 2020SK2104, National Students' Platform for Innovation and Entrepreneurship Training Program grant number 202210542038.

Conflicts of Interest: The authors declare no conflicts of interest.

References

1. Pate, N.; Jawed, S.; Nigar, N.; Junaid, F.; Wadood, A.A.; Abdullah, F. Frequency and pattern of congenital heart defects in a tertiary care cardiac hospital of Karachi. *Pak. J. Med. Sci.* **2016**, *32*, 79–84. [PubMed]
2. Hoffman, J.I.; Kaplan, S. The incidence of congenital heart disease. *J. Am. Coll. Cardiol.* **2002**, *39*, 1890–1900. [CrossRef] [PubMed]
3. Writing Group Members; Mozaffarian, D.; Benjamin, E.J.; Go, A.S.; Arnett, D.K.; Blaha, M.J.; Cushman, M.; Das, S.R.; de Ferranti, S.; Despres, J.P.; et al. Heart Disease and Stroke Statistics-2016 Update: A Report from the American Heart Association. *Circulation* **2016**, *133*, e38–e360.
4. Howe, K.; Clark, M.D.; Torroja, C.F.; Torrance, J.; Berthelot, C.; Muffato, M.; Collins, J.E.; Humphray, S.; McLaren, K.; Matthews, L.; et al. The zebrafish reference genome sequence and its relationship to the human genome. *Nature* **2013**, *496*, 498–503. [CrossRef] [PubMed]
5. González-Rosa, J.M. Zebrafish Models of Cardiac Disease: From Fortuitous Mutants to Precision Medicine. *Circ. Res.* **2022**, *130*, 1803–1826. [CrossRef] [PubMed]
6. von der Heyde, B.; Emmanouilidou, A.; Mazzaferro, E.; Vicenzi, S.; Hoijer, I.; Klingstrom, T.; Jumaa, S.; Dethlefsen, O.; Snieder, H.; de Geus, E.; et al. Translating GWAS-identified loci for cardiac rhythm and rate using an in vivo image- and CRISPR/Cas9-based approach. *Sci. Rep.* **2020**, *10*, 11831. [CrossRef] [PubMed]
7. Milan, D.J.; Jones, I.L.; Ellinor, P.T.; MacRae, C.A. In vivo recording of adult zebrafish electrocardiogram and assessment of drug-induced QT prolongation. *Am. J. Physiol. Heart Circ. Physiol.* **2006**, *291*, H269–H273. [CrossRef]
8. Lin, M.H.; Chou, H.C.; Chen, Y.F.; Liu, W.; Lee, C.C.; Liu, L.Y.; Chuang, Y.J. Development of a rapid and economic in vivo electrocardiogram platform for cardiovascular drug assay and electrophysiology research in adult zebrafish. *Sci. Rep.* **2018**, *8*, 15986. [CrossRef]
9. Liu, C.C.; Li, L.; Lam, Y.W.; Siu, C.W.; Cheng, S.H. Improvement of surface ECG recording in adult zebrafish reveals that the value of this model exceeds our expectation. *Sci. Rep.* **2016**, *6*, 25073. [CrossRef]
10. Nemtsas, P.; Wettwer, E.; Christ, T.; Weidinger, G.; Ravens, U. Adult zebrafish heart as a model for human heart? An electrophysiological study. *J. Mol. Cell Cardiol.* **2010**, *48*, 161–171. [CrossRef]
11. Porrello, E.R.; Mahmoud, A.I.; Simpson, E.; Hill, J.A.; Richardson, J.A.; Olson, E.N.; Sadek, H.A. Transient regenerative potential of the neonatal mouse heart. *Science* **2011**, *331*, 1078–1080. [CrossRef] [PubMed]
12. Andersen, D.C.; Jensen, C.H.; Baun, C.; Hvidsten, S.; Zebrowski, D.C.; Engel, F.B.; Sheikh, S.P. Persistent scarring and dilated cardiomyopathy suggest incomplete regeneration of the apex resected neonatal mouse myocardium—A 180 days follow up study. *J. Mol. Cell Cardiol.* **2016**, *90*, 47–52. [CrossRef] [PubMed]
13. Papaioannou, V.E.; Behringer, R.R. Early embryonic lethality in genetically engineered mice: Diagnosis and phenotypic analysis. *Vet. Pathol.* **2012**, *49*, 64–70. [CrossRef]
14. Bevan, L.; Lim, Z.W.; Venkatesh, B.; Riley, P.R.; Martin, P.; Richardson, R.J. Specific macrophage populations promote both cardiac scar deposition and subsequent resolution in adult zebrafish. *Cardiovasc. Res.* **2020**, *116*, 1357–1371. [CrossRef] [PubMed]
15. Simoes, F.C.; Cahill, T.J.; Kenyon, A.; Gavriouchkina, D.; Vieira, J.M.; Sun, X.; Pezzolla, D.; Ravaud, C.; Masmanian, E.; Weinberger, M.; et al. Macrophages directly contribute collagen to scar formation during zebrafish heart regeneration and mouse heart repair. *Nat. Commun.* **2020**, *11*, 600. [CrossRef] [PubMed]
16. de Preux Charles, A.S.; Bise, T.; Baier, F.; Marro, J.; Jazwinska, A. Distinct effects of inflammation on preconditioning and regeneration of the adult zebrafish heart. *Open Biol.* **2016**, *6*, 160102. [CrossRef]
17. Stainier, D.Y. Zebrafish genetics and vertebrate heart formation. *Nat. Rev. Genet.* **2001**, *2*, 39–48. [CrossRef]
18. Tyser, R.C.; Miranda, A.M.; Chen, C.M.; Davidson, S.M.; Srinivas, S.; Riley, P.R. Calcium handling precedes cardiac differentiation to initiate the first heartbeat. *Elife* **2016**, *5*, e17113. [CrossRef]
19. Bamforth, S.D.; Chaudhry, B.; Bennett, M.; Wilson, R.; Mohun, T.J.; Van Mierop, L.H.; Henderson, D.J.; Anderson, R.H. Clarification of the identity of the mammalian fifth pharyngeal arch artery. *Clin. Anat.* **2013**, *26*, 173–182. [CrossRef]
20. Anderson, M.J.; Pham, V.N.; Vogel, A.M.; Weinstein, B.M.; Roman, B.L. Loss of unc45a precipitates arteriovenous shunting in the aortic arches. *Dev. Biol.* **2008**, *318*, 258–267. [CrossRef]
21. Sun, Y.; Kumar, S.R.; Wong, C.E.D.; Tian, Z.; Bai, H.; Crump, J.G.; Bajpai, R.; Lien, C.L. Craniofacial and cardiac defects in chd7 zebrafish mutants mimic CHARGE syndrome. *Front. Cell Dev. Biol.* **2022**, *10*, 1030587. [CrossRef] [PubMed]

22. Deisseroth, K.; Feng, G.; Majewska, A.K.; Miesenbock, G.; Ting, A.; Schnitzer, M.J. Next-generation optical technologies for illuminating genetically targeted brain circuits. *J. Neurosci.* **2006**, *26*, 10380–10386. [CrossRef] [PubMed]
23. Miesenbock, G. The optogenetic catechism. *Science* **2009**, *326*, 395–399. [CrossRef] [PubMed]
24. Entcheva, E.; Kay, M.W. Cardiac optogenetics: A decade of enlightenment. *Nat. Rev. Cardiol.* **2021**, *18*, 349–367. [CrossRef]
25. Gore, A.V.; Monzo, K.; Cha, Y.R.; Pan, W.; Weinstein, B.M. Vascular development in the zebrafish. *Cold Spring Harb. Perspect. Med.* **2012**, *2*, a006684. [CrossRef]
26. Salgado-Almario, J.; Vicente, M.; Molina, Y.; Martinez-Sielva, A.; Vincent, P.; Domingo, B.; Llopis, J. Simultaneous imaging of calcium and contraction in the beating heart of zebrafish larvae. *Theranostics* **2022**, *12*, 1012–1029. [CrossRef] [PubMed]
27. Vicente, M.; Salgado-Almario, J.; Collins, M.M.; Martinez-Sielva, A.; Minoshima, M.; Kikuchi, K.; Domingo, B.; Llopis, J. Cardioluminescence in Transgenic Zebrafish Larvae: A Calcium Imaging Tool to Study Drug Effects and Pathological Modeling. *Biomedicines* **2021**, *9*, 1294. [CrossRef]
28. Arrenberg, A.B.; Stainier, D.Y.; Baier, H.; Huisken, J. Optogenetic control of cardiac function. *Science* **2010**, *330*, 971–974. [CrossRef] [PubMed]
29. Kirchmaier, B.C.; Poon, K.L.; Schwerte, T.; Huisken, J.; Winkler, C.; Jungblut, B.; Stainier, D.Y.; Brand, T. The Popeye domain containing 2 (popdc2) gene in zebrafish is required for heart and skeletal muscle development. *Dev. Biol.* **2012**, *363*, 438–450. [CrossRef]
30. Baranasic, D.; Hortenhuber, M.; Balwierz, P.J.; Zehnder, T.; Mukarram, A.K.; Nepal, C.; Varnai, C.; Hadzhiev, Y.; Jimenez-Gonzalez, A.; Li, N.; et al. Multiomic atlas with functional stratification and developmental dynamics of zebrafish cis-regulatory elements. *Nat. Genet.* **2022**, *54*, 1037–1050. [CrossRef]
31. Stainier, D.Y.; Fouquet, B.; Chen, J.N.; Warren, K.S.; Weinstein, B.M.; Meiler, S.E.; Mohideen, M.A.; Neuhauss, S.C.; Solnica-Krezel, L.; Schier, A.F.; et al. Mutations affecting the formation and function of the cardiovascular system in the zebrafish embryo. *Development* **1996**, *123*, 285–292. [CrossRef] [PubMed]
32. Weinstein, B.M.; Stemple, D.L.; Driever, W.; Fishman, M.C. Gridlock, a localized heritable vascular patterning defect in the zebrafish. *Nat. Med.* **1995**, *1*, 1143–1147. [CrossRef] [PubMed]
33. Bournele, D.; Beis, D. Zebrafish models of cardiovascular disease. *Heart Fail. Rev.* **2016**, *21*, 803–813. [CrossRef] [PubMed]
34. Lu, S.; Hu, M.; Wang, Z.; Liu, H.; Kou, Y.; Lyu, Z.; Tian, J. Generation and Application of the Zebrafish heg1 Mutant as a Cardiovascular Disease Model. *Biomolecules* **2020**, *10*, 1542. [CrossRef] [PubMed]
35. Harrison, M.R.; Bussmann, J.; Huang, Y.; Zhao, L.; Osorio, A.; Burns, C.G.; Burns, C.E.; Sucov, H.M.; Siekmann, A.F.; Lien, C.L. Chemokine-guided angiogenesis directs coronary vasculature formation in zebrafish. *Dev. Cell* **2015**, *33*, 442–454. [CrossRef] [PubMed]
36. Cotter, P.A.; Han, A.J.; Everson, J.J.; Rodnick, K.J. Cardiac hemodynamics of the rainbow trout (*Oncorhynchus mykiss*) using simultaneous Doppler echocardiography and electrocardiography. *J. Exp. Zool. Part. A Ecol. Genet. Physiol.* **2008**, *309*, 243–254. [CrossRef] [PubMed]
37. Bovo, E.; Dvornikov, A.V.; Mazurek, S.R.; de Tombe, P.P.; Zima, A.V. Mechanisms of Ca^{2+} handling in zebrafish ventricular myocytes. *Pflug. Arch. Eur. J. Physiol.* **2013**, *465*, 1775–1784. [CrossRef] [PubMed]
38. Haverinen, J.; Hassinen, M.; Dash, S.N.; Vornanen, M. Expression of calcium channel transcripts in the zebrafish heart: Dominance of T-type channels. *J. Exp. Biol.* **2018**, *221 Pt 10*, jeb179226. [CrossRef] [PubMed]
39. Vornanen, M.; Hassinen, M. Zebrafish heart as a model for human cardiac electrophysiology. *Channels* **2016**, *10*, 101–110. [CrossRef]
40. Gaborit, N.; Le Bouter, S.; Szuts, V.; Varro, A.; Escande, D.; Nattel, S.; Demolombe, S. Regional and tissue specific transcript signatures of ion channel genes in the non-diseased human heart. *J. Physiol.* **2007**, *582 Pt 2*, 675–693. [CrossRef]
41. Pieske, B.; Trost, S.; Schütt, K.; Minami, K.; Just, H.; Hasenfuss, G. Influence of forskolin on the force-frequency behavior in nonfailing and end-stage failing human myocardium. *Basic. Res. Cardiol.* **1998**, *93* (Suppl. S1), 66–75. [CrossRef] [PubMed]
42. Haustein, M.; Hannes, T.; Trieschmann, J.; Verhaegh, R.; Köster, A.; Hescheler, J.; Brockmeier, K.; Adelmann, R.; Khalil, M. Excitation-contraction coupling in zebrafish ventricular myocardium is regulated by trans-sarcolemmal Ca^{2+} influx and sarcoplasmic reticulum Ca^{2+} release. *PLoS ONE* **2015**, *10*, e0125654. [CrossRef] [PubMed]
43. Arnaout, R.; Ferrer, T.; Huisken, J.; Spitzer, K.; Stainier, D.Y.; Tristani-Firouzi, M.; Chi, N.C. Zebrafish model for human long QT syndrome. *Proc. Natl. Acad. Sci. USA* **2007**, *104*, 11316–11321. [CrossRef] [PubMed]
44. Thorsen, K.; Dam, V.S.; Kjaer-Sorensen, K.; Pedersen, L.N.; Skeberdis, V.A.; Jurevicius, J.; Treinys, R.; Petersen, I.M.B.S.; Nielsen, M.S.; Oxvig, C.; et al. Loss-of-activity-mutation in the cardiac chloride-bicarbonate exchanger AE3 causes short QT syndrome. *Nat. Commun.* **2017**, *8*, 1696. [CrossRef] [PubMed]
45. Verkerk, A.O.; Remme, C.A. Zebrafish: A novel research tool for cardiac (patho)electrophysiology and ion channel disorders. *Front. Physiol.* **2012**, *3*, 255. [CrossRef] [PubMed]
46. Sanchez-Iranzo, H.; Galardi-Castilla, M.; Sanz-Morejon, A.; Gonzalez-Rosa, J.M.; Costa, R.; Ernst, A.; Sainz de Aja, J.; Langa, X.; Mercader, N. Transient fibrosis resolves via fibroblast inactivation in the regenerating zebrafish heart. *Proc. Natl. Acad. Sci. USA* **2018**, *115*, 4188–4193. [CrossRef] [PubMed]
47. Kossack, M.; Hein, S.; Juergensen, L.; Siragusa, M.; Benz, A.; Katus, H.A.; Most, P.; Hassel, D. Induction of cardiac dysfunction in developing and adult zebrafish by chronic isoproterenol stimulation. *J. Mol. Cell Cardiol.* **2017**, *108*, 95–105. [CrossRef] [PubMed]

48. Huttner, I.G.; Wang, L.W.; Santiago, C.F.; Horvat, C.; Johnson, R.; Cheng, D.; von Frieling-Salewsky, M.; Hillcoat, K.; Bemand, T.J.; Trivedi, G.; et al. A-Band Titin Truncation in Zebrafish Causes Dilated Cardiomyopathy and Hemodynamic Stress Intolerance. *Circ. Genom. Precis. Med.* **2018**, *11*, e002135. [CrossRef] [PubMed]
49. Meyer, A.; Schartl, M. Gene and genome duplications in vertebrates: The one-to-four (-to-eight in fish) rule and the evolution of novel gene functions. *Curr. Opin. Cell Biol.* **1999**, *11*, 699–704. [CrossRef]
50. Hoegg, S.; Brinkmann, H.; Taylor, J.S.; Meyer, A. Phylogenetic timing of the fish-specific genome duplication correlates with the diversification of teleost fish. *J. Mol. Evol.* **2004**, *59*, 190–203. [CrossRef]
51. Duong, T.B.; Ravisankar, P.; Song, Y.C.; Gafranek, J.T.; Rydeen, A.B.; Dohn, T.E.; Barske, L.A.; Crump, J.G.; Waxman, J.S. Nr2f1a balances atrial chamber and atrioventricular canal size via BMP signaling-independent and -dependent mechanisms. *Dev. Biol.* **2018**, *434*, 7–14. [CrossRef] [PubMed]
52. Halabi, R.; Cechmanek, P.B.; Hehr, C.L.; McFarlane, S. Semaphorin3f as a cardiomyocyte derived regulator of heart chamber development. *Cell Commun. Signal* **2022**, *20*, 126. [CrossRef] [PubMed]
53. Montague, T.G.; Gagnon, J.A.; Schier, A.F. Conserved regulation of Nodal-mediated left-right patterning in zebrafish and mouse. *Development* **2018**, *145*, dev171090. [CrossRef] [PubMed]
54. Long, S.; Ahmad, N.; Rebagliati, M. The zebrafish nodal-related gene southpaw is required for visceral and diencephalic left-right asymmetry. *Development* **2003**, *130*, 2303–2316. [CrossRef] [PubMed]
55. Schottenfeld, J.; Sullivan-Brown, J.; Burdine, R.D. Zebrafish curly up encodes a Pkd2 ortholog that restricts left-side-specific expression of southpaw. *Development* **2007**, *134*, 1605–1615. [CrossRef] [PubMed]
56. Caron, A.; Xu, X.; Lin, X. Wnt/beta-catenin signaling directly regulates Foxj1 expression and ciliogenesis in zebrafish Kupffer's vesicle. *Development* **2012**, *139*, 514–524. [CrossRef] [PubMed]
57. Smith, K.A.; Chocron, S.; von der Hardt, S.; de Pater, E.; Soufan, A.; Bussmann, J.; Schulte-Merker, S.; Hammerschmidt, M.; Bakkers, J. Rotation and asymmetric development of the zebrafish heart requires directed migration of cardiac progenitor cells. *Dev. Cell* **2008**, *14*, 287–297. [CrossRef]
58. Vermot, J.; Forouhar, A.S.; Liebling, M.; Wu, D.; Plummer, D.; Gharib, M.; Fraser, S.E. Reversing blood flows act through klf2a to ensure normal valvulogenesis in the developing heart. *PLoS Biol.* **2009**, *7*, e1000246. [CrossRef]
59. Juan, T.; Ribeiro da Silva, A.; Cardoso, B.; Lim, S.; Charteau, V.; Stainier, D.Y.R. Multiple pkd and piezo gene family members are required for atrioventricular valve formation. *Nat. Commun.* **2023**, *14*, 214. [CrossRef] [PubMed]
60. Cai, C.; Sang, C.; Du, J.; Jia, H.; Tu, J.; Wan, Q.; Bao, B.; Xie, S.; Huang, Y.; Li, A.; et al. Knockout of tnni1b in zebrafish causes defects in atrioventricular valve development via the inhibition of the myocardial wnt signaling pathway. *FASEB J.* **2019**, *33*, 696–710. [CrossRef]
61. Gunawan, F.; Gentile, A.; Gauvrit, S.; Stainier, D.Y.R.; Bensimon-Brito, A. Nfatc1 Promotes Interstitial Cell Formation during Cardiac Valve Development in Zebrafish. *Circ. Res.* **2020**, *126*, 968–984. [CrossRef]
62. Patel, D.A.; Lavie, C.J.; Milani, R.V.; Shah, S.; Gilliland, Y. Clinical implications of left atrial enlargement: A review. *Ochsner J.* **2009**, *9*, 191–196.
63. Mishra-Gorur, K.; Barak, T.; Kaulen, L.D.; Henegariu, O.; Jin, S.C.; Aguilera, S.M.; Yalbir, E.; Goles, G.; Nishimura, S.; Miyagishima, D.; et al. Pleiotropic role of TRAF7 in skull-base meningiomas and congenital heart disease. *Proc. Natl. Acad. Sci. USA* **2023**, *120*, e2214997120. [CrossRef]
64. Kelly, M.L.; Astsaturov, A.; Rhodes, J.; Chernoff, J. A Pak1/Erk signaling module acts through Gata6 to regulate cardiovascular development in zebrafish. *Dev. Cell* **2014**, *29*, 350–359. [CrossRef]
65. Sidhwani, P.; Leerberg, D.M.; Boezio, G.L.M.; Capasso, T.L.; Yang, H.; Chi, N.C.; Roman, B.L.; Stainier, D.Y.R.; Yelon, D. Cardiac function modulates endocardial cell dynamics to shape the cardiac outflow tract. *Development* **2020**, *147*, dev185900. [CrossRef]
66. Shi, Y.; Li, Y.; Wang, Y.; Zhu, P.; Chen, Y.; Wang, H.; Yue, S.; Xia, X.; Chen, J.; Jiang, Z.; et al. BVES downregulation in non-syndromic tetralogy of fallot is associated with ventricular outflow tract stenosis. *Sci. Rep.* **2020**, *10*, 14167. [CrossRef]
67. Hao, L.; Ma, J.; Wu, F.; Ma, X.; Qian, M.; Sheng, W.; Yan, T.; Tang, N.; Jiang, X.; Zhang, B.; et al. WDR62 variants contribute to congenital heart disease by inhibiting cardiomyocyte proliferation. *Clin. Transl. Med.* **2022**, *12*, e941. [CrossRef]
68. Bonetti, M.; Rodriguez-Martinez, V.; Paardekooper Overman, J.; Overvoorde, J.; van Eekelen, M.; Jopling, C.; Hertog, J. Distinct and overlapping functions of ptpn11 genes in Zebrafish development. *PLoS ONE* **2014**, *9*, e94884. [CrossRef]
69. Solman, M.; Blokzijl-Franke, S.; Piques, F.; Yan, C.; Yang, Q.; Strullu, M.; Kamel, S.M.; Ak, P.; Bakkers, J.; Langenau, D.M.; et al. Inflammatory response in hematopoietic stem and progenitor cells triggered by activating SHP2 mutations evokes blood defects. *Elife* **2022**, *11*, e73040. [CrossRef]
70. Zhang, D.; Gates, K.P.; Barske, L.; Wang, G.; Lancman, J.J.; Zeng, X.I.; Groff, M.; Wang, K.; Parsons, M.J.; Crump, J.G.; et al. Endoderm Jagged induces liver and pancreas duct lineage in zebrafish. *Nat. Commun.* **2017**, *8*, 769. [CrossRef]
71. Zhao, C.; Matalonga, J.; Lancman, J.J.; Liu, L.; Xiao, C.; Kumar, S.; Gates, K.P.; He, J.; Graves, A.; Huisken, J.; et al. Regenerative failure of intrahepatic biliary cells in Alagille syndrome rescued by elevated Jagged/Notch/Sox9 signaling. *Proc. Natl. Acad. Sci. USA* **2022**, *119*, e2201097119. [CrossRef]
72. Prendergast, A.; Ziganshin, B.A.; Papanikolaou, D.; Zafar, M.A.; Nicoli, S.; Mukherjee, S.; Elefteriades, J.A. Phenotyping Zebrafish Mutant Models to Assess Candidate Genes Associated with Aortic Aneurysm. *Genes* **2022**, *13*, 123. [CrossRef]
73. Mitchell, E.; Gilbert, M.; Loomes, K.M. Alagille Syndrome. *Clin. Liver Dis.* **2018**, *22*, 625–641. [CrossRef]

74. Yue, Y.; Jiang, M.; He, L.; Zhang, Z.; Zhang, Q.; Gu, C.; Liu, M.; Li, N.; Zhao, Q. The transcription factor Foxc1a in zebrafish directly regulates expression of nkx2.5, encoding a transcriptional regulator of cardiac progenitor cells. *J. Biol. Chem.* **2018**, *293*, 638–650. [CrossRef]
75. Chrystal, P.W.; French, C.R.; Jean, F.; Havrylov, S.; van Baarle, S.; Peturson, A.M.; Xu, P.; Crump, J.G.; Pilgrim, D.B.; Lehmann, O.J.; et al. The Axenfeld-Rieger Syndrome Gene FOXC1 Contributes to Left-Right Patterning. *Genes* **2021**, *12*, 170. [CrossRef]
76. Xu, P.; Balczerski, B.; Ciozda, A.; Louie, K.; Oralova, V.; Huysseune, A.; Crump, J.G. Fox proteins are modular competency factors for facial cartilage and tooth specification. *Development* **2018**, *145*, dev165498. [CrossRef]
77. Skarie, J.M.; Link, B.A. FoxC1 is essential for vascular basement membrane integrity and hyaloid vessel morphogenesis. *Investig. Ophthalmol. Vis. Sci.* **2009**, *50*, 5026–5034. [CrossRef]
78. Collins, M.M.; Ahlberg, G.; Hansen, C.V.; Guenther, S.; Marin-Juez, R.; Sokol, A.M.; El-Sammak, H.; Piesker, J.; Hellsten, Y.; Olesen, M.S.; et al. Early sarcomere and metabolic defects in a zebrafish pitx2c cardiac arrhythmia model. *Proc. Natl. Acad. Sci. USA* **2019**, *116*, 24115–24121. [CrossRef]
79. Al Turki, S.; Manickaraj, A.K.; Mercer, C.L.; Gerety, S.S.; Hitz, M.P.; Lindsay, S.; D'Alessandro, L.C.; Swaminathan, G.J.; Bentham, J.; Arndt, A.K.; et al. Rare variants in NR2F2 cause congenital heart defects in humans. *Am. J. Hum. Genet.* **2014**, *94*, 574–585. [CrossRef]
80. Schott, J.J.; Benson, D.W.; Basson, C.T.; Pease, W.; Silberbach, G.M.; Moak, J.P.; Maron, B.J.; Seidman, C.E.; Seidman, J.G. Congenital heart disease caused by mutations in the transcription factor NKX2-5. *Science* **1998**, *281*, 108–111. [CrossRef]
81. Marques, S.R.; Yelon, D. Differential requirement for BMP signaling in atrial and ventricular lineages establishes cardiac chamber proportionality. *Dev. Biol.* **2009**, *328*, 472–482. [CrossRef]
82. Meilhac, S.M.; Esner, M.; Kelly, R.G.; Nicolas, J.F.; Buckingham, M.E. The clonal origin of myocardial cells in different regions of the embryonic mouse heart. *Dev. Cell* **2004**, *6*, 685–698. [CrossRef]
83. Kelly, R.G.; Brown, N.A.; Buckingham, M.E. The arterial pole of the mouse heart forms from Fgf10-expressing cells in pharyngeal mesoderm. *Dev. Cell* **2001**, *1*, 435–440. [CrossRef]
84. Cai, C.L.; Liang, X.; Shi, Y.; Chu, P.H.; Pfaff, S.L.; Chen, J.; Evans, S. Isl1 identifies a cardiac progenitor population that proliferates prior to differentiation and contributes a majority of cells to the heart. *Dev. Cell* **2003**, *5*, 877–889. [CrossRef]
85. Buckingham, M.; Meilhac, S.; Zaffran, S. Building the mammalian heart from two sources of myocardial cells. *Nat. Rev. Genet.* **2005**, *6*, 826–835. [CrossRef]
86. Lin, F.J.; Qin, J.; Tang, K.; Tsai, S.Y.; Tsai, M.J. Coup d'Etat: An orphan takes control. *Endocr. Rev.* **2011**, *32*, 404–421. [CrossRef]
87. Laursen, K.B.; Mongan, N.P.; Zhuang, Y.; Ng, M.M.; Benoit, Y.D.; Gudas, L.J. Polycomb recruitment attenuates retinoic acid-induced transcription of the bivalent NR2F1 gene. *Nucleic Acids Res.* **2013**, *41*, 6430–6443. [CrossRef]
88. Pereira, F.A.; Tsai, M.J.; Tsai, S.Y. COUP-TF orphan nuclear receptors in development and differentiation. *Cell Mol. Life Sci.* **2000**, *57*, 1388–1398. [CrossRef]
89. Li, G.; Xu, A.; Sim, S.; Priest, J.R.; Tian, X.; Khan, T.; Quertermous, T.; Zhou, B.; Tsao, P.S.; Quake, S.R.; et al. Transcriptomic Profiling Maps Anatomically Patterned Subpopulations among Single Embryonic Cardiac Cells. *Dev. Cell* **2016**, *39*, 491–507. [CrossRef]
90. Devalla, H.D.; Schwach, V.; Ford, J.W.; Milnes, J.T.; El-Haou, S.; Jackson, C.; Gkatzis, K.; Elliott, D.A.; Chuva de Sousa Lopes, S.M.; Mummery, C.L.; et al. Atrial-like cardiomyocytes from human pluripotent stem cells are a robust preclinical model for assessing atrial-selective pharmacology. *EMBO Mol. Med.* **2015**, *7*, 394–410. [CrossRef]
91. Epstein, J.A.; Aghajanian, H.; Singh, M.K. Semaphorin signaling in cardiovascular development. *Cell Metab.* **2015**, *21*, 163–173. [CrossRef]
92. Alto, L.T.; Terman, J.R. Semaphorins and their Signaling Mechanisms. *Methods Mol. Biol.* **2017**, *1493*, 1–25.
93. Jongbloets, B.C.; Pasterkamp, R.J. Semaphorin signalling during development. *Development* **2014**, *141*, 3292–3297. [CrossRef]
94. Sanchez-Castro, M.; Pichon, O.; Briand, A.; Poulain, D.; Gournay, V.; David, A.; Le Caignec, C. Disruption of the SEMA3D gene in a patient with congenital heart defects. *Hum. Mutat.* **2015**, *36*, 30–33. [CrossRef]
95. Silversides, C.K.; Lionel, A.C.; Costain, G.; Merico, D.; Migita, O.; Liu, B.; Yuen, T.; Rickaby, J.; Thiruvahindrapuram, B.; Marshall, C.R.; et al. Rare copy number variations in adults with tetralogy of Fallot implicate novel risk gene pathways. *PLoS Genet.* **2012**, *8*, e1002843. [CrossRef]
96. Degenhardt, K.; Rychik, J. Fetal Situs, Isomerism, Heterotaxy Syndrome: Diagnostic Evaluation and Implication for Postnatal Management. *Curr. Treat. Opt. Cardiovasc. Med.* **2016**, *18*, 77. [CrossRef]
97. Taketazu, M.; Lougheed, J.; Yoo, S.J.; Lim, J.S.; Hornberger, L.K. Spectrum of cardiovascular disease, accuracy of diagnosis, and outcome in fetal heterotaxy syndrome. *Am. J. Cardiol.* **2006**, *97*, 720–724. [CrossRef]
98. Desgrange, A.; Le Garrec, J.F.; Meilhac, S.M. Left-right asymmetry in heart development and disease: Forming the right loop. *Development* **2018**, *145*, dev162776. [CrossRef]
99. Shiraishi, I.; Ichikawa, H. Human heterotaxy syndrome—From molecular genetics to clinical features, management, and prognosis. *Circ. J.* **2012**, *76*, 2066–2075. [CrossRef]
100. Bakkers, J.; Verhoeven, M.C.; Abdelilah-Seyfried, S. Shaping the zebrafish heart: From left-right axis specification to epithelial tissue morphogenesis. *Dev. Biol.* **2009**, *330*, 213–220. [CrossRef]
101. Essner, J.J.; Amack, J.D.; Nyholm, M.K.; Harris, E.B.; Yost, H.J. Kupffer's vesicle is a ciliated organ of asymmetry in the zebrafish embryo that initiates left-right development of the brain, heart and gut. *Development* **2005**, *132*, 1247–1260. [CrossRef]

102. Essner, J.J.; Vogan, K.J.; Wagner, M.K.; Tabin, C.J.; Yost, H.J.; Brueckner, M. Conserved function for embryonic nodal cilia. *Nature* **2002**, *418*, 37–38. [CrossRef] [PubMed]
103. Djenoune, L.; Mahamdeh, M.; Truong, T.V.; Nguyen, C.T.; Fraser, S.E.; Brueckner, M.; Howard, J.; Yuan, S. Cilia function as calcium-mediated mechanosensors that instruct left-right asymmetry. *Science* **2023**, *379*, 71–78. [CrossRef]
104. Yuan, S.; Zhao, L.; Brueckner, M.; Sun, Z. Intraciliary calcium oscillations initiate vertebrate left-right asymmetry. *Curr. Biol.* **2015**, *25*, 556–567. [CrossRef]
105. Hashimoto, H.; Rebagliati, M.; Ahmad, N.; Muraoka, O.; Kurokawa, T.; Hibi, M.; Suzuki, T. The Cerberus/Dan-family protein Charon is a negative regulator of Nodal signaling during left-right patterning in zebrafish. *Development* **2004**, *131*, 1741–1753. [CrossRef] [PubMed]
106. Smith, K.A.; Noel, E.; Thurlings, I.; Rehmann, H.; Chocron, S.; Bakkers, J. Bmp and nodal independently regulate lefty1 expression to maintain unilateral nodal activity during left-right axis specification in zebrafish. *PLoS Genet.* **2011**, *7*, e1002289. [CrossRef]
107. Branford, W.W.; Yost, H.J. Lefty-dependent inhibition of Nodal- and Wnt-responsive organizer gene expression is essential for normal gastrulation. *Curr. Biol.* **2002**, *12*, 2136–2141. [CrossRef] [PubMed]
108. Rogers, K.W.; Lord, N.D.; Gagnon, J.A.; Pauli, A.; Zimmerman, S.; Aksel, D.C.; Reyon, D.; Tsai, S.Q.; Joung, J.K.; Schier, A.F. Nodal patterning without Lefty inhibitory feedback is functional but fragile. *Elife* **2017**, *6*, e28785. [CrossRef]
109. Wang, X.; Yost, H.J. Initiation and propagation of posterior to anterior (PA) waves in zebrafish left-right development. *Dev. Dyn.* **2008**, *237*, 3640–3647. [CrossRef]
110. Liu, S.; Wei, W.; Wang, P.; Liu, C.; Jiang, X.; Li, T.; Li, F.; Wu, Y.; Chen, S.; Sun, K.; et al. LOF variants identifying candidate genes of laterality defects patients with congenital heart disease. *PLoS Genet.* **2022**, *18*, e1010530. [CrossRef]
111. LaHaye, S.; Lincoln, J.; Garg, V. Genetics of valvular heart disease. *Curr. Cardiol. Rep.* **2014**, *16*, 487. [CrossRef] [PubMed]
112. MacGrogan, D.; Luxan, G.; Driessen-Mol, A.; Bouten, C.; Baaijens, F.; de la Pompa, J.L. How to make a heart valve: From embryonic development to bioengineering of living valve substitutes. *Cold Spring Harb. Perspect. Med.* **2014**, *4*, a013912. [CrossRef] [PubMed]
113. Lin, C.J.; Lin, C.Y.; Chen, C.H.; Zhou, B.; Chang, C.P. Partitioning the heart: Mechanisms of cardiac septation and valve development. *Development* **2012**, *139*, 3277–3299. [CrossRef] [PubMed]
114. Huisken, J.; Stainier, D.Y. Selective plane illumination microscopy techniques in developmental biology. *Development* **2009**, *136*, 1963–1975. [CrossRef] [PubMed]
115. Butcher, J.T.; Markwald, R.R. Valvulogenesis: The moving target. *Philos. Trans. R. Soc. Lond. B Biol. Sci.* **2007**, *362*, 1489–1503. [CrossRef] [PubMed]
116. Steed, E.; Boselli, F.; Vermot, J. Hemodynamics driven cardiac valve morphogenesis. *Biochim. Biophys. Acta* **2016**, *1863 Pt B*, 1760–1766. [CrossRef]
117. Armstrong, E.J.; Bischoff, J. Heart valve development: Endothelial cell signaling and differentiation. *Circ. Res.* **2004**, *95*, 459–470. [CrossRef]
118. Fontana, F.; Haack, T.; Reichenbach, M.; Knaus, P.; Puceat, M.; Abdelilah-Seyfried, S. Antagonistic Activities of Vegfr3/Flt4 and Notch1b Fine-tune Mechanosensitive Signaling during Zebrafish Cardiac Valvulogenesis. *Cell Rep.* **2020**, *32*, 107883. [CrossRef]
119. Heckel, E.; Boselli, F.; Roth, S.; Krudewig, A.; Belting, H.G.; Charvin, G.; Vermot, J. Oscillatory Flow Modulates Mechanosensitive klf2a Expression through trpv4 and trpp2 during Heart Valve Development. *Curr. Biol.* **2015**, *25*, 1354–1361. [CrossRef]
120. Liu, Y.; Bai, F.; Tang, Z.; Liu, N.; Liu, Q. Integrative transcriptomic, proteomic, and machine learning approach to identifying feature genes of atrial fibrillation using atrial samples from patients with valvular heart disease. *BMC Cardiovasc. Disord.* **2021**, *21*, 52. [CrossRef]
121. Abdul-Sater, Z.; Yehya, A.; Beresian, J.; Salem, E.; Kamar, A.; Baydoun, S.; Shibbani, K.; Soubra, A.; Bitar, F.; Nemer, G. Two heterozygous mutations in NFATC1 in a patient with Tricuspid Atresia. *PLoS ONE* **2012**, *7*, e49532. [CrossRef] [PubMed]
122. Ferese, R.; Bonetti, M.; Consoli, F.; Guida, V.; Sarkozy, A.; Lepri, F.R.; Versacci, P.; Gambardella, S.; Calcagni, G.; Margiotti, K.; et al. Heterozygous missense mutations in NFATC1 are associated with atrioventricular septal defect. *Hum. Mutat.* **2018**, *39*, 1428–1441. [CrossRef] [PubMed]
123. Gu, H.; Gong, J.; Qiu, W.; Cao, H.; Xu, J.; Chen, S.; Chen, Y. Association of a tandem repeat polymorphism in NFATc1 with increased risk of perimembranous ventricular septal defect in a Chinese population. *Biochem. Genet.* **2011**, *49*, 592–600. [CrossRef] [PubMed]
124. Bajolle, F.; Zaffran, S.; Kelly, R.G.; Hadchouel, J.; Bonnet, D.; Brown, N.A.; Buckingham, M.E. Rotation of the myocardial wall of the outflow tract is implicated in the normal positioning of the great arteries. *Circ. Res.* **2006**, *98*, 421–428. [CrossRef] [PubMed]
125. Neeb, Z.; Lajiness, J.D.; Bolanis, E.; Conway, S.J. Cardiac outflow tract anomalies. *Wiley Interdiscip. Rev. Dev. Biol.* **2013**, *2*, 499–530. [CrossRef] [PubMed]
126. Pierpont, M.E.; Brueckner, M.; Chung, W.K.; Garg, V.; Lacro, R.V.; McGuire, A.L.; Mital, S.; Priest, J.R.; Pu, W.T.; Roberts, A.; et al. Genetic Basis for Congenital Heart Disease: Revisited: A Scientific Statement from the American Heart Association. *Circulation* **2018**, *138*, e653–e711. [CrossRef]
127. George, R.M.; Maldonado-Velez, G.; Firulli, A.B. The heart of the neural crest: Cardiac neural crest cells in development and regeneration. *Development* **2020**, *147*, dev188706. [CrossRef] [PubMed]

128. Kodo, K.; Nishizawa, T.; Furutani, M.; Arai, S.; Yamamura, E.; Joo, K.; Takahashi, T.; Matsuoka, R.; Yamagishi, H. GATA6 mutations cause human cardiac outflow tract defects by disrupting semaphorin-plexin signaling. *Proc. Natl. Acad. Sci. USA* **2009**, *106*, 13933–13938. [CrossRef] [PubMed]
129. Li, J.; Yang, S.; Pu, Z.; Dai, J.; Jiang, T.; Du, F.; Jiang, Z.; Cheng, Y.; Dai, G.; Wang, J.; et al. Whole-exome sequencing identifies SGCD and ACVRL1 mutations associated with total anomalous pulmonary venous return (TAPVR) in Chinese population. *Oncotarget* **2017**, *8*, 27812–27819. [CrossRef]
130. Haarman, M.G.; Kerstjens-Frederikse, W.S.; Vissia-Kazemier, T.R.; Breeman, K.T.N.; Timens, W.; Vos, Y.J.; Roofthooft, M.T.R.; Hillege, H.L.; Berger, R.M.F. The Genetic Epidemiology of Pediatric Pulmonary Arterial Hypertension. *J. Pediatr.* **2020**, *225*, 65–73.e5. [CrossRef]
131. van der Linde, D.; Konings, E.E.; Slager, M.A.; Witsenburg, M.; Helbing, W.A.; Takkenberg, J.J.; Roos-Hesselink, J.W. Birth prevalence of congenital heart disease worldwide: A systematic review and meta-analysis. *J. Am. Coll. Cardiol.* **2011**, *58*, 2241–2247. [CrossRef] [PubMed]
132. Matos-Nieves, A.; Yasuhara, J.; Garg, V. Another Notch in the Genetic Puzzle of Tetralogy of Fallot. *Circ. Res.* **2019**, *124*, 462–464. [CrossRef] [PubMed]
133. Apitz, C.; Webb, G.D.; Redington, A.N. Tetralogy of Fallot. *Lancet* **2009**, *374*, 1462–1471. [CrossRef] [PubMed]
134. Sommer, R.J.; Hijazi, Z.M.; Rhodes, J.F. Pathophysiology of congenital heart disease in the adult: Part III: Complex congenital heart disease. *Circulation* **2008**, *117*, 1340–1350. [CrossRef] [PubMed]
135. Goldmuntz, E.; Clark, B.J.; Mitchell, L.E.; Jawad, A.F.; Cuneo, B.F.; Reed, L.; McDonald-McGinn, D.; Chien, P.; Feuer, J.; Zackai, E.H.; et al. Frequency of 22q11 deletions in patients with conotruncal defects. *J. Am. Coll. Cardiol.* **1998**, *32*, 492–498. [CrossRef] [PubMed]
136. Benson, D.W.; Silberbach, G.M.; Kavanaugh-McHugh, A.; Cottrill, C.; Zhang, Y.; Riggs, S.; Smalls, O.; Johnson, M.C.; Watson, M.S.; Seidman, J.G.; et al. Mutations in the cardiac transcription factor NKX2.5 affect diverse cardiac developmental pathways. *J. Clin. Investig.* **1999**, *104*, 1567–1573. [CrossRef] [PubMed]
137. Olson, E.N. Gene regulatory networks in the evolution and development of the heart. *Science* **2006**, *313*, 1922–1927. [CrossRef] [PubMed]
138. Reese, D.E.; Zavaljevski, M.; Streiff, N.L.; Bader, D. bves: A novel gene expressed during coronary blood vessel development. *Dev. Biol.* **1999**, *209*, 159–171. [CrossRef] [PubMed]
139. Amunjela, J.N.; Swan, A.H.; Brand, T. The Role of the Popeye Domain Containing Gene Family in Organ Homeostasis. *Cells* **2019**, *8*, 1594. [CrossRef]
140. Wu, M.; Li, Y.; He, X.; Shao, X.; Yang, F.; Zhao, M.; Wu, C.; Zhang, C.; Zhou, L. Mutational and functional analysis of the BVES gene coding region in Chinese patients with non-syndromic tetralogy of Fallot. *Int. J. Mol. Med.* **2013**, *31*, 899–903. [CrossRef]
141. Shi, Y.; Li, Y.; Wang, Y.; Zhuang, J.; Wang, H.; Hu, M.; Mo, X.; Yue, S.; Chen, Y.; Fan, X.; et al. The Functional Polymorphism R129W in the BVES Gene Is Associated with Sporadic Tetralogy of Fallot in the Han Chinese Population. *Genet. Test. Mol. Biomark.* **2019**, *23*, 601–609. [CrossRef] [PubMed]
142. Gingold-Belfer, R.; Bergman, M.; Alcalay, Y.; Schlesinger, H.; Aravot, D.; Berman, M.; Salman, H.; Brand, T.; Kessler-Icekson, G. Popeye domain-containing 1 is down-regulated in failing human hearts. *Int. J. Mol. Med.* **2011**, *27*, 25–31. [PubMed]
143. Andrée, B.; Fleige, A.; Arnold, H.H.; Brand, T. Mouse Pop1 is required for muscle regeneration in adult skeletal muscle. *Mol. Cell. Biol.* **2002**, *22*, 1504–1512. [CrossRef] [PubMed]
144. Smith, T.K.; Bader, D.M. Characterization of Bves expression during mouse development using newly generated immunoreagents. *Dev. Dyn. Off. Publ. Am. Assoc. Anat.* **2006**, *235*, 1701–1708. [CrossRef] [PubMed]
145. Froese, A.; Breher, S.S.; Waldeyer, C.; Schindler, R.F.; Nikolaev, V.O.; Rinné, S.; Wischmeyer, E.; Schlueter, J.; Becher, J.; Simrick, S.; et al. Popeye domain containing proteins are essential for stress-mediated modulation of cardiac pacemaking in mice. *J. Clin. Investig.* **2012**, *122*, 1119–1130. [CrossRef]
146. Ward, C.; Stadt, H.; Hutson, M.; Kirby, M.L. Ablation of the secondary heart field leads to tetralogy of Fallot and pulmonary atresia. *Dev. Biol.* **2005**, *284*, 72–83. [CrossRef] [PubMed]
147. Yelbuz, T.M.; Waldo, K.L.; Kumiski, D.H.; Stadt, H.A.; Wolfe, R.R.; Leatherbury, L.; Kirby, M.L. Shortened outflow tract leads to altered cardiac looping after neural crest ablation. *Circulation* **2002**, *106*, 504–510. [CrossRef] [PubMed]
148. Abu-Issa, R.; Kirby, M.L. Patterning of the heart field in the chick. *Dev. Biol.* **2008**, *319*, 223–233. [CrossRef] [PubMed]
149. Janssen, N.; Bergman, J.E.; Swertz, M.A.; Tranebjaerg, L.; Lodahl, M.; Schoots, J.; Hofstra, R.M.; van Ravenswaaij-Arts, C.M.; Hoefsloot, L.H. Mutation update on the CHD7 gene involved in CHARGE syndrome. *Hum. Mutat.* **2012**, *33*, 1149–1160. [CrossRef]
150. Qin, Z.; Su, J.; Li, M.; Yang, Q.; Yi, S.; Zheng, H.; Zhang, Q.; Chen, F.; Yi, S.; Lu, W.; et al. Clinical and Genetic Analysis of CHD7 Expands the Genotype and Phenotype of CHARGE Syndrome. *Front. Genet.* **2020**, *11*, 592. [CrossRef]
151. Zentner, G.E.; Layman, W.S.; Martin, D.M.; Scacheri, P.C. Molecular and phenotypic aspects of CHD7 mutation in CHARGE syndrome. *Am. J. Med. Genet. A* **2010**, *152A*, 674–686. [CrossRef]
152. Corsten-Janssen, N.; Kerstjens-Frederikse, W.S.; du Marchie Sarvaas, G.J.; Baardman, M.E.; Bakker, M.K.; Bergman, J.E.; Hove, H.D.; Heimdal, K.R.; Rustad, C.F.; Hennekam, R.C.; et al. The cardiac phenotype in patients with a CHD7 mutation. *Circ. Cardiovasc. Genet.* **2013**, *6*, 248–254. [CrossRef] [PubMed]

153. Zaidi, S.; Choi, M.; Wakimoto, H.; Ma, L.; Jiang, J.; Overton, J.D.; Romano-Adesman, A.; Bjornson, R.D.; Breitbart, R.E.; Brown, K.K.; et al. De novo mutations in histone-modifying genes in congenital heart disease. *Nature* **2013**, *498*, 220–223. [CrossRef] [PubMed]
154. Yan, S.; Thienthanasit, R.; Chen, D.; Engelen, E.; Brühl, J.; Crossman, D.K.; Kesterson, R.; Wang, Q.; Bouazoune, K.; Jiao, K. CHD7 regulates cardiovascular development through ATP-dependent and -independent activities. *Proc. Natl. Acad. Sci. USA* **2020**, *117*, 28847–28858. [CrossRef] [PubMed]
155. Morrell, C.H.; Brant, L.J. Modelling hearing thresholds in the elderly. *Stat. Med.* **1991**, *10*, 1453–1464. [CrossRef] [PubMed]
156. Hurd, E.A.; Capers, P.L.; Blauwkamp, M.N.; Adams, M.E.; Raphael, Y.; Poucher, H.K.; Martin, D.M. Loss of Chd7 function in gene-trapped reporter mice is embryonic lethal and associated with severe defects in multiple developing tissues. *Mamm. Genome Off. J. Int. Mamm. Genome Soc.* **2007**, *18*, 94–104. [CrossRef] [PubMed]
157. Layman, W.S.; Hurd, E.A.; Martin, D.M. Chromodomain proteins in development: Lessons from CHARGE syndrome. *Clin. Genet.* **2010**, *78*, 11–20. [CrossRef]
158. Copp, A.J. Death before birth: Clues from gene knockouts and mutations. *Trends Genet. TIG* **1995**, *11*, 87–93. [CrossRef]
159. Prykhozhij, S.V.; Steele, S.L.; Razaghi, B.; Berman, J.N. A rapid and effective method for screening, sequencing and reporter verification of engineered frameshift mutations in zebrafish. *Dis. Models Mech.* **2017**, *10*, 811–822. [CrossRef]
160. Jamadagni, P.; Breuer, M.; Schmeisser, K.; Cardinal, T.; Kassa, B.; Parker, J.A.; Pilon, N.; Samarut, E.; Patten, S.A. Chromatin remodeller CHD7 is required for GABAergic neuron development by promoting PAQR3 expression. *EMBO Rep.* **2021**, *22*, e50958. [CrossRef]
161. Cloney, K.; Steele, S.L.; Stoyek, M.R.; Croll, R.P.; Smith, F.M.; Prykhozhij, S.V.; Brown, M.M.; Midgen, C.; Blake, K.; Berman, J.N. Etiology and functional validation of gastrointestinal motility dysfunction in a zebrafish model of CHARGE syndrome. *FEBS J.* **2018**, *285*, 2125–2140. [CrossRef] [PubMed]
162. Asad, Z.; Pandey, A.; Babu, A.; Sun, Y.; Shevade, K.; Kapoor, S.; Ullah, I.; Ranjan, S.; Scaria, V.; Bajpai, R.; et al. Rescue of neural crest-derived phenotypes in a zebrafish CHARGE model by Sox10 downregulation. *Hum. Mol. Genet.* **2016**, *25*, 3539–3554. [CrossRef] [PubMed]
163. Blake, K.; MacCuspie, J.; Hartshorne, T.S.; Roy, M.; Davenport, S.L.; Corsten, G. Postoperative airway events of individuals with CHARGE syndrome. *Int. J. Pediatr. Otorhinolaryngol.* **2009**, *73*, 219–226. [CrossRef] [PubMed]
164. MacLean, J.E.; Wertman, J.N.; Prykhozhij, S.V.; Chedrawe, E.; Langley, S.; Steele, S.L.; Ban, K.; Blake, K.; Berman, J.N. phox2ba: The Potential Genetic Link behind the Overlap in the Symptomatology between CHARGE and Central Congenital Hypoventilation Syndromes. *Genes* **2023**, *14*, 1086. [CrossRef] [PubMed]
165. Nagashimada, M.; Ohta, H.; Li, C.; Nakao, K.; Uesaka, T.; Brunet, J.F.; Amiel, J.; Trochet, D.; Wakayama, T.; Enomoto, H. Autonomic neurocristopathy-associated mutations in PHOX2B dysregulate Sox10 expression. *J. Clin. Investig.* **2012**, *122*, 3145–3158. [CrossRef] [PubMed]
166. Bishara, J.; Keens, T.G.; Perez, I.A. The genetics of congenital central hypoventilation syndrome: Clinical implications. *Appl. Clin. Genet.* **2018**, *11*, 135–144. [CrossRef] [PubMed]
167. Marino, B.; Digilio, M.C.; Toscano, A.; Giannotti, A.; Dallapiccola, B. Congenital heart diseases in children with Noonan syndrome: An expanded cardiac spectrum with high prevalence of atrioventricular canal. *J. Pediatr.* **1999**, *135*, 703–706. [CrossRef] [PubMed]
168. Calcagni, G.; Limongelli, G.; D'Ambrosio, A.; Gesualdo, F.; Digilio, M.C.; Baban, A.; Albanese, S.B.; Versacci, P.; De Luca, E.; Ferrero, G.B.; et al. Cardiac defects, morbidity and mortality in patients affected by RASopathies. CARNET study results. *Int. J. Cardiol.* **2017**, *245*, 92–98. [CrossRef]
169. Prendiville, T.W.; Gauvreau, K.; Tworog-Dube, E.; Patkin, L.; Kucherlapati, R.S.; Roberts, A.E.; Lacro, R.V. Cardiovascular disease in Noonan syndrome. *Arch. Dis. Child.* **2014**, *99*, 629–634. [CrossRef]
170. Digilio, M.C.; Romana Lepri, F.; Dentici, M.L.; Henderson, A.; Baban, A.; Roberti, M.C.; Capolino, R.; Versacci, P.; Surace, C.; Angioni, A.; et al. Atrioventricular canal defect in patients with RASopathies. *Eur. J. Hum. Genet.* **2013**, *21*, 200–204. [CrossRef]
171. Neel, B.G.; Gu, H.; Pao, L. The 'Shp'ing news: SH2 domain-containing tyrosine phosphatases in cell signaling. *Trends Biochem. Sci.* **2003**, *28*, 284–293. [CrossRef] [PubMed]
172. Yang, W.; Klaman, L.D.; Chen, B.; Araki, T.; Harada, H.; Thomas, S.M.; George, E.L.; Neel, B.G. An Shp2/SFK/Ras/Erk signaling pathway controls trophoblast stem cell survival. *Dev. Cell* **2006**, *10*, 317–327. [CrossRef] [PubMed]
173. van Eekelen, M.; Overvoorde, J.; van Rooijen, C.; den Hertog, J. Identification and expression of the family of classical protein-tyrosine phosphatases in zebrafish. *PLoS ONE* **2010**, *5*, e12573. [CrossRef] [PubMed]
174. Saxton, T.M.; Henkemeyer, M.; Gasca, S.; Shen, R.; Rossi, D.J.; Shalaby, F.; Feng, G.S.; Pawson, T. Abnormal mesoderm patterning in mouse embryos mutant for the SH2 tyrosine phosphatase Shp-2. *EMBO J.* **1997**, *16*, 2352–2364. [CrossRef] [PubMed]
175. Feng, G.S. Shp-2 tyrosine phosphatase: Signaling one cell or many. *Exp. Cell Res.* **1999**, *253*, 47–54. [CrossRef] [PubMed]
176. Bonetti, M.; Paardekooper Overman, J.; Tessadori, F.; Noel, E.; Bakkers, J.; den Hertog, J. Noonan and LEOPARD syndrome Shp2 variants induce heart displacement defects in zebrafish. *Development* **2014**, *141*, 1961–1970. [CrossRef] [PubMed]
177. Schubbert, S.; Shannon, K.; Bollag, G. Hyperactive Ras in developmental disorders and cancer. *Nat. Rev. Cancer* **2007**, *7*, 295–308. [CrossRef] [PubMed]
178. Matozaki, T.; Murata, Y.; Saito, Y.; Okazawa, H.; Ohnishi, H. Protein tyrosine phosphatase SHP-2: A proto-oncogene product that promotes Ras activation. *Cancer Sci.* **2009**, *100*, 1786–1793. [CrossRef] [PubMed]

179. Andelfinger, G.; Marquis, C.; Raboisson, M.J.; Theoret, Y.; Waldmuller, S.; Wiegand, G.; Gelb, B.D.; Zenker, M.; Delrue, M.A.; Hofbeck, M. Hypertrophic Cardiomyopathy in Noonan Syndrome Treated by MEK-Inhibition. *J. Am. Coll. Cardiol.* **2019**, *73*, 2237–2239. [CrossRef]
180. Noyes, P.D.; Haggard, D.E.; Gonnerman, G.D.; Tanguay, R.L. Advanced morphological—Behavioral test platform reveals neurodevelopmental defects in embryonic zebrafish exposed to comprehensive suite of halogenated and organophosphate flame retardants. *Toxicol. Sci.* **2015**, *145*, 177–195. [CrossRef]
181. Padilla, S.; Corum, D.; Padnos, B.; Hunter, D.L.; Beam, A.; Houck, K.A.; Sipes, N.; Kleinstreuer, N.; Knudsen, T.; Dix, D.J.; et al. Zebrafish developmental screening of the ToxCast Phase I chemical library. *Reprod. Toxicol.* **2012**, *33*, 174–187. [CrossRef]
182. Truong, L.; Reif, D.M.; St Mary, L.; Geier, M.C.; Truong, H.D.; Tanguay, R.L. Multidimensional in vivo hazard assessment using zebrafish. *Toxicol. Sci.* **2014**, *137*, 212–233. [CrossRef] [PubMed]
183. Balcombe, J.P.; Barnard, N.D.; Sandusky, C. Laboratory routines cause animal stress. *Contemp. Top. Lab. Anim. Sci.* **2004**, *43*, 42–51.
184. Li, D.; March, M.E.; Gutierrez-Uzquiza, A.; Kao, C.; Seiler, C.; Pinto, E.; Matsuoka, L.S.; Battig, M.R.; Bhoj, E.J.; Wenger, T.L.; et al. ARAF recurrent mutation causes central conducting lymphatic anomaly treatable with a MEK inhibitor. *Nat. Med.* **2019**, *25*, 1116–1122. [CrossRef]
185. Siebert, J.; Schneider, M.; Reuter-Schmitt, D.; Wurtemberger, J.; Neubuser, A.; Driever, W.; Hettmer, S.; Kapp, F.G. Rhabdomyosarcoma xenotransplants in zebrafish embryos. *Pediatr. Blood Cancer* **2023**, *70*, e30053. [CrossRef]
186. Ki, D.H.; He, S.; Rodig, S.; Look, A.T. Overexpression of PDGFRA cooperates with loss of NF1 and p53 to accelerate the molecular pathogenesis of malignant peripheral nerve sheath tumors. *Oncogene* **2017**, *36*, 1058–1068. [CrossRef]
187. Parasido, E.; Avetian, G.S.; Naeem, A.; Graham, G.; Pishvaian, M.; Glasgow, E.; Mudambi, S.; Lee, Y.; Ihemelandu, C.; Choudhry, M.; et al. The Sustained Induction of c-MYC Drives Nab-Paclitaxel Resistance in Primary Pancreatic Ductal Carcinoma Cells. *Mol. Cancer Res.* **2019**, *17*, 1815–1827. [CrossRef]
188. Akagi, K.; Tanaka, T.; Baba, S. Successful living donor liver transplantation after stent implantation in a patient with Alagille syndrome and severe bilateral pulmonary artery stenosis. *Cardiol. Young* **2018**, *28*, 1465–1467. [CrossRef] [PubMed]
189. Emerick, K.M.; Rand, E.B.; Goldmuntz, E.; Krantz, I.D.; Spinner, N.B.; Piccoli, D.A. Features of Alagille syndrome in 92 patients: Frequency and relation to prognosis. *Hepatology* **1999**, *29*, 822–829. [CrossRef] [PubMed]
190. McElhinney, D.B.; Krantz, I.D.; Bason, L.; Piccoli, D.A.; Emerick, K.M.; Spinner, N.B.; Goldmuntz, E. Analysis of cardiovascular phenotype and genotype-phenotype correlation in individuals with a JAG1 mutation and/or Alagille syndrome. *Circulation* **2002**, *106*, 2567–2574. [CrossRef]
191. Gilbert, M.A.; Bauer, R.C.; Rajagopalan, R.; Grochowski, C.M.; Chao, G.; McEldrew, D.; Nassur, J.A.; Rand, E.B.; Krock, B.L.; Kamath, B.M.; et al. Alagille syndrome mutation update: Comprehensive overview of JAG1 and NOTCH2 mutation frequencies and insight into missense variant classification. *Hum. Mutat.* **2019**, *40*, 2197–2220. [CrossRef]
192. Zhao, C.; Lancman, J.J.; Yang, Y.; Gates, K.P.; Cao, D.; Barske, L.; Matalonga, J.; Pan, X.; He, J.; Graves, A.; et al. Intrahepatic cholangiocyte regeneration from an Fgf-dependent extrahepatic progenitor niche in a zebrafish model of Alagille Syndrome. *Hepatology* **2022**, *75*, 567–583. [CrossRef] [PubMed]
193. Schroeter, E.H.; Kisslinger, J.A.; Kopan, R. Notch-1 signalling requires ligand-induced proteolytic release of intracellular domain. *Nature* **1998**, *393*, 382–386. [CrossRef] [PubMed]
194. Guo, B.; McMillan, B.J.; Blacklow, S.C. Structure and function of the Mind bomb E3 ligase in the context of Notch signal transduction. *Curr. Opin. Struct. Biol.* **2016**, *41*, 38–45. [CrossRef] [PubMed]
195. Berndt, J.D.; Aoyagi, A.; Yang, P.; Anastas, J.N.; Tang, L.; Moon, R.T. Mindbomb 1, an E3 ubiquitin ligase, forms a complex with RYK to activate Wnt/beta-catenin signaling. *J. Cell Biol.* **2011**, *194*, 737–750. [CrossRef] [PubMed]
196. Honkanen, R.A.; Nishimura, D.Y.; Swiderski, R.E.; Bennett, S.R.; Hong, S.; Kwon, Y.H.; Stone, E.M.; Sheffield, V.C.; Alward, W.L. A family with Axenfeld-Rieger syndrome and Peters Anomaly caused by a point mutation (Phe112Ser) in the FOXC1 gene. *Am. J. Ophthalmol.* **2003**, *135*, 368–375. [CrossRef] [PubMed]
197. Gripp, K.W.; Hopkins, E.; Jenny, K.; Thacker, D.; Salvin, J. Cardiac anomalies in Axenfeld-Rieger syndrome due to a novel FOXC1 mutation. *Am. J. Med. Genet. A* **2013**, *161A*, 114–119. [CrossRef] [PubMed]
198. Du, R.F.; Huang, H.; Fan, L.L.; Li, X.P.; Xia, K.; Xiang, R. A Novel Mutation of FOXC1 (R127L) in an Axenfeld-Rieger Syndrome Family with Glaucoma and Multiple Congenital Heart Diseases. *Ophthalmic Genet.* **2016**, *37*, 111–115. [PubMed]
199. Seifi, M.; Walter, M.A. Axenfeld-Rieger syndrome. *Clin. Genet.* **2018**, *93*, 1123–1130. [CrossRef]
200. D'Haene, B.; Meire, F.; Claerhout, I.; Kroes, H.Y.; Plomp, A.; Arens, Y.H.; de Ravel, T.; Casteels, I.; De Jaegere, S.; Hooghe, S.; et al. Expanding the spectrum of FOXC1 and PITX2 mutations and copy number changes in patients with anterior segment malformations. *Investig. Ophthalmol. Vis. Sci.* **2011**, *52*, 324–333. [CrossRef]
201. Semina, E.V.; Reiter, R.; Leysens, N.J.; Alward, W.L.; Small, K.W.; Datson, N.A.; Siegel-Bartelt, J.; Bierke-Nelson, D.; Bitoun, P.; Zabel, B.U.; et al. Cloning and characterization of a novel bicoid-related homeobox transcription factor gene, RIEG, involved in Rieger syndrome. *Nat. Genet.* **1996**, *14*, 392–399. [CrossRef] [PubMed]
202. Topczewska, J.M.; Topczewski, J.; Solnica-Krezel, L.; Hogan, B.L. Sequence and expression of zebrafish foxc1a and foxc1b, encoding conserved forkhead/winged helix transcription factors. *Mech. Dev.* **2001**, *100*, 343–347. [CrossRef] [PubMed]
203. Ferre-Fernandez, J.J.; Sorokina, E.A.; Thompson, S.; Collery, R.F.; Nordquist, E.; Lincoln, J.; Semina, E.V. Disruption of foxc1 genes in zebrafish results in dosage-dependent phenotypes overlapping Axenfeld-Rieger syndrome. *Hum. Mol. Genet.* **2020**, *29*, 2723–2735. [CrossRef]

204. Zhao, C.M.; Peng, L.Y.; Li, L.; Liu, X.Y.; Wang, J.; Zhang, X.L.; Yuan, F.; Li, R.G.; Qiu, X.B.; Yang, Y.Q. PITX2 Loss-of-Function Mutation Contributes to Congenital Endocardial Cushion Defect and Axenfeld-Rieger Syndrome. *PLoS ONE* **2015**, *10*, e0124409. [CrossRef]
205. Strungaru, M.H.; Dinu, I.; Walter, M.A. Genotype-phenotype correlations in Axenfeld-Rieger malformation and glaucoma patients with FOXC1 and PITX2 mutations. *Investig. Ophthalmol. Vis. Sci.* **2007**, *48*, 228–237. [CrossRef]
206. Ji, Y.; Buel, S.M.; Amack, J.D. Mutations in zebrafish pitx2 model congenital malformations in Axenfeld-Rieger syndrome but do not disrupt left-right placement of visceral organs. *Dev. Biol.* **2016**, *416*, 69–81. [CrossRef]
207. Landrigan, P.J.; De Garbino, J.P.; Newman, B. Framing the future in light of the past: Living in a chemical world. *Ann. N. Y Acad. Sci.* **2006**, *1076*, 657–659. [CrossRef] [PubMed]
208. Scott-Drechsel, D.E.; Rugonyi, S.; Marks, D.L.; Thornburg, K.L.; Hinds, M.T. Hyperglycemia slows embryonic growth and suppresses cell cycle via cyclin D1 and p21. *Diabetes* **2013**, *62*, 234–242. [CrossRef]
209. Yang, P.; Chen, X.; Kaushal, S.; Reece, E.A.; Yang, P. High glucose suppresses embryonic stem cell differentiation into cardiomyocytes: High glucose inhibits ES cell cardiogenesis. *Stem Cell Res. Ther.* **2016**, *7*, 187. [CrossRef]
210. Sankar, S.; Jayabalan, M.; Venkatesh, S.; Ibrahim, M. Effect of hyperglycemia on tbx5a and nppa gene expression and its correlation to structural and functional changes in developing zebrafish heart. *Cell Biol. Int.* **2022**, *46*, 2173–2184. [CrossRef]
211. Jin, Y.M.; Zhao, S.Z.; Zhang, Z.L.; Chen, Y.; Cheng, X.; Chuai, M.; Liu, G.S.; Lee, K.K.H.; Yang, X. High Glucose Level Induces Cardiovascular Dysplasia During Early Embryo Development. *Exp. Clin. Endocrinol. Diabetes* **2019**, *127*, 590–597. [CrossRef]
212. Garnham, E.A.; Beck, F.; Clarke, C.A.; Stanisstreet, M. Effects of glucose on rat embryos in culture. *Diabetologia* **1983**, *25*, 291–295. [CrossRef] [PubMed]
213. Liang, J.; Gui, Y.; Wang, W.; Gao, S.; Li, J.; Song, H. Elevated glucose induces congenital heart defects by altering the expression of tbx5, tbx20, and has2 in developing zebrafish embryos. *Birth Defects Res. A Clin. Mol. Teratol.* **2010**, *88*, 480–486. [CrossRef]
214. Basson, C.T.; Bachinsky, D.R.; Lin, R.C.; Levi, T.; Elkins, J.A.; Soults, J.; Grayzel, D.; Kroumpouzou, E.; Traill, T.A.; Leblanc-Straceski, J.; et al. Mutations in human TBX5 [corrected] cause limb and cardiac malformation in Holt-Oram syndrome. *Nat. Genet.* **1997**, *15*, 30–35. [CrossRef]
215. Lu, J.; Tsai, T.; Choo, S.; Yeh, S.; Tang, R.; Yang, A.; Lee, H.; Lu, J. Induction of apoptosis and inhibition of cell growth by tbx5 knockdown contribute to dysmorphogenesis in Zebrafish embryos. *J. Biomed. Sci.* **2011**, *18*, 73. [CrossRef]
216. King-Heiden, T.C.; Mehta, V.; Xiong, K.M.; Lanham, K.A.; Antkiewicz, D.S.; Ganser, A.; Heideman, W.; Peterson, R.E. Reproductive and developmental toxicity of dioxin in fish. *Mol. Cell Endocrinol.* **2012**, *354*, 121–138. [CrossRef]
217. Lanham, K.A.; Peterson, R.E.; Heideman, W. Sensitivity to dioxin decreases as zebrafish mature. *Toxicol. Sci.* **2012**, *127*, 360–370. [CrossRef]
218. Antkiewicz, D.S.; Burns, C.G.; Carney, S.A.; Peterson, R.E.; Heideman, W. Heart malformation is an early response to TCDD in embryonic zebrafish. *Toxicol. Sci.* **2005**, *84*, 368–377. [CrossRef]
219. Henry, T.R.; Spitsbergen, J.M.; Hornung, M.W.; Abnet, C.C.; Peterson, R.E. Early life stage toxicity of 2,3,7,8-tetrachlorodibenzo-p-dioxin in zebrafish (*Danio rerio*). *Toxicol. Appl. Pharmacol.* **1997**, *142*, 56–68. [CrossRef]
220. Zodrow, J.M.; Stegeman, J.J.; Tanguay, R.L. Histological analysis of acute toxicity of 2,3,7,8-tetrachlorodibenzo-p-dioxin (TCDD) in zebrafish. *Aquat. Toxicol.* **2004**, *66*, 25–38. [CrossRef]
221. Lanham, K.A.; Plavicki, J.; Peterson, R.E.; Heideman, W. Cardiac myocyte-specific AHR activation phenocopies TCDD-induced toxicity in zebrafish. *Toxicol. Sci.* **2014**, *141*, 141–154. [CrossRef] [PubMed]
222. Dlugos, C.A.; Rabin, R.A. Structural and functional effects of developmental exposure to ethanol on the zebrafish heart. *Alcohol. Clin. Exp. Res.* **2010**, *34*, 1013–1021. [CrossRef] [PubMed]
223. Sarmah, S.; Muralidharan, P.; Marrs, J.A. Embryonic Ethanol Exposure Dysregulates BMP and Notch Signaling, Leading to Persistent Atrio-Ventricular Valve Defects in Zebrafish. *PLoS ONE* **2016**, *11*, e0161205. [CrossRef] [PubMed]
224. Staudt, D.; Stainier, D. Uncovering the molecular and cellular mechanisms of heart development using the zebrafish. *Annu. Rev. Genet.* **2012**, *46*, 397–418. [CrossRef]
225. Palencia-Desai, S.; Rost, M.S.; Schumacher, J.A.; Ton, Q.V.; Craig, M.P.; Baltrunaite, K.; Koenig, A.L.; Wang, J.; Poss, K.D.; Chi, N.C.; et al. Myocardium and BMP signaling are required for endocardial differentiation. *Development* **2015**, *142*, 2304–2315. [CrossRef] [PubMed]
226. Liu, J.; Stainier, D.Y. Zebrafish in the study of early cardiac development. *Circ. Res.* **2012**, *110*, 870–874. [CrossRef]
227. Delot, E.C.; Bahamonde, M.E.; Zhao, M.; Lyons, K.M. BMP signaling is required for septation of the outflow tract of the mammalian heart. *Development* **2003**, *130*, 209–220. [CrossRef] [PubMed]
228. High, F.A.; Epstein, J.A. The multifaceted role of Notch in cardiac development and disease. *Nat. Rev. Genet.* **2008**, *9*, 49–61. [CrossRef] [PubMed]
229. Niessen, K.; Karsan, A. Notch signaling in cardiac development. *Circ. Res.* **2008**, *102*, 1169–1181. [CrossRef]
230. Timmerman, L.A.; Grego-Bessa, J.; Raya, A.; Bertran, E.; Perez-Pomares, J.M.; Diez, J.; Aranda, S.; Palomo, S.; McCormick, F.; Izpisua-Belmonte, J.C.; et al. Notch promotes epithelial-mesenchymal transition during cardiac development and oncogenic transformation. *Genes Dev.* **2004**, *18*, 99–115. [CrossRef]
231. Cadena, P.G.; Cadena, M.R.S.; Sarmah, S.; Marrs, J.A. Folic acid reduces the ethanol-induced morphological and behavioral defects in embryonic and larval zebrafish (*Danio rerio*) as a model for fetal alcohol spectrum disorder (FASD). *Reprod. Toxicol.* **2020**, *96*, 249–257. [CrossRef] [PubMed]

232. Sarmah, S.; Marrs, J.A. Complex cardiac defects after ethanol exposure during discrete cardiogenic events in zebrafish: Prevention with folic acid. *Dev. Dyn. Off. Publ. Am. Assoc. Anat.* **2013**, *242*, 1184–1201. [CrossRef] [PubMed]
233. Mitchell, J.J.; Paiva, M.; Heaton, M.B. The antioxidants vitamin E and beta-carotene protect against ethanol-induced neurotoxicity in embryonic rat hippocampal cultures. *Alcohol* **1999**, *17*, 163–168. [CrossRef] [PubMed]
234. Joshi, R.; Adhikari, S.; Patro, B.S.; Chattopadhyay, S.; Mukherjee, T. Free radical scavenging behavior of folic acid: Evidence for possible antioxidant activity. *Free Radic. Biol. Med.* **2001**, *30*, 1390–1399. [CrossRef] [PubMed]
235. Rao, K.S.; Kameswaran, V.; Bruneau, B.G. Modeling congenital heart disease: Lessons from mice, hPSC-based models, and organoids. *Genes Dev.* **2022**, *36*, 652–663. [CrossRef] [PubMed]
236. Tambi, R.; Zehra, B.; Nandkishore, S.; Sharafat, S.; Kader, F.; Nassir, N.; Mohamed, N.; Ahmed, A.; Abdel Hameid, R.; Alasrawi, S.; et al. Single-cell reconstruction and mutation enrichment analysis identifies dysregulated cardiomyocyte and endothelial cells in congenital heart disease. *Physiol. Genom.* **2023**, *55*, 634–646. [CrossRef]
237. Cornean, A.; Gierten, J.; Welz, B.; Mateo, J.L.; Thumberger, T.; Wittbrodt, J. Precise in vivo functional analysis of DNA variants with base editing using ACEofBASEs target prediction. *Elife* **2022**, *11*, e72124. [CrossRef]

International Journal of
Molecular Sciences

MDPI

Article

Ace Deficiency Induces Intestinal Inflammation in Zebrafish

Mingxia Wei [1,†], Qinqing Yu [1,†], Enguang Li [1], Yibing Zhao [1], Chen Sun [1,2], Hongyan Li [1,2], Zhenhui Liu [1,2,*] and Guangdong Ji [1,2,*]

1 College of Marine Life Sciences, Ocean University of China, Qingdao 266003, China; weimingxia@stu.ouc.edu.cn (M.W.); yuqinqing@stu.ouc.edu.cn (Q.Y.)

2 Key Laboratory of Evolution & Marine Biodiversity (Ministry of Education), Institute of Evolution & Marine Biodiversity, Ocean University of China, Qingdao 266003, China

* Correspondence: zhenhuiliu@ouc.edu.cn (Z.L.); jamesdong@ouc.edu.cn (G.J.); Tel.: +86-0532-82032170 (Z.L.); +86-0532-82031665 (G.J.)

† These authors contributed equally to this work.

Abstract: Inflammatory bowel disease (IBD) is a nonspecific chronic inflammatory disease resulting from an immune disorder in the intestine that is prone to relapse and incurable. The understanding of the pathogenesis of IBD remains unclear. In this study, we found that *ace* (angiotensin-converting enzyme), expressed abundantly in the intestine, plays an important role in IBD. The deletion of *ace* in zebrafish caused intestinal inflammation with increased expression of the inflammatory marker genes interleukin 1 beta (*il1b*), matrix metallopeptidase 9 (*mmp9*), myeloid-specific peroxidase (*mpx*), leukocyte cell-derived chemotaxin-2-like (*lect2l*), and chemokine (C-X-C motif) ligand 8b (*cxcl8b*). Moreover, the secretion of mucus in the $ace^{-/-}$ mutants was significantly higher than that in the wild-type zebrafish, validating the phenotype of intestinal inflammation. This was further confirmed by the IBD model constructed using dextran sodium sulfate (DSS), in which the mutant zebrafish had a higher susceptibility to enteritis. Our study reveals the role of *ace* in intestinal homeostasis, providing a new target for potential therapeutic interventions.

Keywords: teleost; angiotensin-converting enzyme; inflammatory bowel disease; enteritis

1. Introduction

Chronic intestinal inflammation is a result of a breakdown in intestinal homeostasis. Intestinal inflammation occurs with many gastrointestinal diseases, among which inflammatory bowel disease (IBD) is typical. IBD includes Crohn's disease (CD) and ulcerative colitis (UC) [1,2]. IBD occurs primarily in the mucosa of the large intestine, leading to debilitating conditions including diarrhea, rectal bleeding, and weight loss [3–5]. Animals with IBD have abnormal expressions of proinflammatory molecules such as IL-6, IL-1b, and TNFa and immunoregulatory cytokines such as TGFb, IL-10, and IL-35 [6,7]. Various undetermined environmental and genetic factors, even an inappropriate immune response to gut bacteria, contribute to the disease [8–11]. However, there is currently no clear pathogenesis of IBD. A widely used model to investigate the pathogenesis of IBD is the dextran sulfate sodium (DSS)-induced colitis model in mice [12–15].

Zebrafish (*Danio rerio*) have been used as animal models for human intestinal inflammation diseases including IBD [14,16–18]. Many genes that function in inflammatory responses are conserved between zebrafish and mammals. In mammals, for example, NOD1 and NOD2 are both involved in the detection of bacteria and contribute to gastrointestinal inflammation. In zebrafish, Nod1 and Nod2, which are expressed in intestinal epithelial cells and neutrophils, are also IBD-susceptible genes [19,20]. Other genes such as Il-1β and Sst3 also have a similar function in the innate immune response between zebrafish and mammals [21–23]. Therefore, zebrafish have proven to be effective models for studying intestinal damage.

The angiotensin-converting enzyme (ACE) is an essential enzyme in the renin–angiotensin–aldosterone system (RAAS), which regulates blood pressure by the cascading of enzyme proteolysis [24]. Two isoforms of mammalian ACE are recognized: the somatic form (sACE) and the germinal form (gACE) [25–28]. ACE was implicated in the pathological processes of brain ischemic injury, cardiovascular disorders, lung injury, and other processes [29]. In mice lacking *ace*, manifestations of hypotension, renal vascular thickening, and impaired urine concentration were observed [30–33]. It was reported that the apoptosis and proliferation of intestinal epithelial cells within the intestinal epithelium are compromised in mice with ACE deficiency [34–36]. Thus, potential correlations of ACE with inflammation in the gastrointestinal tract were demonstrated [37–39]. Nevertheless, the exact mechanism by which ACE exerts its regulatory effects on the gastrointestinal tract remains unclear.

Based on this, the aim of this study is (1) to understand the evolutionary conservation of zebrafish Ace among vertebrates and its specific expression in the intestine; (2) to investigate the role of Ace in the development and progression of IBD; (3) to analyze the effects of *ace* deletion in zebrafish on intestinal inflammation; and (4) to confirm the role of Ace in intestinal inflammation using a dextran sodium sulfate (DSS)-induced IBD model.

2. Results

2.1. Ace Is Evolutionarily Conserved in Vertebrates

The zebrafish *ace* gene was cloned based on the gene information from Ensembl database (Ensembl ID: ENSDARG00000079166). The Ace protein in zebrafish was identified to have two metalloproteinase domains and a transmembrane region, which are conserved among vertebrates (Figure 1A). Also, they have a similar 3-D structure (Figure 1B). The phylogenetic analysis showed that zebrafish Ace proteins are clustered with those from other vertebrates (Figure 1C), and there is syntenic conservation of *ace* between zebrafish and flameback cichlids; however, no conservation was observed among zebrafish *ace* genomic neighborhoods and those of humans and xenopus (Figure 1D), suggesting that genomic rearrangements, such as inversions or translocations, may have occurred between ray-finned fishes and tetrapods over evolutionary time, leading to the difference in gene order and synteny.

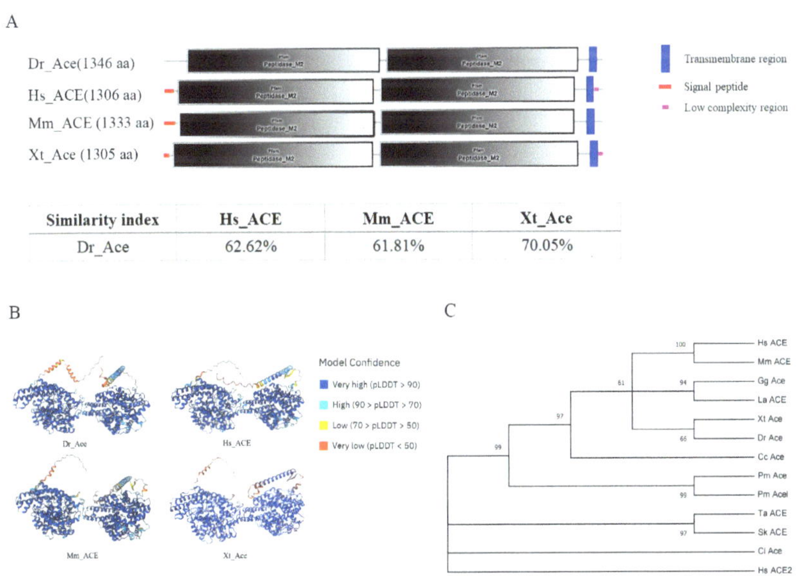

Figure 1. *Cont.*

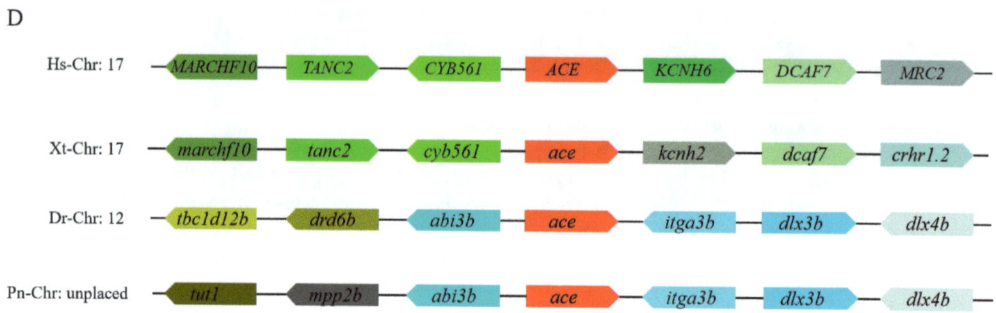

Figure 1. Homology comparison of Ace homologues in different species. (**A**) The secondary structures of Ace homologues in *D. rerio*, *H. sapiens*, *M. musculus*, and *X. tropicalis*. This diagram was generated using SMART online software (http://smart.embl-heidelberg.de/, accessed on 1 December 2023). Red squares, signal peptides; blue rectangles, transmembrane regions; M2, metallopeptidase 2 domain. The similarity index was derived by DNAMAN software (9.0.1.116). (**B**) The three-dimensional structures of *D. rerio* ACE, *H. sapiens* ACE, *M. musculus* ACE, and *X. tropicalis* ACE. This diagram was generated using the Alphafold online server (https://colab.research.google.com/github/deepmind/alphafold/blob/main/notebooks/AlphaFold.ipynb accessed on 5 December 2023). AlphaFold produces a per-residue model confidence score (pLDDT, predicted local distance difference test) between 0 and 100. Some regions below 50 pLDDT may be unstructured in isolation. (**C**) A phylogenetic tree of Ace homologues of various species was constructed with IQ-tree using the maximum likelihood method. Hs: *H. sapiens* (NP_000780), Mm: *M. musculus* (NP_997507), Gg: *G. gallus* (NP_001161204), La: *L. agilis* (XP_033024785), Xt: *X. tropicalis* (NP_001116882), Cc: *C. carcharias* (XP_041029546), Dr: *D. rerio* (XP_694336), Pn: *P. nyererei* (XP_005721704), Lo: *L. oculatus* (XP_015217378), Pm: *P. marinus* (XP_032821231, XP_032807619), Ci: *C. intestinalis* (XP_026693950), Sk: *S. kowalevskii* (XP_002741143), and Ta: *T. adhaerens* (XP_002111333). Each node was bootstrapped with 1000 replications to estimate its reliability. (**D**) Syntenic analysis among zebrafish *ace* genomic neighborhoods and those of humans, frogs, and flameback cichlids. Different colors indicate different genes. Orthologs of these genes in other species are shown in corresponding colors.

2.2. Zebrafish ace Expressed in Intestines during Early Development

The relative transcript levels of *ace* were examined in different tissues of adult zebrafish. We observed that the highest expression of *ace* was detected in the intestinal tissue of adult zebrafish. During the early stages of development, *ace* expression became detectable starting from 14 hpf. As the development progressed, the expression of this gene gradually increased (Figure 2C).

To further explore the spatio-temporal expression pattern of *ace* during development, WISH was performed. The results indicated that the *ace* gene was specifically expressed in the intestine of zebrafish at 4 dpf and 5 dpf (Figure 2D). This finding was corroborated by sections of the larvae following in situ hybridization (Figure 2E).

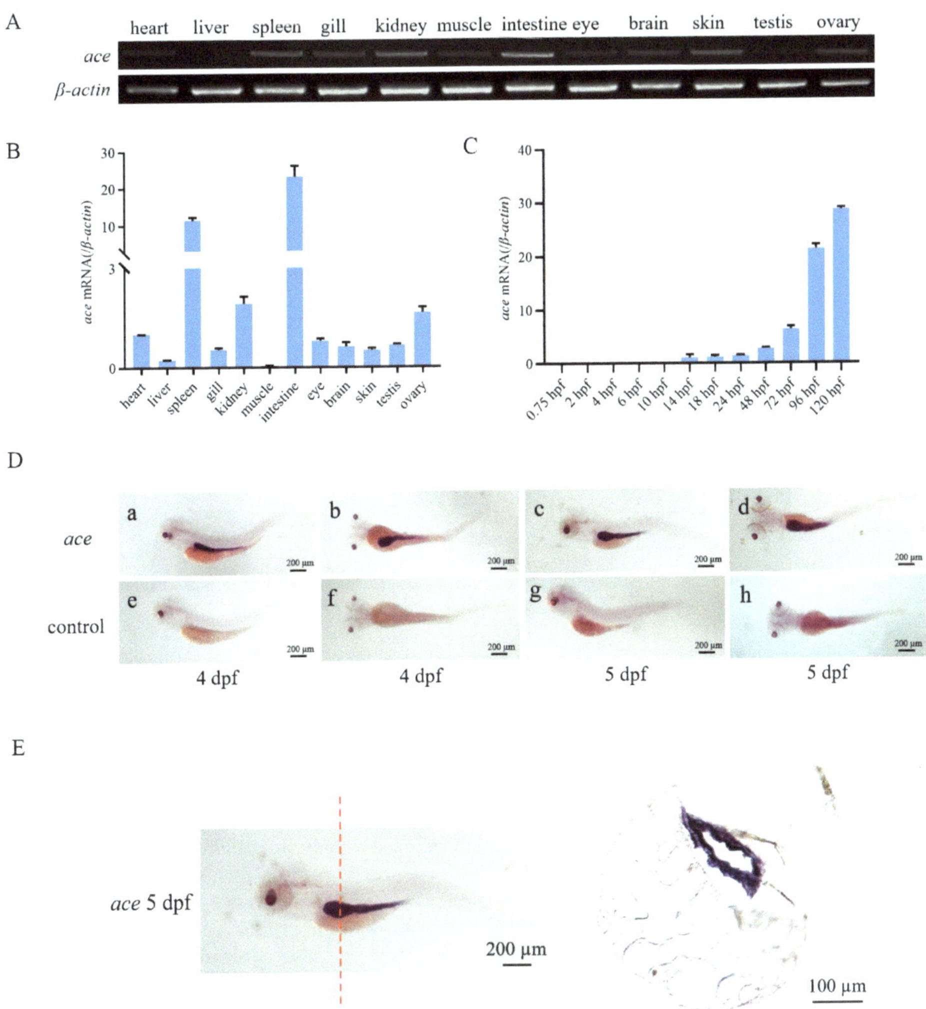

Figure 2. Expression patterns of *ace* mRNA in zebrafish. (**A**) Transcriptional expression patterns of *ace* mRNA in different tissues (heart, liver, spleen, gill, kidney, muscle, intestine, eye, brain, skin, testis, and ovary) were detected by RT-PCR. *β-actin* was used as an internal control. (**B**,**C**) Relative transcriptional expression of *ace* in different tissues and different developmental stages as measured by RT-qPCR. hpf: hours post-fertilization. *β-actin* was used as an internal control. (**D**) Spatio-temporal expression patterns of *ace* in embryos detected by WISH. "a–d" represents the results of anti-sense probe hybridization, while "e–h" represents the results of sense probe hybridization (negative control). dpf: days post-fertilization. (**E**) The intestinal region of the larvae at 5 dpf after WISH was frozen and sectioned (8 µm). The red line indicates the position of the frozen section.

2.3. Ace Localized on the Cell Membrane

To investigate the subcellular localization of Ace, HEK293 T cells were transfected with either *pcDNA3.1/V5/ace/eGFP* or *pcDNA3.1/V5/eGFP* (control). Following transfection, the cells were stained with DAPI. In comparison to the control, which displayed a uniform distribution of eGFP throughout the cells, Ace exhibited distinct green fluorescence on the cell membrane and endoplasmic reticulum (Figure 3). This observation may be attributed to the transmembrane region of Ace [40].

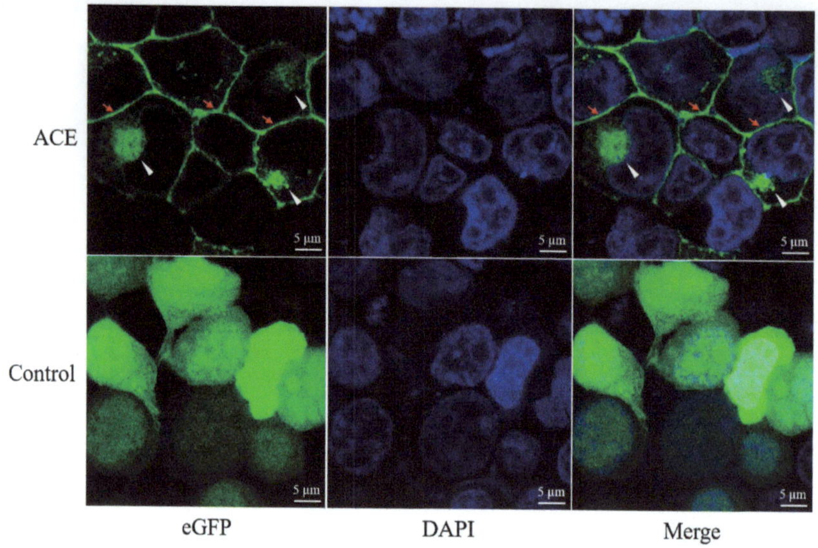

Figure 3. Subcellular localization of Ace in HEK293T cells. Recombinant plasmids *pcDNA3.1/V5/Ace/eGFP* and *pcDNA3.1/V5/eGFP* (control) were transiently transfected into HET293T cells. DAPI stained the nucleus. Red arrows point to the cell membrane and white arrowheads point to the endoplasmic reticulum.

2.4. Ace Deficiency Does Not Result in Intestine Defects in Zebrafish

To investigate the function of *ace*, we generated an *ace* knockout ($ace^{-/-}$) zebrafish by the CRISPR/Cas9 approach. The mutation induced a frameshift in the protein-coding region, leading to premature termination of translation (Figure 4A). As expected, mRNA levels in $ace^{-/-}$ mutants were significantly reduced compared to those in the wild type (Figure 4B). Surprisingly, during development, the *ace*-deficient zebrafish appeared normal, exhibiting survival to adulthood and fertility without any observable morphological or developmental abnormalities. To explore whether the absence of *ace* impacts intestinal tube development, we synthesized the intestinal marker *fabp2* to probe intestinal development. The WISH analysis did not indicate any noticeable abnormalities, suggesting that the intestinal development program is not evidently affected at 5 dpf (Figure 4B). This was further confirmed by a histological examination (HE staining) of intestinal sections (Figure 4C). Thus, our results suggest that *ace* deficiency does not significantly alter intestinal development.

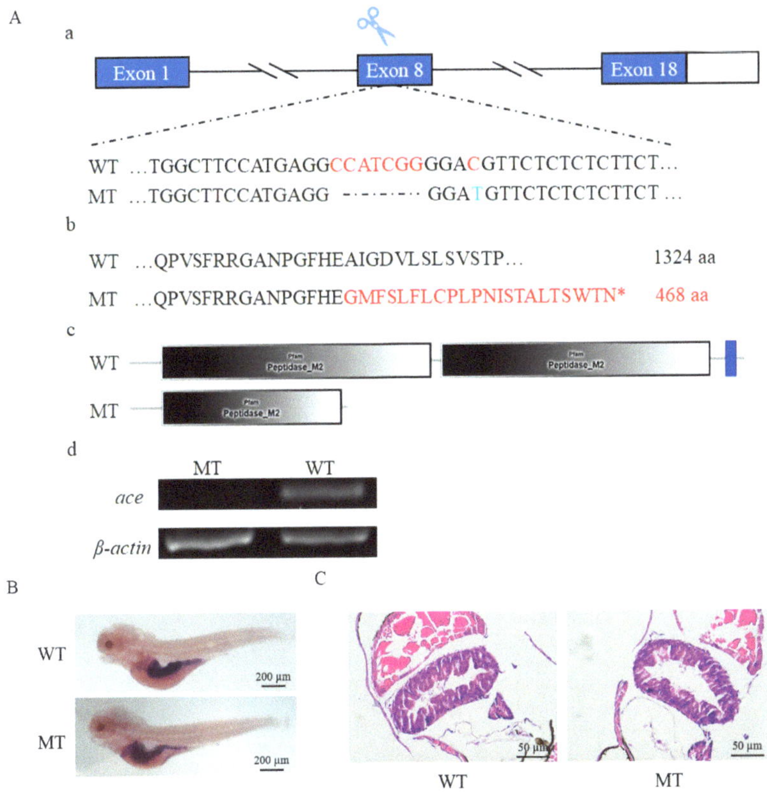

Figure 4. Knockout of ace and morphological examination of ace$^{-/-}$ mutant larvae. (**A**) The generation of ace mutations in zebrafish through CRISPR/Cas9 technology. (**a**). The targeted exon 8 of ace containing the knockout site. Exons are indicated by boxes, blue box means coding region of exon, while white box means non-coding region of exon, introns are represented by fold lines. The red letter in the wild type (WT) means the position some changes will happen, while the dotted line and blue letter corresponded to red letters means deletion and transition happened in the mutant (MT), respectively. (**b**,**c**). In comparison to the wild type, the mutants had a deletion of 7 nucleotides in exon 8, leading to skipped mutation (red letters), and premature termination of translation at the 468th amino acid. The asterisk in the figure indicates the position of protein translation termination. (**d**). To assess the impact of ace knockout, RT-PCR was performed to analyze ace expression in both ace$^{-/-}$ and wild-type samples (**B**) Expression patterns of fabp2 in ace$^{-/-}$ mutants and wild-type larvae at 5 dpf, as detected by WISH. (**C**) The morphology of the intestinal epithelium in ace$^{-/-}$ mutants and wild-type larvae at 5 dpf, as shown by HE staining.

2.5. Ace Deficiency Induces Intestinal Inflammation

To investigate the differential expressed genes (DEGs) between wild-type and *ace*$^{-/-}$ mutant larvae at 5 dpf, an RNA-seq analysis was performed. The results of the Gene Ontology (GO) analysis indicated an upregulation of endopeptidase activity in the absence of *ace* (Figure 5A). Additionally, Kyoto Encyclopedia of Genes and Genomes (KEGG) analyses revealed an enrichment of DEGs in the phagosome signaling pathway, suggesting a potential involvement in immunity (Figure 5B). Notably, the DEGs also included genes expressed in the intestinal epithelium and associated with inflammation (Figure 5C). Further, *ace* is mainly expressed in leukocytes in *fxr*$^{-/-}$ zebrafish larva intestine at 6 dpf [41], which is in line with the RNA-seq results above, indicating that *ace* plays an important role in intestinal immunity (Figure 5D).

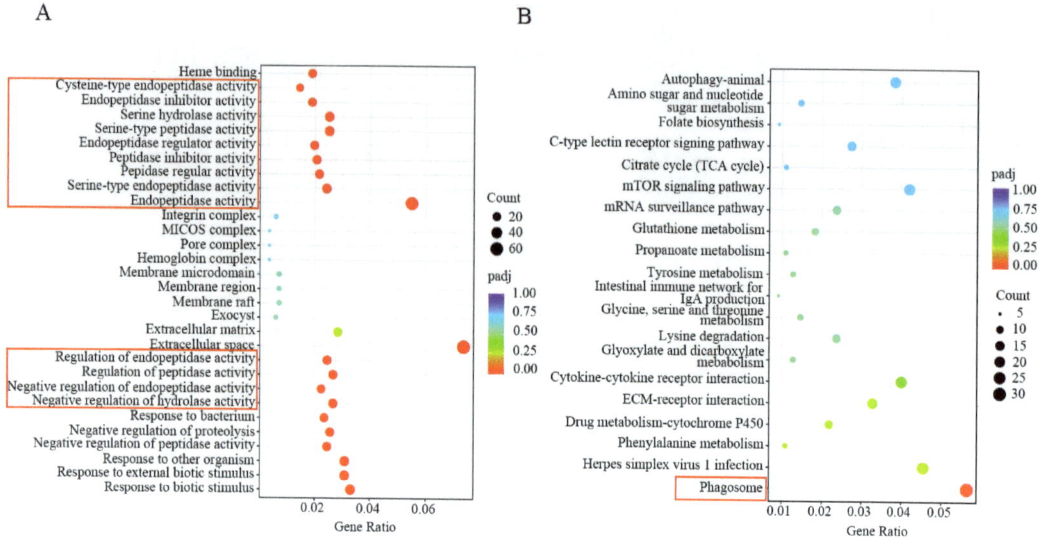

C

Gene ID	Log2FoldChange	Pvalue	Gene name	Gene chr	Gene description
ENSDARG00000014031	7.50335555016198	4.66×10^{-31}	*abcc2*	13	ATP-binding cassette, sub-family C (CFTR/MRP), member 2
ENSDARG00000114451	6.89471776976595	2.64×10^{-25}	*mmp13a*	10	matrix metallopeptidase 13a
ENSDARG00000042816	4.56163500506937	2.21×10^{-20}	*mmp9*	8	matrix metallopeptidase 9
ENSDARG00000012076	−4.25755706746673	6.84×10^{-19}	*apoa1a*	5	apolipoprotein A-Ia
ENSDARG00000043406	−5.63009142383027	3.29×10^{-18}	*slc8a1b*	17	solute carrier family 8 (sodium/calcium exchanger), member 1b
ENSDARG00000098700	4.01385587230294	1.24×10^{-11}	*il1b*	10	interleukin 1, beta
ENSDARG00000102299	3.54935545659283	1.05×10^{-9}	*cxcl8b.1*	7	chemokine (C-X-C motif) ligand 8b, duplicate 1
ENSDARG00000019521	2.12085638345123	1.73×10^{-6}	*mpx*	10	myeloid-specific peroxidase

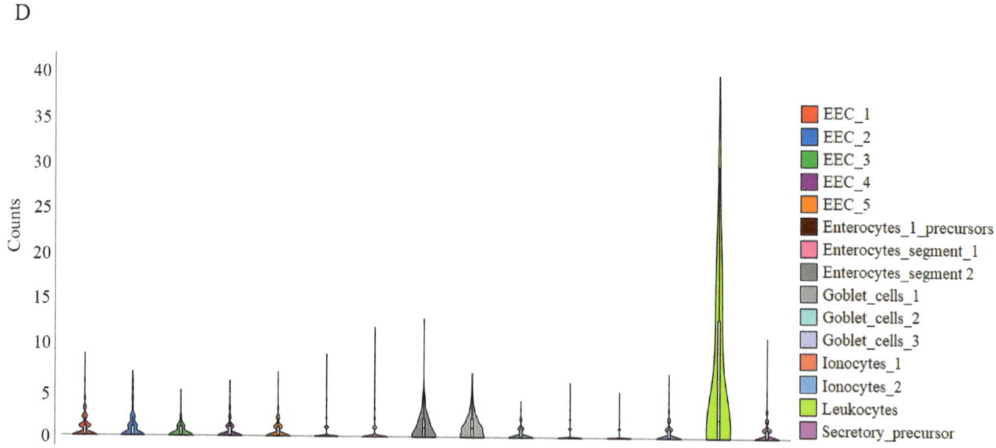

Figure 5. RNA-seq analysis of $ace^{-/-}$ mutants and wild type at 5 dpf. (**A**,**B**) GO and KEGG analysis of the pathways through the enrichment of DEGs between $ace^{-/-}$ mutants and wild type. The red box signifies the peptide enzyme or immune-related signaling pathways. (**C**) The table shows some DEGs with significant changes, including genes highly expressed in intestinal epithelium and some immune genes. (**D**) Single-cell profile of ACE in $fxr^{-/-}$ zebrafish larvae at 6 dpf.

We then assessed the disparity in inflammatory cytokine expression between wild-type and *ace*$^{-/-}$ zebrafish larvae by real-time PCR. The results clearly demonstrate that the absence of *ace* substantially augments the expression of interleukin 1 beta (*il1b*), matrix metallopeptidase 9 (*mmp9*), leukocyte cell-derived chemotaxin-2-like (*lect2l*), chemokine (C-X-C motif) ligand 8b (*cxcl8b*), and the neutrophil marker myeloid-specific peroxidase (*mpx*) (Figure 6A), as well as some other validated members of the MMP families, such as *mmp9*, *mmp13a*, *mmp14b*, and *mmp30* (Figure 6B). The heightened levels of pro-inflammatory cytokines in *ace*$^{-/-}$ zebrafish indicate the occurrence of an inflammatory response. It is known that goblet cells secrete mucus to safeguard the intestines against infection, primarily concentrated in the mid- and posterior intestine regions. The abundance of intestinal mucus was further evaluated through AB-PAS staining. Interestingly, *ace*$^{-/-}$ mutants, but not the wild type, exhibited enriched mucus in their digestive tracts at 5 dpf (Figure 6C,D), indicating that goblet cells and mucus could promote intestinal defense and homeostasis. The observation of heightened mucus secretion and elevated expression of pro-inflammatory cytokines in *ace*-deficient zebrafish suggests that the deletion of *ace* may contribute to the induction of intestinal inflammation and defense.

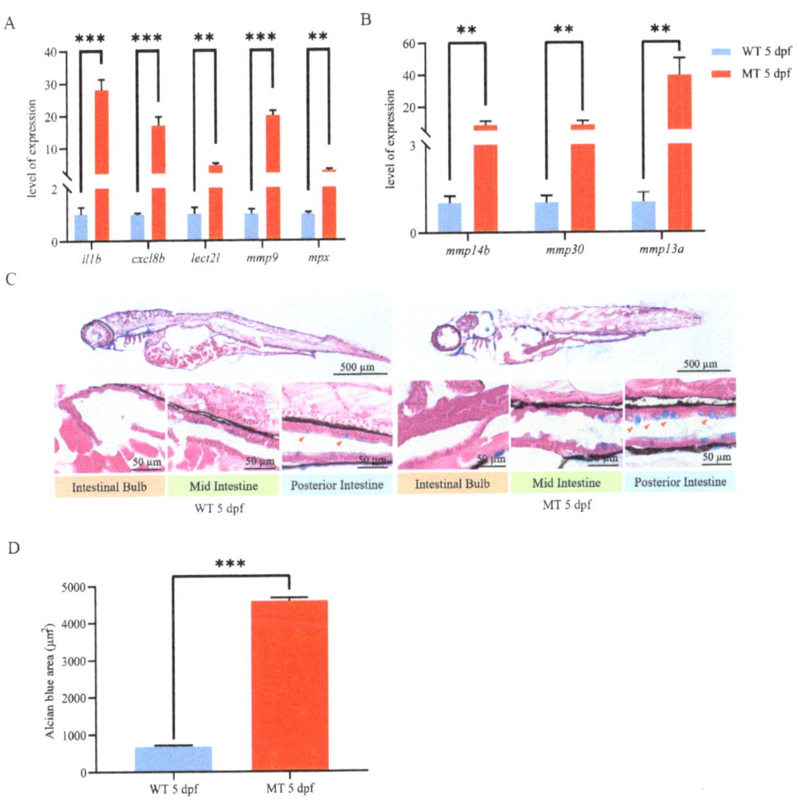

Figure 6. Inflammatory response evaluation for *ace*$^{-/-}$ mutants and wild-type larvae at 5 dpf. (**A**,**B**) qRT-PCR analysis of selected inflammation-related genes. (**C**) AB-PAS staining sections label mucus in the intestine of both *ace*$^{-/-}$ mutants and wild-type larvae at 5 dpf (upper panel). The lower panels provide enlarged photographs of the intestinal bulb and mid- and posterior intestine regions. The red arrowhead in the figure indicates the staining signal. (**D**) Cell statistics for Alcian blue staining in the digestive tract of *ace*$^{-/-}$ mutants and wild-type larvae at 5 dpf. Data are presented as the mean ± SD of 10 fish. ** $p < 0.01$, *** $p < 0.001$. The experiments were independently repeated three times.

2.6. *Ace-Deficient Zebrafish Are Susceptible to IBD*

To determine whether *ace*-deficient zebrafish are susceptible to IBD, we first used DSS to construct an intestinal infection model. Three days after drinking DSS, the expression of pro-inflammatory factors, such as *il1b*, *lect21*, *cxcl8b*, *mmp9*, and *mpx*, was increased, indicating that the intestinal infection model was successfully generated (Figure 7A,B). When zebrafish larvae were exposed to DSS at a dose between 0.5 and 1% (w/v), the survival rate of *ace*$^{-/-}$ mutants was lower than that of the wild type (Figure 7C). By HE staining on intestinal sections, we found that the intestinal tract of the *ace*$^{-/-}$ mutant larvae was significantly smaller, with a thickened wall and a smaller diameter (Figure 7D).

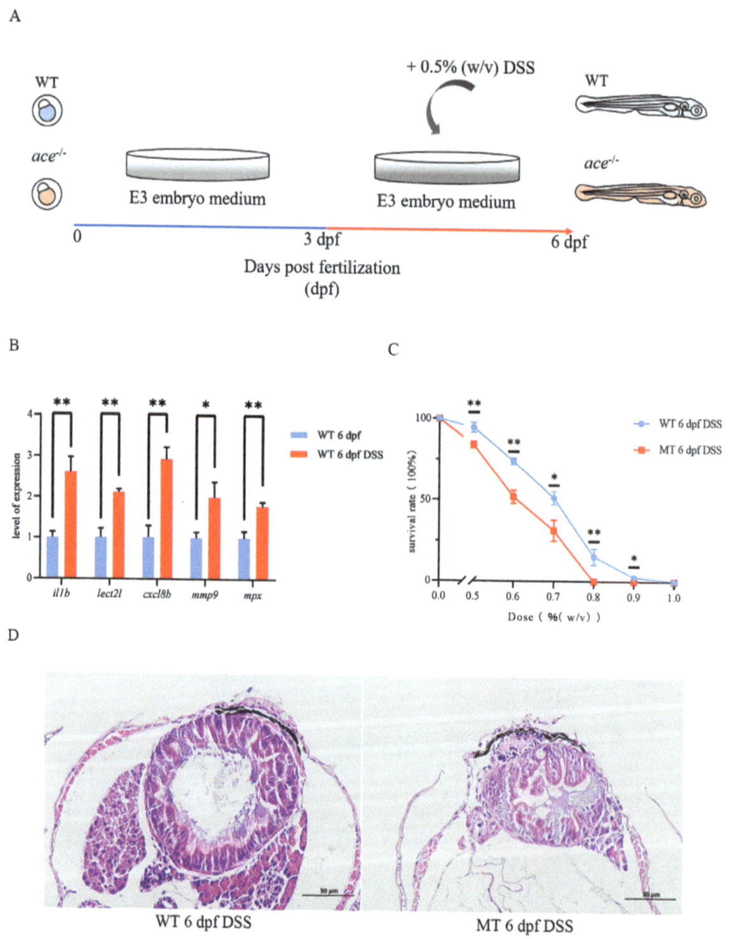

Figure 7. The IBD model was successfully constructed using DSS. (**A**) A diagram of IBD model development. (**B**) Changes in DSS-treated and untreated wild-type larvae were examined by qRT-PCR for several pro-inflammatory factors: *ilb*, *lect2l*, *cxcl8b*, *mmp9*, and *mpx*. Data represent the mean ± SD. * $p < 0.05$, ** $p < 0.01$. Three independent biological replicates were performed. (**C**) Line chart of survival rates comparing *ace*$^{-/-}$ mutants and wild-type larvae treated as the DSS dose (% (w/v)) increased from 0% to 1% (n = 50). The significance of differences is annotated above the nodes in the line chart. * $p < 0.05$, ** $p < 0.01$. (**D**) HE staining shows a more severely disorganized intestinal epithelium in *ace*$^{-/-}$ mutants compared with the wild type at 6 dpf following DSS treatment.

Furthermore, AB-PAS staining showed more mucus in the *ace*$^{-/-}$ mutants compared to wild-type zebrafish (Figure 8A). Additionally, we compared the expression of inflammatory factors between *ace*$^{-/-}$ mutants and wild-type larvae treated with DSS. Compared to that in wild-type zebrafish, the expression of *il1b*, *mmp9*, *lect2l*, *cxcl8b*, and *mpx* was significantly upregulated in *ace*-deficient zebrafish (Figure 8B). Collectively, our data indicate that *ace*-deficient zebrafish are highly susceptible to IBD.

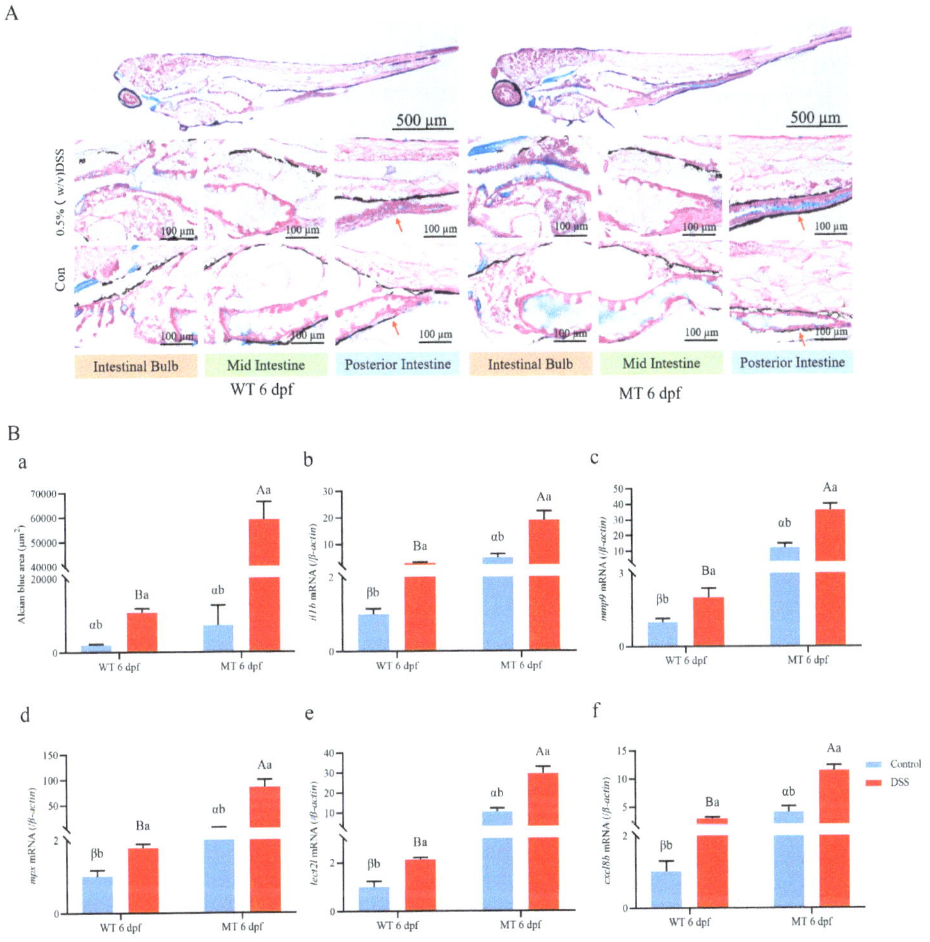

Figure 8. Histopathological assessment comparing *ace*$^{-/-}$ mutant and wild-type larvae treated with DSS. (**A**) AB-PAS staining sections highlighting mucus in the intestine of DSS-treated and untreated *ace*$^{-/-}$ mutants and wild-type larvae (upper panel). The lower panels show enlarged photographs of the intestinal bulb and the mid- and posterior intestine regions. At the bottom, the control group exhibits AB-PAS staining sections that did not undergo DSS treatment. The red arrow in the figure indicates the staining signal. (**B**) (**a**). Quantification of Alcian blue-stained cells in the digestive tract of *ace*$^{-/-}$ mutants and wild-type larvae at 6 dpf. Data represent the mean ± SD of 10 fish. (**b**–**f**). The expression of several pro-inflammatory factors in *ace*$^{-/-}$ mutants and wild-type larvae examined by qRT-PCR. Data are presented as the mean ± SD. Different lowercase letters indicate significant differences ($p < 0.05$) within the group; different uppercase letters indicate significant differences ($p < 0.05$) between the groups after DSS treatment; different Latin letters indicate significant differences ($p < 0.05$) between the groups before DSS treatment.

3. Discussion

Angiotensin-converting enzyme (ACE), a zinc-dependent dipeptidyl carboxypeptidase composed of two metalloproteinase domains, plays a vital role in the renin–angiotensin–aldosterone system (RAAS) and is involved in immune regulation [42,43]. In mice, *ace* deficiency has been reported to affect intestinal epithelial renewal, but its precise function in intestinal inflammation remains unexplored [34]. In this study, we characterized zebrafish *ace* and found that its deletion induced intestinal inflammation, thereby expanding our understanding of ACE's functions.

In humans, ACE contains sACE and gACE isoforms. sACE is significantly expressed in various tissues such as the small intestine, duodenum, lungs, kidneys, choroid plexus, and placenta [44,45], while gACE is specifically expressed in testes and is associated with male fertility [46,47]. Notably, the highest expression level of sACE was observed in the small intestine [48]. This is in line with our finding in zebrafish that the expression of the *ace* gene is highest in the intestine. However, an analysis of tissue expression revealed a lack of notable expression in the testes in zebrafish, and we also found that both female and male *ace* mutants' reproduction was not affected when compared to the wild type; this is different from the gACE detected in human, suggesting that there is no functional gACE in zebrafish.

Although the absence of *ace* appears to have no discernible impact on the zebrafish's intestinal development as indicated by its morphology, transcripts of the *ace* gene were detected as early as 14 hpf and continued to be expressed throughout development in different adult tissues including the spleen and intestine. The zebrafish begins to form a digestive tract in a segmental fashion at the mid-somite stages (18 hpf) and completes gut tube morphogenesis at 34 hpf; after the onset of exogenous feeding (5 dpf), there is a dramatic increase in the size of the intestine and the appearance of epithelial cells and other cell types [49]. In accordance with this process, from 4 to 5 dpf, *ace* specifically localized in the intestine. Moreover, a single-cell transcriptome analysis of zebrafish intestine at 6 dpf showed that *ace* expressed in many cell types, including the highest expression in leukocytes and moderate expression in enteroendocrine cells (EECs), enterocytes, goblet cells, ionocytes, and secretory precursors [41], suggesting that *ace* may play an important role in intestinal homeostasis and immunity.

More mucus was secreted in the mid- and posterior intestine of $ace^{-/-}$ mutants compared to the wild type at 5 dpf, indicating the occurrence of intestinal inflammation. Further, the transcriptome sequencing of 5 dpf larvae revealed that the absence of *ace* had a significant impact on immune pathways, leading to a notable increase in the expression of proinflammatory factors including *il1b*, *lect21*, *cxcl8b*, *mpx*, and *mmp9*, as well as several other members of the *mmp* gene family, such as *mmp9*, *mmp13a*, *mmp14b*, and *mmp30*. It has been observed that some of the MMP family members are associated with various inflammatory responses [50,51]. Thus, there is a potential correlation between intestinal inflammation and the deletion of *ace*. This correlation was further supported by the constructed IBD model, where intestinal inflammation induced by DSS showed a similar expression pattern of proinflammatory factors to that observed in $ace^{-/-}$ mutants. Moreover, compared to the wild-type larvae, the $ace^{-/-}$ mutants exhibited a significantly lower survival rate, increased mucus secretion in the intestine, and a notable upregulation of various inflammatory factors in this IBD model, suggesting that both *ace* deletion and DSS induction synergistically contribute to the occurrence and progression of inflammation. Although ACE in wild-type larvae is mildly upregulated in a dextran sodium sulfate colitis model [52], it is possible that *ace* deficiency impedes the ability of larvae to mount an appropriate immune response and increases their vulnerability to DSS stimuli, ultimately resulting in earlier mortality compared to the control group.

Many studies have documented the relationship between ACE and inflammation in mammals, and some results seem controversial. For example, ACE overexpression in myeloid-derived cells has been shown to increase the production of pro-inflammatory cytokines, such as IL-12β, TNF, or nitric oxide, while ACE overexpression in neutrophils has been shown to increase resistance to infections with MRSA, Klebsiella pneumoniae, and Pseudomonas aeruginosa [40]. *Ace*$^{-/-}$ mice have a less vigorous immune response to MRSA infection [40]. However, ACE-overexpressing macrophages have been shown to attenuate neuropathology and neuroinflammation [51]. Although ACE inhibitors have been shown to reduce vascular inflammation, there is no convincing evidence indicating that ACE inhibitors reduce plasma levels of major inflammatory markers in hypertension models [53]. This controversy is partly attributed to the possibility that inflammatory responses may be controlled by local rather than global immune, vascular, and inflammatory cell responses to infection or injury [54], with other localized factors involved in the inflammation process.

In this study, ACE was specifically localized in the intestine and was highly expressed by leukocytes present in intestinal tissue. A deficiency of ACE induced a mild inflammatory response in the intestine of zebrafish, indicating a potential role for ACE in immune regulation and gut health. It appears that other factors also participate in the *ace*-deficiency-induced inflammatory response, which are likely to depend on immune cells and molecules at the site of tissue damage or infection.

4. Materials and Methods

4.1. Ethics Statement

Embryos were produced through natural mating. All zebrafish studies were conducted according to the Animal Care and Use Committee of the Ocean University of China (SD2007695).

4.2. Zebrafish Strains and Mutants

Zebrafish (*D. rerio*) from the AB strain were kept at 28 °C and fed twice daily during dark periods of 10 h and 14 h. The fragments of *ace* were amplified using the specific primers S1 and AS1 by PCR (Table 1). Using CRISPR/Cas9 technology, *ace*$^{-/-}$ mutant lines were derived from the AB line of zebrafish. This study used 5′-TGGCTTCCATGAGGCCATCG-3′ in exon 8 as the knockout target. Mixtures of Cas9 mRNA and targeting gRNA were microinjected into one-cell-stage zebrafish embryos. A comparison with the wild-type zebrafish sequence confirmed the mutation sites. Fish from the F1 generation carrying a 7 bp deletion were crossed to obtain the F_2 generation [55,56].

Table 1. Primers used in this study.

Name	Sequence (5′-3′)	Sequence Information
S1	CGCAACCAGTAACTACGCATT	For *ace* knockout (*ace*)
AS1	CTCTCCACTGAACACACTCCAT	
S2	CTCCACAGCATTGACCTCCT	For WISH (*ace*)
AS2	GTAAGCGTTCCAGACCTCCT	
S3	CGCTCTACCTCAGCGTTCAT	For RT-qPCR (*ace*)
AS3	GCCCACATGTTTCCAAGCAG	
S4	GGTATTGTGATGGACTCTGGTGAT	For RT-qPCR (*β-actin*)
AS4	TCGGCTGTGGTGGTGAAG	
S5	CCGGAATTCATGAACAGAGGGAAAAGGGAGAG	For recombinant (*ace*)
AS5	CCGGATATCCCTTGAGCTCCATCTGAGACATGG	
S6	GATCCGCTTGCAATGAGCTAC	For RT-qPCR (*il1b*)
AS6	TCAGGGCGATGATGACGTTC	
S7	TCAGGGCGATGATGACGTTC	For RT-qPCR (*mmp9*)
AS7	TAGCGGGTTTGAATGGCTGG	

Table 1. *Cont.*

Name	Sequence (5′-3′)	Sequence Information
S8	CCAGAACCAGTGAGCCTGAG	For RT-qPCR (*mpx*)
AS8	ACTCTCTTCTTCTGCCCCCA	
S9	TAGCTTGAGTGGAGGAGGTCT	For RT-qPCR (*lect2l*)
AS9	CATGGGAAGTGATGCCAGGA	
S10	GTGCGCCAATGAGGGTGAA	For RT-qPCR (*cxcl8b*)
AS10	ACCCACGTCTCGGTAGGATT	
S11	CGCCAACAACCAGGTTTACAGTTAT	For RT-qPCR (*mmp13a*)
AS11	TCAGGACGCGTAACAGCTTG	
S12	ACAATGCGTTCTGTCGATGC	For RT-qPCR (*mmp14b*)
AS12	AGCGGGAGAATACAGACGTT	
S13	GCAGCTCTCATCCTTGTGGT	For RT-qPCR (*mmp30a*)
AS13	ACTGCGACAAATACGCCTCT	

4.3. Bioinformatics Analysis

A set of 12 ACE sequences was primarily collected from National Center for Biotechnology Information (NCBI) website (http://www.ncbi.nlm.nih.gov/). Protein domains were predicted by the SMART website (http://smart.embl.de/); protein 3-D structures were predicted by Alphafold (https://alphafold.com/). A phylogenetic tree was constructed using IQ-tree with the maximum-likelihood algorithm, the L+G4 model, and 1000 bootstrap replications.

4.4. Whole-Mount In Situ Hybridization (WISH) in Zebrafish Larvae

The zebrafish embryos for WISH were cultured from 12 h post-fertilization (hpf) in E3 embryo medium containing 0.004% PTU (Sigema, P6148, Fukushima, Japan). The fragments of *ace* were amplified using the specific primers S2 and AS2 by PCR (Table 1). The fragments were digested with Sph I (Takara Bio, 1246S, Kusatsu, Japan), and Sp6 RNA polymerase (Thermo Scientific, EP0131) (Waltham, MA, USA) was used to synthesize an antisense probe labelled with digoxigenin (DIG) (Sigema, 11277073910). WISH followed the protocol in its procedures [57]. Stereomicroscopy (Nikon, SMZ1270/1270i, Tokyo, Japan) was used to observe and photograph stained embryos.

4.5. RNA Extraction and RT-qPCR from Zebrafish Samples

Zebrafish total RNA was extracted using Total RNA Kit I (Omega, R6834-02) (Biel/Bienne, Switzerland). DNase was used to treat the RNA, and PrimeScript^{TM} RT reagent with gDNA Eraser (TaKaRa, RP047A) was used to synthesize the cDNA. As controls, reverse transcriptase-free samples were added. Quantitative PCR was performed on an ABI 7500 machine with ChamQ SYBR Color qPCR Master Mix (Vazyme, Q431-02) (Nanjing, China). To normalize the data, the *β-actin* gene was used as the internal reference gene. A comparative Ct method ($2^{-\Delta\Delta Ct}$ method) was used to calculate relative expression levels. All quantitative PCR experiments were conducted in triplicate. This study used the primers shown in Table 1.

4.6. AB-PAS Staining and HE Staining in Zebrafish Larvae

Ice-cold acetone was used to fix zebrafish embryos. Wuhan Servicebio Technology Company performed microtomy and staining of the embryos. In each section (8 μm) after AB-PAS staining, mucus-containing areas of the staining in the mid-intestine and posterior intestine were quantified using ImageJ software (1.48v). The image processing followed the protocol in its procedures [58].

4.7. RNA-Seq Analysis

RNA was extracted from both wild-type and *ace*$^{-/-}$ larvae at 5 days post-fertilization (dpf). The same batch of samples was sequenced for both control and knockout groups. The transcriptome was sequenced using Novogene (Beijing, China). The gene abundance was calculated and normalized using RPKM (reads per kb per million reads). DEGs (differentially expressed genes) between groups were determined using the EdgeR package (http://www.r-project.org/). Differential gene expression analysis was conducted with a fold change criterion of ≥2 or ≤0.5, and a false discovery rate (FDR) of <0.05 was considered significant. Subsequently, the KEGG pathways (Kyoto Encyclopedia of Genes and Genomes) were analyzed for enrichment. The single-cell profile of ACE in *fxr*$^{-/-}$ zebrafish larvae at 6 dpf was generated using the Single Cell Portal website (https://singlecell.broadinstitute.org/single_cell/study/SCP1675/zebrafish-intestinal-epithelial-cells-wt-and-fxr?genes=ace&tab=distribution#study-visualize, accessed on 1 December 2023), and the raw data can be accessed from the NCBI GEO Database: GSE173570 [41].

4.8. DSS Treatment

Analyses of DSS (Yeasen Biotech, 60316ES25) effects were conducted on wild-type and *ace*$^{-/-}$ larvae. All embryos (n = 26–32 larvae per concentration) were cultured to 3 dpf in E3 embryo medium before the tests. From a 10% DSS stock solution (w/v), dilutions were prepared (0.5%, 0.6%, 0.7%, 0.8%, 0.9% w/v, 1.0%) with the medium replaced daily. After 3 days of treatment, the survival rate was calculated. Later experiments (0.5% DSS) used the lowest toxicity dosage. From 3 dpf to 6 dpf, the wild-type and *ace*$^{-/-}$ DSS groups were cultured in 0.5% DSS in E3 embryo medium, while the corresponding control groups were cultured in parallel in E3 embryo medium only. The RNA was isolated from half of the larvae at 6 dpf, and the other half were used for histopathological analysis [23,59].

4.9. Subcellular Localization

To examine the Ace protein localization in subcellular compartments, a subcellular localization assay was conducted. The *pcDNA3.1/V5-His A* plasmid was used to create the *pcDNA3.1/V5/eGFP* vector by cloning eGFP. Next, with primers that contain EcoR I and EcoR V restriction enzyme cutting sites, the complete coding region of *ace* was amplified by PCR using specific primers (S5, AS5). Recombinant plasmids were constructed by inserting the fragments upstream of *pcDNA3.1/V5/eGFP*. To investigate the subcellular localization of Ace in HEK293 T cells, the *pcDNA3.1/V5/ace/eGFP* and *pcDNA3.1/V5/eGFP* recombinant plasmids were separately transfected. DAPI staining was performed 24 h post-transfection, and observations were made using confocal microscopes after washing the samples with PBS [60].

4.10. Statistical Analysis

Statistical analyses were conducted using Graphpad Prism 9.0.0 software. The data are presented as the mean ± standard deviation. Student's t-test was used for statistical analysis. $p < 0.05$ was considered statistically significant.

5. Conclusions

In summary, we demonstrated the evolutionary conservation of zebrafish *ace* among vertebrates, with specific expression observed in the intestine. Interestingly, deficiency in *ace* does not lead to intestinal defects in zebrafish but rather triggers intestinal inflammation. Furthermore, zebrafish lacking *ace* show heightened susceptibility to inflammatory bowel disease (IBD) phenotypes. Future studies are warranted to identify the particular cell types impacted by Ace deficiency, along with the specific signaling pathways in which Ace is involved in innate immunity. This understanding holds significant potential for advancing novel therapeutic strategies for inflammatory bowel diseases in human patients.

Author Contributions: Investigation, methodology, visualization, validation, writing—original draft: M.W. and Q.Y.; visualization, validation, writing—review and editing: E.L., Y.Z., C.S., and H.L.; conceptualization, formal analysis, methodology, visualization, validation, writing—review and editing, project administration, funding acquisition: Z.L. and G.J. All authors have read and agreed to the published version of the manuscript.

Funding: This work was funded by grants from the Natural Science Foundation of China (NSFC) (grant number: 31970429), Science & Technology Innovation Project of Laoshan Laboratory (grant number: LSKJ202203204), National Key Research and Development Project of the Ministry of Science and Technology (grant number: 2023YFE0199500), and Shandong Provincial Natural Science Foundation (grant number: ZR2020MC050).

Institutional Review Board Statement: The usage of fish was in strict accordance with the recommendations in the Guidelines for the Animal Care and Use Committee of the Ocean University of China (SD2007695). All possible endeavors were made to minimize suffering and maintain animal welfare.

Informed Consent Statement: Not applicable.

Data Availability Statement: The original contributions presented in this study are included in the article. The raw RNA-Seq data were deposited into the NCBI data bank with the accession number PRJNA1082072. Further inquiries can be directed to the corresponding authors.

Conflicts of Interest: The authors declare no conflicts of interest.

References

1. Bai, L.; Scott, M.K.D.; Steinberg, E.; Kalesinskas, L.; Habtezion, A.; Shah, N.H.; Khatri, P. Computational drug repositioning of atorvastatin for ulcerative colitis. *J. Am. Med. Inf. Assoc.* **2021**, *28*, 2325–2335. [CrossRef] [PubMed]
2. Molodecky, N.A.; Soon, I.S.; Rabi, D.M.; Ghali, W.A.; Ferris, M.; Chernoff, G.; Benchimol, E.I.; Panaccione, R.; Ghosh, S.; Barkema, H.W.; et al. Increasing incidence and prevalence of the inflammatory bowel diseases with time, based on systematic review. *Gastroenterology* **2012**, *142*, 46–54. [CrossRef] [PubMed]
3. Chachu, K.A.; Osterman, M.T. How to diagnose and treat IBD mimics in the refractory ibd patient who does not have IBD. *Inflamm. Bowel Dis.* **2016**, *22*, 1262–1274. [CrossRef] [PubMed]
4. Abraham, C.; Cho, J.H. Inflammatory bowel disease. *N. Engl. J. Med.* **2009**, *361*, 2066–2078. [CrossRef] [PubMed]
5. Kozlowski, C.; Jeet, S.; Beyer, J.; Guerrero, S.; Lesch, J.; Wang, X.; Devoss, J.; Diehl, L. An entirely automated method to score dss-induced colitis in mice by digital image analysis of pathology slides. *Dis. Model. Mech.* **2013**, *6*, 855–865. [CrossRef] [PubMed]
6. Levite, M. Neuropeptides, by direct interaction with T cells, induce cytokine secretion and break the commitment to a distinct T helper phenotype. *Proc. Natl. Acad. Sci. USA* **1998**, *95*, 12544–12549. [CrossRef] [PubMed]
7. Blum, A.M.; Metwali, A.; Mathew, R.C.; Cook, G.; Elliott, D.; Weinstock, J.V. Granuloma t lymphocytes in murine schistosomiasis mansoni have somatostatin receptors and respond to somatostatin with decreased ifn-gamma secretion. *J. Immunol.* **1992**, *149*, 3621–3626. [CrossRef] [PubMed]
8. Takagi, T.; Homma, T.; Fujii, J.; Shirasawa, N.; Yoriki, H.; Hotta, Y.; Higashimura, Y.; Mizushima, K.; Hirai, Y.; Katada, K.; et al. Elevated er stress exacerbates dextran sulfate sodium-induced colitis in prdx4-knockout mice. *Free. Radic. Biol. Med.* **2019**, *134*, 153–164. [CrossRef] [PubMed]
9. de Souza, H.S.P.; Fiocchi, C. Immunopathogenesis of IBD: Current state of the art. *Nat. Rev. Gastroenterol. Hepatol.* **2016**, *13*, 13–27. [CrossRef]
10. Keshteli, A.H. Hyperhomocysteinemia as a potential contributor of colorectal cancer development in inflammatory bowel diseases: A review. *World J. Gastroenterol.* **2015**, *21*, 1081. [CrossRef]
11. Fonseca-Camarillo, G.; Yamamoto-Furusho, J.K. Immunoregulatory pathways involved in inflammatory bowel disease. *Inflamm. Bowel Dis.* **2015**, *21*, 2188–2193. [CrossRef] [PubMed]
12. Li, R.; Huang, X.; Yang, L.; Liang, X.; Huang, W.; Lai, K.P.; Zhou, L. Integrated analysis reveals the targets and mechanisms in immunosuppressive effect of mesalazine on ulcerative colitis. *Front. Nutr.* **2022**, *9*, 867692. [CrossRef] [PubMed]
13. Lu, Q.; Zeng, L.; Li, X.; Liu, Y.; Du, X.; Bai, G.; Yan, X. Protective effects of panax notoginseng saponin on dextran sulfate sodium-induced colitis in rats through phosphoinositide-3-kinase protein kinase B signaling pathway inhibition. *World J. Gastroenterol* **2020**, *26*, 1156–1171. [CrossRef]
14. Oehlers, S.H.; Flores, M.V.; Hall, C.J.; Okuda, K.S.; Sison, J.O.; Crosier, K.E.; Crosier, P.S. Chemically induced intestinal damage models in zebrafish larvae. *Zebrafish* **2013**, *10*, 184–193. [CrossRef] [PubMed]
15. Du, C.; Wang, K.; Zhao, Y.; Nan, X.; Chen, R.; Quan, S.; Xiong, B. Supplementation with milk-derived extracellular vesicles shapes the gut microbiota and regulates the transcriptomic landscape in experimental colitis. *Nutrients* **2022**, *14*, 1808. [CrossRef]
16. Oehlers, S.H.; Flores, M.V.; Okuda, K.S.; Hall, C.J.; Crosier, K.E.; Crosier, P.S. A chemical enterocolitis model in zebrafish larvae that is dependent on microbiota and responsive to pharmacological agents. *Dev. Dyn.* **2011**, *240*, 288–298. [CrossRef] [PubMed]

17. Oehlers, S.H.; Flores, M.V.; Hall, C.J.; Wang, L.; Ko, D.C.; Crosier, K.E.; Crosier, P.S. A whole animal chemical screen approach to identify modifiers of intestinal neutrophilic inflammation. *FEBS J.* **2017**, *284*, 402–413. [CrossRef] [PubMed]
18. Okuda, K.S.; Misa, J.P.; Oehlers, S.H.; Hall, C.J.; Ellett, F.; Alasmari, S.; Lieschke, G.J.; Crosier, K.E.; Crosier, P.S.; Astin, J.W. A zebrafish model of inflammatory lymphangiogenesis. *Biol. Open* **2015**, *4*, 1270–1280. [CrossRef] [PubMed]
19. Allen, I.C. A NOD to zebrafish models of inflammatory bowel disease pathogenesis. *Dis. Model. Mech.* **2011**, *4*, 711–712. [CrossRef]
20. Oehlers, S.H.; Flores, M.V.; Hall, C.J.; Swift, S.; Crosier, K.E.; Crosier, P.S. The inflammatory bowel disease (IBD) susceptibility genes NOD1 and NOD2 have conserved anti-bacterial roles in zebrafish. *Dis. Model. Mech.* **2011**, *4*, 832–841. [CrossRef]
21. van der Vaart, M.; van Soest, J.J.; Spaink, H.P.; Meijer, A.H. Functional analysis of a zebrafish myd88 mutant identifies key transcriptional components of the innate immune system. *Dis. Model. Mech.* **2013**, *6*, 841–854. [CrossRef]
22. Ogryzko, N.V.; Hoggett, E.E.; Solaymani-Kohal, S.; Tazzyman, S.; Chico, T.J.A.; Renshaw, S.A.; Wilson, H.L. Zebrafish tissue injury causes upregulation of interleukin-1 and caspase-dependent amplification of the inflammatory response. *Dis. Model. Mech.* **2014**, *7*, 259–264. [CrossRef] [PubMed]
23. Ma, J.; Chen, J.; Louro, B.; Martins, R.S.T.; Canario, A.V.M. Somatostatin 3 loss of function impairs the innate immune response to intestinal inflammation. *Aquac. Fish.* **2021**, *6*, 548–557. [CrossRef]
24. Li, C.; Liu, K.; Chen, S.; Han, L.; Han, W. Gaussian accelerated molecular dynamics simulations investigation on the mechanism of angiotensin-converting enzyme (ACE) C-domain inhibition by dipeptides. *Foods* **2022**, *11*, 327. [CrossRef]
25. Burnham, S.; Smith, J.A.; Lee, A.J.; Isaac, R.E.; Shirras, A.D. The angiotensin-converting enzyme (*ace*) gene family of anopheles gambiae. *BMC Genomics* **2005**, *6*, 172. [CrossRef] [PubMed]
26. Lucero, H.A.; Kintsurashvili, E.; Marketou, M.E.; Gavras, H. Cell signaling, internalization, and nuclear localization of the angiotensin converting enzyme in smooth muscle and endothelial cells. *J. Biol. Chem.* **2010**, *285*, 5555–5568. [CrossRef]
27. Fagyas, M.; Uri, K.; Siket, I.M.; Darago, A.; Boczan, J.; Banyai, E.; Edes, I.; Papp, Z.; Toth, A. New perspectives in the renin-angiotensin-aldosterone system (RAAS) III: Endogenous inhibition of angiotensin converting enzyme (ACE) provides protection against cardiovascular diseases. *PLoS ONE* **2014**, *9*, e93719. [CrossRef]
28. Lubbe, L.; Sturrock, E.D. Interacting cogs in the machinery of the renin angiotensin system. *Biophys Rev.* **2019**, *11*, 583–589. [CrossRef]
29. Ou, Z.; Tao, M.; Gao, Q.; Zhang, X.; Yang, Y.; Zhou, J.; Zhang, Y. Up-regulation of angiotensin-converting enzyme in response to acute ischemic stroke via ERK/NF-κB pathway in spontaneously hypertensive rats. *Oncotarget* **2017**, *8*, 97041–97051. [CrossRef]
30. Esther, C.R.; Marino, E.M.; Howard, T.E.; Machaud, A.; Corvol, P.; Capecchi, M.R.; Bernstein, K.E. The critical role of tissue angiotensin-converting enzyme as revealed by gene targeting in mice. *J. Clin. Investig.* **1997**, *99*, 2375–2385. [CrossRef]
31. Cole, J.; Le Quach, D.; Sundaram, K.; Corvol, P.; Capecchi, M.R.; Bernstein, K.E. Mice lacking endothelial angiotensin-converting enzyme have a normal blood pressure. *Circ. Res.* **2002**, *90*, 87–92. [CrossRef] [PubMed]
32. Kessler, S.P.; Senanayake, P.D.; Gaughan, C.; Sen, G.C. Vascular expression of germinal *ace* fails to maintain normal blood pressure in $ACE^{-/-}$ mice. *FASEB. J.* **2007**, *21*, 156–166. [CrossRef] [PubMed]
33. Esther, C.R.; Howard, T.E.; Marino, E.M.; Goddard, J.M.; Capecchi, M.R.; Bernstein, K.E. Mice lacking angiotensin-converting enzyme have low blood pressure, renal pathology, and reduced male fertility. *Lab. Investig.* **1996**, *74*, 953–965. [PubMed]
34. Haxhija, E.Q.; Yang, H.; Spencer, A.U.; Koga, H.; Sun, X.; Teitelbaum, D.H. Modulation of mouse intestinal epithelial cell turnover in the absence of angiotensin converting enzyme. *Am. J. Physiol.-Gastroint. Liver Physiol.* **2008**, *295*, G88–G98. [CrossRef] [PubMed]
35. Wildhaber, B.E.; Yang, H.; Haxhija, E.Q.; Spencer, A.U.; Teitelbaum, D.H. Intestinal intraepithelial lymphocyte derived angiotensin converting enzyme modulates epithelial cell apoptosis. *Apoptosis* **2005**, *10*, 1305–1315. [CrossRef] [PubMed]
36. Garg, M.; Burrell, L.M.; Velkoska, E.; Griggs, K.; Angus, P.W.; Gibson, P.R.; Lubel, J.S. Upregulation of circulating components of the alternative renin-angiotensin system in inflammatory bowel disease: A pilot study. *J. Renin-Angiotensin-Aldosterone Syst.* **2015**, *16*, 559–569. [CrossRef] [PubMed]
37. Pacurari, M.; Kafoury, R.; Tchounwou, P.B.; Ndebele, K.; Cavaillon, J. The Renin-Angiotensin-aldosterone system in vascular inflammation and remodeling. *Int. J. Inflamm.* **2014**, *2014*, 689360. [CrossRef] [PubMed]
38. Heilmann, R.M.M.; Csukovich, G.; Burgener, I.A.A.; Dengler, F. Time to eraase chronic inflammation: Current advances and future perspectives on renin-angiotensin-aldosterone-system and chronic intestinal inflammation in dogs and humans. *Front. Vet. Sci.* **2023**, *10*, 1180125. [CrossRef] [PubMed]
39. Zhang, L.; Zetter, M.A.; Guerra, E.C.; Hernández, V.S.; Mahata, S.K.; Eiden, L.E. Ace2 in the second act of COVID-19 syndrome: Peptide dysregulation and possible correction with oestrogen. *J. Neuroendocrinol.* **2021**, *33*, e12935. [CrossRef]
40. Bernstein, K.E.; Khan, Z.; Giani, J.F.; Cao, D.Y.; Bernstein, E.A.; Shen, X.Z. Angiotensin-converting enzyme in innate and adaptive immunity. *Nat. Rev. Nephrol.* **2018**, *14*, 325–336. [CrossRef]
41. Wen, J.; Mercado, G.P.; Volland, A.; Doden, H.L.; Lickwar, C.R.; Crooks, T.; Kakiyama, G.; Kelly, C.; Cocchiaro, J.L.; Ridlon, J.M.; et al. Fxr signaling and microbial metabolism of bile salts in the zebrafish intestine. *Sci. Adv.* **2021**, *7*, eabg1371. [CrossRef] [PubMed]
42. Oosthuizen, D.; Sturrock, E.D. Exploring the impact of ACE inhibition in immunity and disease. *J. Renin-Angiotensin-Aldosterone Syst.* **2022**, *2022*, 9028969. [CrossRef]

43. Duarte, T.; Silva, M.M.; Michelotti, P.; Barbosa, N.; Feltes, B.C.; Dorn, M.; Rocha, J.; Dalla, C.C. The drosophila melanogaster ACE2 ortholog genes are differently expressed in obesity/diabetes and aging models: Implications for COVID-19 pathology. *Biochim. Biophys. Acta-Mol. Basis Dis.* **2022**, *1868*, 166551. [CrossRef] [PubMed]
44. Metzger, R.; Franke, F.E.; Bohle, R.M.; Alhenc-Gelas, F.; Danilov, S.M. Heterogeneous distribution of angiotensin i-converting enzyme (CD143) in the human and rat vascular systems: Vessel, organ and species specificity. *Microvasc. Res.* **2011**, *81*, 206–215. [CrossRef]
45. Bernstein, K.E.; Ong, F.S.; Blackwell, W.L.; Shah, K.H.; Giani, J.F.; Gonzalez-Villalobos, R.A.; Shen, X.Z.; Fuchs, S.; Touyz, R.M. A modern understanding of the traditional and nontraditional biological functions of angiotensin-converting enzyme. *Pharmacol. Rev.* **2013**, *65*, 1–46. [CrossRef]
46. Huang, Q.; Luo, L.; Alamdar, A.; Zhang, J.; Liu, L.; Tian, M.; Eqani, S.A.; Shen, H. Integrated proteomics and metabolomics analysis of rat testis: Mechanism of arsenic-induced male reproductive toxicity. *Sci. Rep.* **2016**, *6*, 32518. [CrossRef] [PubMed]
47. Belleannee, C.; Da, S.N.; Shum, W.W.; Marsolais, M.; Laprade, R.; Brown, D.; Breton, S. Segmental expression of the bradykinin type 2 receptor in rat efferent ducts and epididymis and its role in the regulation of aquaporin 9. *Biol. Reprod.* **2009**, *80*, 134–143. [CrossRef]
48. Fagerberg, L.; Hallstrom, B.M.; Oksvold, P.; Kampf, C.; Djureinovic, D.; Odeberg, J.; Habuka, M.; Tahmasebpoor, S.; Danielsson, A.; Edlund, K.; et al. Analysis of the human tissue-specific expression by genome-wide integration of transcriptomics and antibody-based proteomics. *Mol. Cell. Proteomics* **2014**, *13*, 397–406. [CrossRef]
49. Wallace, K.N.; Akhter, S.; Smith, E.M.; Lorent, K.; Pack, M. Intestinal growth and differentiation in zebrafish. *Mech. Dev.* **2005**, *122*, 157–173. [CrossRef]
50. Nagai, T.; Kanasaki, M.; Srivastava, S.P.; Nakamura, Y.; Ishigaki, Y.; Kitada, M.; Shi, S.; Kanasaki, K.; Koya, D. N-acetyl-seryl-aspartyl-lysyl-proline inhibits diabetes-associated kidney fibrosis and endothelial-mesenchymal transition. *Biomed Res. Int.* **2014**, *2014*, 696475. [CrossRef]
51. Razavian, M.; Bordenave, T.; Georgiadis, D.; Beau, F.; Zhang, J.; Golestani, R.; Toczek, J.; Jung, J.J.; Ye, Y.; Kim, H.Y.; et al. Optical imaging of mmp-12 active form in inflammation and aneurysm. *Sci. Rep.* **2016**, *6*, 38345. [CrossRef] [PubMed]
52. Sueyoshi, R.; Ignatoski, K.M.W.; Daignault, S.; Okawada, M.; Teitelbaum, D.H. Angiotensin converting enzyme-inhibitor reduces colitis severity in an il-10 knockout model. *Dig. Dis. Sci.* **2013**, *58*, 3165–3177. [CrossRef] [PubMed]
53. Di Raimondo, D.; Tuttolomondo, A.; Butta, C.; Miceli, S.; Licata, G.; Pinto, A. Effects of ACE-inhibitors and angiotensin receptor blockers on inflammation. *Curr. Pharm. Des.* **2012**, *18*, 4385–4413. [CrossRef]
54. Chen, L.; Deng, H.; Cui, H.; Fang, J.; Zuo, Z.; Deng, J.; Li, Y.; Wang, X.; Zhao, L. Inflammatory responses and inflammation-associated diseases in organs. *Oncotarget* **2018**, *9*, 7204–7218. [CrossRef] [PubMed]
55. Jin, Q.; Gao, Y.; Shuai, S.; Chen, Y.; Wang, K.; Chen, J.; Peng, J.; Gao, C. Cdx1b protects intestinal cell fate by repressing signaling networks for liver specification. *J. Genet. Genomics* **2022**, *49*, 1101–1113. [CrossRef]
56. Li, X.; Guo, R.; Yang, S.; Zhang, X.; Yin, X.; Teng, L.; Zhang, S.; Ji, G.; Li, H. Cd248a and Cd248b in zebrafish participate in innate immune responses. *Front. Immunol.* **2022**, *13*, 970626. [CrossRef]
57. Thisse, C.; Thisse, B. High-resolution in situ hybridization to whole-mount zebrafish embryos. *Nat. Protoc.* **2008**, *3*, 59–69. [CrossRef]
58. Grove, C.; Jerram, D.A. jPOR: An imagej macro to quantify total optical porosity from blue-stained thin sections. *Comput. Geosci.* **2011**, *37*, 1850–1859. [CrossRef]
59. Flores, E.; Dutta, S.; Bosserman, R.; van Hoof, A.; Krachler, A. Colonization of larval zebrafish (*Danio rerio*) with adherent-invasive *Escherichia coli* prevents recovery of the intestinal mucosa from drug-induced enterocolitis. *mSphere* **2023**, *8*, e00512-23. [CrossRef]
60. Li, X.L.; Pongor, L.; Tang, W.; Das, S.; Muys, B.R.; Jones, M.F.; Lazar, S.B.; Dangelmaier, E.A.; Hartford, C.C.; Grammatikakis, I.; et al. A small protein encoded by a putative lncRNA regulates apoptosis and tumorigenicity in human colorectal cancer cells. *Elife* **2020**, *9*, e53734. [CrossRef]

International Journal of
Molecular Sciences

MDPI

Article

Mitral Cell Dendritic Morphology in the Adult Zebrafish Olfactory Bulb following Growth, Injury and Recovery

John P. Rozofsky [1], Joanna M. Pozzuto [2] and Christine A. Byrd-Jacobs [1,*]

[1] Department of Biological Sciences, Western Michigan University, 1903 W Michigan Ave., Kalamazoo, MI 49009, USA; john.p.rozofsky@wmich.edu

[2] Department of Biology, Kalamazoo Valley Community College, 6767 W O Ave., Kalamazoo, MI 49009, USA; jpozzuto@kvcc.edu

* Correspondence: christine.byrd@wmich.edu

Abstract: The role of afferent target interactions in dendritic plasticity within the adult brain remains poorly understood. There is a paucity of data regarding the effects of deafferentation and subsequent dendritic recovery in adult brain structures. Moreover, although adult zebrafish demonstrate ongoing growth, investigations into the impact of growth on mitral cell (MC) dendritic arbor structure and complexity are lacking. Leveraging the regenerative capabilities of the zebrafish olfactory system, we conducted a comprehensive study to address these gaps. Employing an eight-week reversible deafferentation injury model followed by retrograde labeling, we observed substantial morphological alterations in MC dendrites. Our hypothesis posited that cessation of injury would facilitate recovery of MC dendritic arbor structure and complexity, potentially influenced by growth dynamics. Statistical analyses revealed significant changes in MC dendritic morphology following growth and recovery periods, indicating that MC total dendritic branch length retained significance after 8 weeks of deafferentation injury when normalized to individual fish physical characteristics. This suggests that regeneration of branch length could potentially function relatively independently of growth-related changes. These findings underscore the remarkable plasticity of adult dendritic arbor structures in a sophisticated model organism and highlight the efficacy of zebrafish as a vital implement for studying neuroregenerative processes.

Keywords: zebrafish; deafferentation; olfactory bulb; mitral cell; retrograde labeling

1. Introduction

The zebrafish olfactory system provides an excellent platform for the study of brain plasticity due to the well-known regenerative abilities of zebrafish [1] and the inherent constitutive and regenerative processes of the olfactory system [2,3]. As in other animals, zebrafish exhibit life-long plasticity in both the olfactory epithelium and the olfactory bulb [4,5]. Several methods have been developed in adult zebrafish to cause damage to the olfactory organ that results in deafferentation of the olfactory bulb [6–10]. Degeneration and regeneration of the zebrafish olfactory bulb following deafferentation has been shown to result in numerous changes to bulb morphology and neurochemistry [6,7]. These effects are accompanied by social behavioral deficits that return to normal when given time to recover [8,11]. Of note for the current study is the ability to perform repeated chemical ablation to the olfactory epithelium to cause long-term, chronic, partial deafferentation of the olfactory bulb [7], resulting in morphological, pharmacological, and behavioral effects. With cessation of treatment, the olfactory bulb recovers, showing that this is a reversible deafferentation model [11]. Calvo-Ochoa and Byrd-Jacobs [5] performed an extensive review of deafferentation-induced changes in the zebrafish olfactory bulb that outlines additional results and reinforces the use of zebrafish as a model system for neural plasticity.

The teleost olfactory bulb is organized in a diffusely ordered laminar structure, with output neurons found throughout the glomerular layer and mitral cell (MC) dendrites synapsing with olfactory sensory neuron axons [12]. Zebrafish MCs are morphologically heterogeneous and possess both multi- and uni-dendritic arbors [12]. Upon receiving sensory information from the olfactory epithelium, MCs serve as the primary relay neurons within the olfactory bulb propagating the signal from the olfactory bulb to higher order brain structures via the medial and lateral olfactory tracts [13–15]. The role of afferent-target interactions in maintenance of dendritic morphology in the adult brain is not well understood, and a reversible deafferentation technique is effective in the exploration of this process.

Dendritic trees form a foundational network in the nervous system, and plasticity of dendritic structures relates to injury and recovery processes within the adult brain. Numerous animal models have been used to study dendritic plasticity. In mice, exposure to physical and social stress is associated with increased dendritic spine density in the nucleus accumbens, with adolescents showing different responses to social stress compared to adults [16]. Additionally in mice, whisker trimming induces functional reorganization of intracortical circuits within the barrel cortex resulting in destabilization of persistent spines and stabilization of new spines in pyramidal neurons [17]. Dendrites of optic tectal neurons in young *Xenopus* substantially increase in complexity when exposed to tyrosine hydroxylase [18]. It is rarer to find studies of dendrite plasticity in adult animals, but zebrafish provide a good model for this. For example, in the retina of adult zebrafish, dendritic field size and complexity of bipolar neurons is restored after chemical damage, showing recovery of dendritic morphology in an adult [19].

The olfactory system's inherent adaptability makes it a good model for this area of study. In the *Xenopus* olfactory bulb, dendritic plasticity is influenced by olfactory stimulation and deprivation. Single-cell electroporation of granule cells with green fluorescent protein allowed for short and long-term imaging of synaptic structure plasticity with odor stimulation resulting in increased stability of dendrites [20]. Another study in *Xenopus* used sparse cell electroporation of a dextran to show that mitral/tufted cell dendritic tufts are less complex and smaller following olfactory nerve transection [21]. This deafferentation effect is transient and dendritic complexity returns with reinnervation. In rodents, large scale plasticity of dendritic dynamics has been shown in numerous cell types in the olfactory bulb. Periglomerular neurons in the mouse olfactory bulb imaged using two photon microscopy under odor-enriched environments demonstrate accelerated dendritic development [22]. Adult-born mouse granule cell dendrites are regulated in early maturational stages through the activation of NMDA receptors [23]. Information specific to zebrafish dendritic plasticity is lacking, despite that model organism's prevalence in regeneration studies.

MC dendritic morphology is impacted by loss of afferent input. Permanent ablation of the olfactory epithelium in adult zebrafish disrupts glomerular structures after 3 weeks, with fluorescently labeled phallotoxin showing loss of axonal and dendritic components [24]. Following 8 weeks of permanent deafferentation, phallotoxin labeling shows indistinguishable glomeruli and decreased bulb volume. In addition, retrograde labeling of MCs reveals significant reductions in total length of major dendritic branches, area of the dendritic field, and optical density measures of the fine processes in the dendritic tuft [24]. Similarly, 8 weeks of chronic, partial deafferentation results in numerous morphological changes in MC dendrites and significant reductions in olfactory bulb MC dendritic arborization measures including modified Sholl analysis, number of major branches, total length of major dendritic branches, area of dendritic field, and distribution of fine processes [24]. Given that this chronic, partial deafferentation method in zebrafish is reversible, it is possible to explore the ability of MC dendrites to recover from this deafferentation-induced damage.

Building upon the results of Pozzuto and colleagues [24] demonstrating the loss of MC dendritic structures following injury in zebrafish, this study examined the effects of cessation of treatment after long-term chronic deafferentation in order to explore dendritic plasticity during recovery from damage. Our hypothesis was that the cessation of deafferentation would allow for the recovery of MC dendritic arbor structure and complexity.

Additionally, considering that zebrafish exhibit continued growth with age [25] and these experiments included survival times of up to four months, we hypothesized that there would be changes in dendritic arbor structures due to growth.

2. Results

2.1. Recovery of Mitral Cell Dendritic Arbors

Our previous study showed that 8 weeks of repeated chemical ablation of the olfactory epithelium causes significant effects on MC dendritic arbors (Figure 1A,B), including diminished secondary branches, reduction in the number of major dendritic branches, significantly reduced total length of major dendritic branches, decreased relative size of dendritic arbor, and reduction in the optical density of the distribution of fine processes within the dendritic arbor [24]. To explore the ability of MC dendrites to recover from these deafferentation-induced effects, dendritic arbors were analyzed for complexity after various time periods following reinnervation, using our reversible deafferentation model [7]. Morphological analysis included the previously mentioned measures, as well as modified Sholl analysis, performed 3 and 8 weeks after cessation of treatment (Figure 1C–F).

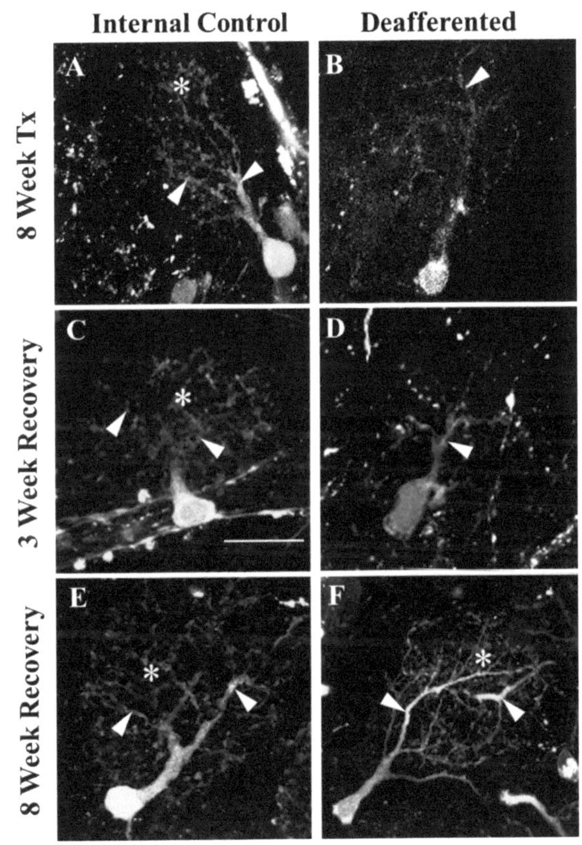

Figure 1. The ability of mitral cell dendritic arbor morphology to recover was examined at 3 and 8 weeks after cessation of treatment following 8 weeks of chronic, partial deafferentation. (**A**) Control mitral cell dendritic arbors had numerous branches (arrowheads) and a dense tuft of fine processes (*). With deafferentation (**B**), there was an obvious reduction in fine processes and branches. When allowed 3 weeks of recovery, internal control mitral cells (**C**) exhibited control morphologies. Previously deafferented mitral cells (**D**) still lacked smaller branches and fine processes typical of control cells,

though major branches were present (arrowheads) and were more abundant than in deafferented cells. Internal control cells (**E**) following 8 weeks of recovery exhibited control morphologies and appeared to be larger in size than control cells at previous time points. Following 8 weeks of chronic partial deafferentation and 8 weeks of recovery, mitral cell dendritic arbors appeared to return to control morphology, with abundant primary, major, and fine dendritic branches (**F**). Scale = 20 μm, for all.

There was a significant reduction in the number of major branches in MC dendritic arbors at 8 weeks post-deafferentation as reported previously [24]. When allowed 3 weeks of recovery, the number of major branches of previously deafferented cells returned to control levels, with no significant differences when compared to internal control cells and unlesioned control cells (Figure 2A). Additionally, following 8 weeks of recovery the total number of major branches within MC dendritic arbors in internal control cells was increased significantly compared to unlesioned control cells (Figure 2A).

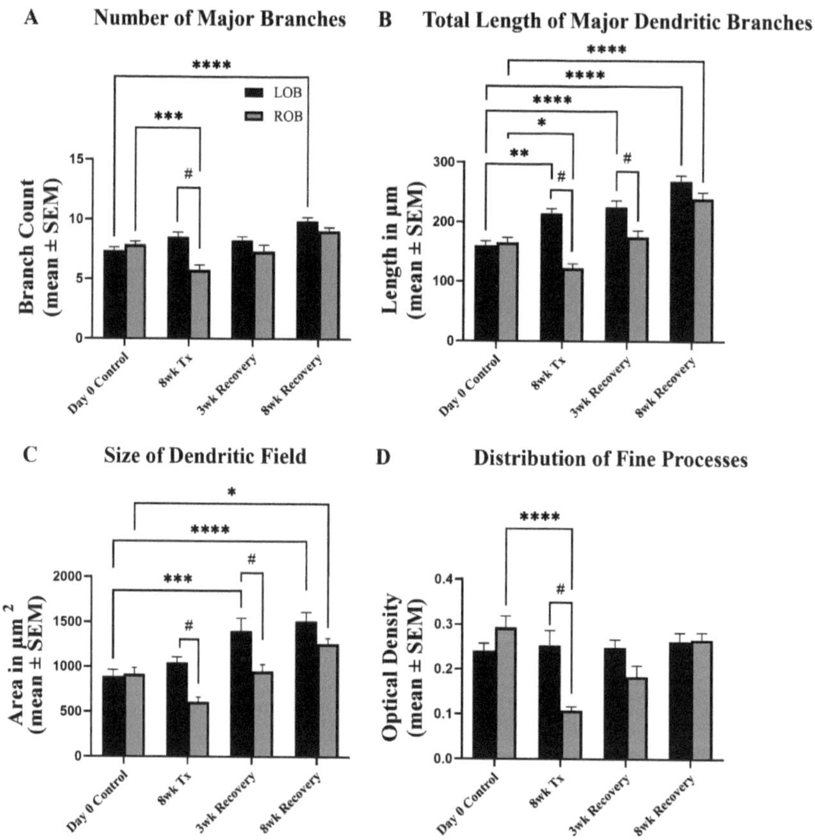

Figure 2. The potential of mitral cell dendritic arbor structures to recover following chronic, partial deafferentation was investigated. (**A**) Following 8 weeks of chronic, partial deafferentation there was a significant reduction in the number of major branches compared to internal control and unlesioned day 0 control cells. With 8 weeks of recovery there were significantly more major branches in internal control bulbs compared to day 0 controls. (**B**) Eight weeks of chronic, partial deafferentation results in effects on total length of mitral cell major dendritic branches. With 3 weeks of recovery, significant decreases remained, while the length of major dendritic branches within the internal control bulb were still significantly greater than those in day 0 unlesioned control bulbs. Following 8 weeks of recovery

there was a significant increase compared to day 0 control cells, but no differences were seen between internal control and treated sides. (**C**) There was a significant decrease in the relative size of the dendritic field following 8 weeks of partial chronic deafferentation and these effects remained significantly reduced when allowed 3 weeks of recovery. There were significant increases in internal control cells compared to day 0 unlesioned control cells with 3 weeks of recovery, and in both internal control and previously deafferented cells following 8 weeks of recovery. (**D**) Following 8 weeks of partial chronic deafferentation there was a significant decrease in optical density of the dendritic arbor compared to day 0 unlesioned control cells, and there was a difference between internal control and treated sides. LOB = left, internal-control olfactory bulb, ROB = right, treated olfactory bulb. For day 0 control, n = 5 fish, 26 mitral cells (LOB), 28 mitral cells (ROB). For 8wk TX, n = 4 fish, 21 mitral cells (LOB), 26 mitral cells (ROB). For 3wk Rec, n = 4 fish, 21 mitral cells (LOB), 20 mitral cells (ROB). For 8wk Rec, n = 5 fish, 26 mitral cells (LOB), 26 mitral cells (ROB). Paired t-test compared to internal controls, # = $p < 0.05$; Two-way ANOVA, with Tukey's multiple comparisons test compared to day zero controls, **** = $p < 0.0001$; *** = $p < 0.001$; ** = $p < 0.01$; * = $p < 0.05$. F-value is 16.31 (**A**), 44.22 (**B**), 19.75 (**C**), and 7.58 (**D**). Degrees of freedom is three, for all. 95% Confidence Interval is 0.51 to 1.53 (**A**), 28.66 to 54.65 (**B**), 158.90 to 392.90 (**C**), and 0.01 to 0.07 (**D**).

It was previously shown that following 8 weeks of chronic, partial deafferentation there is a significant decrease in the length of major branches in deafferented cells compared to internal control and unlesioned control cells, and cells in the contralateral internal control bulb at 8 weeks post-deafferentation are significantly longer than those in unlesioned control cells [26]. At 3 weeks following cessation of deafferentation there were significant decreases in the total length of major dendritic branches when compared to internal control cells (Figure 2B). Furthermore, the total length of major dendritic branches of internal control MCs was significantly increased from the lengths of those in unlesioned control fish. With 8 weeks of recovery there were no significant differences in major dendritic branch length between the left and right olfactory bulbs of previously deafferented animals; however, the total dendritic branch lengths of these cells were significantly longer than those of the unlesioned control fish quantified at day 0 (Figure 2B).

Projections of z-stack images were used to estimate the relative size of the dendritic arbor. At 8 weeks post-deafferentation there is a significant decrease in the size of the dendritic field between dendritic arbors of deafferented cells compared to internal control cells and unlesioned control cells, as shown before [24]. With 3 weeks of recovery, there were significant decreases in MC dendritic arbor size between the left and right olfactory bulbs, and the size of the dendritic arbors in the internal control bulb were significantly increased compared to unlesioned control animals (Figure 2C). Following 8 weeks of recovery, MC dendritic arbor size was significantly larger in both the left and right olfactory bulbs compared to unlesioned control cells (Figure 2C).

The optical density for dextran labeling of the dendritic arbor was used to estimate the effects of deafferentation on the fine processes of the dendritic arbor that could not be accurately traced. Pozzuto et al. [24] showed previously that with 8 weeks of partial chronic deafferentation there is a significant decrease in the optical density of dextran labeling in deafferented MC dendritic arbors compared to internal control cells and unlesioned control cells. When allowed 3 and 8 weeks of recovery there were no significant differences in the distribution of fine processes of the dendritic arbor between left and right olfactory bulbs of recovered animals (Figure 2D).

Modified Sholl analyses were used to estimate the recovery of dendritic arbor complexity following deafferentation. Due to dendritic arbor distance from the soma in unidendritic zebrafish MCs, a modified Sholl analysis where the concentric circles started at the base of the arbor was used to examine the overall complexity of the dendritic arbors [24]. Our previous study showed that 8 weeks of partial, chronic deafferentation resulted in a significant decrease in the overall complexity of MC dendritic arbors [24]. Following 3 and 8 weeks of recovery, overall dendritic complexity returned to control levels, with no significant differences in the complexity of MC dendritic arbors between left and right olfactory bulbs of previously deafferented animals (Figure 3A,B).

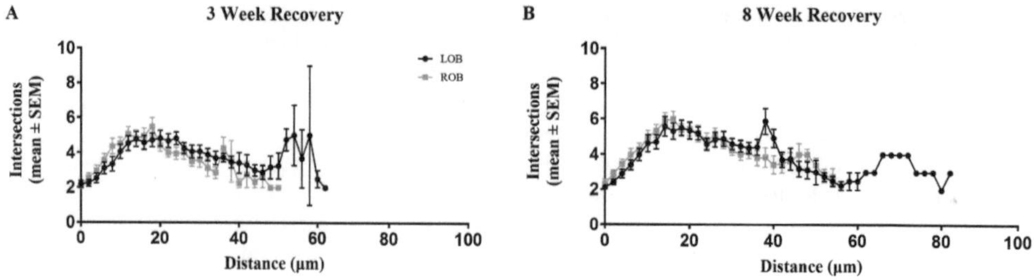

Figure 3. A modified Sholl analysis was used to determine whether the significant decline in overall dendritic complexity that occurs following 8 weeks of chronic, partial deafferentation could be reversed. (**A**) When allowed to recover for 3 weeks, there were no significant differences ($p = 0.63$) in the number of intersections at any distance from the base of the arbor or in overall dendritic complexity, though there is some variability: n = 4 fish, 21 mitral cells (LOB), 20 mitral cells (ROB). (**B**) Following 8 weeks of recovery there were no significant differences ($p = 0.08$) in the number of intersections or in overall dendritic complexity: n = 5 fish, 26 mitral cells (LOB), 26 mitral cells (ROB). Kolmogorov-Smirnov test compared lines formed by the average number of intersections at various distances from the base of the arbor. LOB = left, internal-control olfactory bulb, ROB = right, treated olfactory bulb.

2.2. Growth

Examination of the above data led to the observation that 8 weeks of deafferentation and 8 weeks of recovery, totaling 16 weeks after unlesioned control fish were processed, appeared to allow substantial growth of dendritic arbor structures. To examine the effects of potential growth-related changes to the dendritic arbors of zebrafish MCs over time, a group of cohort control animals were examined alongside the chronic deafferentation and recovery animals. All animals in these studies were taken from the same animal population concurrently with experimental animals, and fish lengths, weights, brain weights, and sex were recorded. In control animals, there was a significant increase in body length at 8 weeks and 16 weeks when compared to the length of control animals at day 0 (Figure 4A). Additionally, animal weights significantly increased at 8 and 16 weeks compared to day 0 control animal weights (Figure 4B). There was also a significant increase in brain weight from day 0 to 16 weeks (Figure 4C).

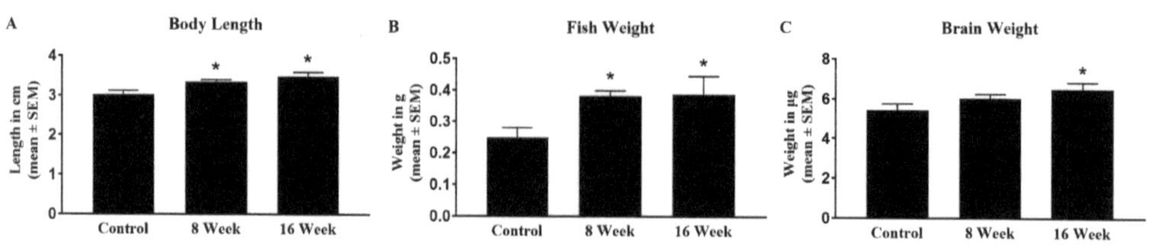

Figure 4. The potential growth of unlesioned control animals over 16 weeks was examined. (**A**) There was a significant increase in body length compared to day 0 control animals (n = 5) at both 8 (n = 4) and 16 (n = 4) weeks. (**B**) After 8 and 16 weeks there were significant increases in animal weight compared to day 0 control animal weights. (**C**) Brain weights of unlesioned cohort control animals at 16 weeks were significantly increased compared to day 0 control animals. Unpaired t-test, * = $p < 0.05$ compared to day 0 unlesioned control animals.

To examine potential growth-related changes in MC dendrites, the total number of major dendritic branches, the length of those branches, the relative area of the dendritic arbor, and the optical densities of dextran labeling of dendritic tufts of MCs were examined in control fish at day 0, 8 weeks, and 16 weeks. There were no significant differences in the total number of major dendritic branches between the right and left olfactory bulbs in day

0, 8-week, and 16-week control MC dendritic arbors (Figure 5A). There was, however, a significant increase in the total number of major branches in MC dendritic arbors of left olfactory bulbs at 8 and 16 weeks when compared to day 0 controls (Figure 5A). The total length of the major dendritic branches was not significantly different between the left and right olfactory bulbs of day 0, 8-week, and 16-week controls (Figure 5B), but at 8 weeks and 16 weeks the total length of major dendritic branches was significantly increased compared to day 0 control cells. Additionally, the total length of major dendritic branches of MCs in the right olfactory bulb at 16 weeks was significantly increased from those in the right olfactory bulb at 8 weeks, indicating substantial growth-related changes to the dendritic arbor over time (Figure 5B).

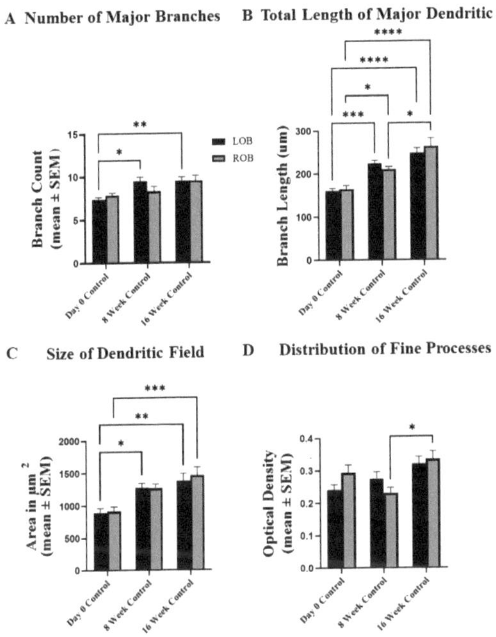

Figure 5. The potential for growth-related changes to dendritic arbor structure was examined in a group of unlesioned cohort control animals over 16 weeks. (**A**) Compared to day 0 control animals, there was a significant increase in the number of major branches in the left olfactory bulbs of animals examined at 8 and 16 weeks. (**B**) At 8 weeks there was a significant increase in the total length of major dendritic branches of mitral cells in the left and right olfactory bulbs compared to day 0 control cells, and this increase was maintained at 16 weeks. Also, at 16 weeks the lengths of major dendritic branches in the right olfactory bulb were significantly longer than those in the right olfactory bulb at 8 weeks. (**C**) There was a significant increase in the relative size of the dendritic field of mitral cells in the left olfactory bulb at 8 weeks compared to day 0 control cells, and at 16 weeks mitral cell dendritic arbors in both the right and left olfactory bulbs were significantly larger than those at day 0. (**D**) The distribution of fine processes in the dendritic arbor remained fairly stable over time, with the only significant increase occurring in cells found in the right olfactory bulb at 16 weeks when compared to those present in the right olfactory bulb at 8 weeks. LOB = left, internal-control olfactory bulb, ROB = right, treated olfactory bulb. For day 0 control, n = 5 fish, 26 mitral cells (LOB), 28 mitral cells (ROB). For 8-week controls, n = 4 fish, 21 mitral cells (LOB), 20 mitral cells (ROB). For 16-week controls, n = 4 fish, 21 mitral cells (LOB), 20 mitral cells (ROB). Two-way ANOVA, with Tukey's multiple comparisons test, **** = $p < 0.0001$; *** = $p < 0.001$; ** = $p < 0.01$; * = $p < 0.05$. F-value is 10.48 (**A**), 41.38 (**B**), 17.98 (**C**), and 6.12 (**D**). Degrees of freedom is 2, for all. 95% Confidence interval is −0.48 to 0.94 (**A**), −19.78 to 14.61 (**B**), −181.10 to 115.90 (**C**), and −0.04 to 0.03 (**D**).

The size of the dendritic field of MCs in the left olfactory bulb at 8 weeks, and both olfactory bulbs at 16 weeks, were significantly larger than the size of the dendritic field of MCs examined at day 0 (Figure 5C), although there were no differences in the size of the dendritic field between MCs in the left and right olfactory bulbs at day 0, 8 weeks, or 16 weeks. Optical density of dextran labeling of MC dendritic arbors showed no differences in the estimate of fine processes between left and right bulbs of animals at day 0, 8 weeks, or 16 weeks, but there was a significant increase in the optical density labeling of MC dendritic arbors in the right olfactory bulb at 16 weeks when compared to that in the right olfactory bulb at 8 weeks, indicating a slight change over time in the distribution of fine processes within the dendritic arbors of MCs (Figure 5D).

The overall dendritic complexity of MCs of untreated, cohort control animals was analyzed at day 0, 8 weeks, and 16 weeks using a modified Sholl analysis. Kolmogorov-Smirnov tests indicated that the dendritic arbors of MCs in the left and right olfactory bulbs of animals at 16 weeks were significantly more complex than MC dendritic arbors at both day 0 and 8-week timepoints, indicating that MC dendritic arbors become more complex with growth over time (Figure 6A,B).

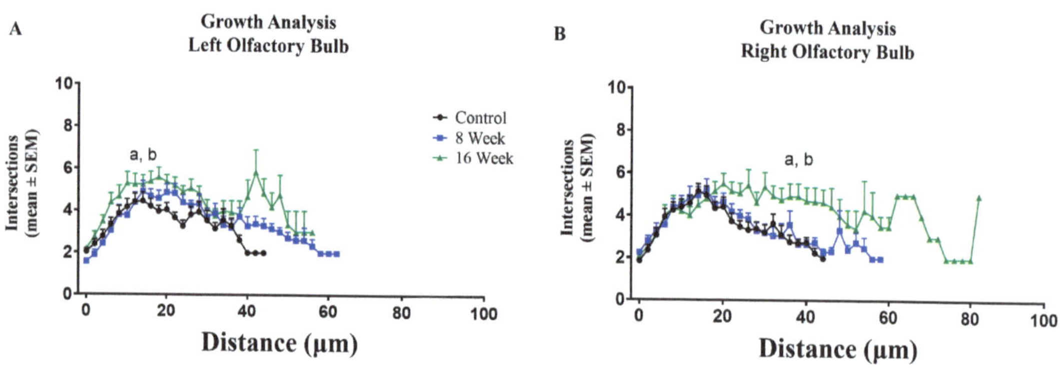

Figure 6. A modified Sholl analysis was used to determine the potential effects of growth on overall dendritic arbor complexity in unlesioned cohort control fish over time. (**A**) In the left olfactory bulb there was a significant increase in overall dendritic complexity at 16 weeks compared to both day 0 ($p = 0.002$) and 8-week ($p = 0.011$) controls. (**B**) In the right olfactory bulb there was a significant increase in overall dendritic complexity at 16 weeks compared to both day 0 ($p = 0.02$) and 8-week ($p = 0.01$) controls. For controls, n = 5 fish, 26 mitral cells (LOB), 28 mitral cells (ROB). For 8-week controls, n = 4 fish, 21 mitral cells (LOB), 20 mitral cells (ROB). For 16-week controls, n = 4 fish, 21 mitral cells (LOB), 20 mitral cells (ROB). Kolmogorov-Smirnov test compared lines formed by the average number of intersections at various distances from the base of the arbor, a = $p < 0.05$ compared to day 0 control cells, b = $p < 0.05$ compared to 8-week control cells.

2.3. Injury and Recovery Normalized to Growth-Related Changes

To distinguish whether recovery of MC dendritic morphology was a measure of regeneration as opposed to an artifact of growth-related changes, the data were normalized to individual values for the physical characteristics of body length, fish weight and brain weight. As shown in Figure 4, these parameters increased significantly with age. MC total dendritic branch length retained significance at the 8-week chronic deafferentation timepoint when normalized to body length, fish weight and brain weight (Figure 7A–C), suggesting that regeneration of branch length could potentially function relatively independently of growth-related changes. Additionally, normalization of the data revealed significant increases in the left olfactory bulb at the 8-week recovery timepoint compared to the day 0 control and in the right olfactory bulb compared to the 8-week deafferentation timepoint (Figure 7A). For all other MC measures when normalized to body length, fish weight and brain weight, the number of major branches, the area of the dendritic arbor and measures of optical density were no longer significant at the 8-week chronic deafferentation timepoint, suggesting that recovery of these characteristics could be a function of growth.

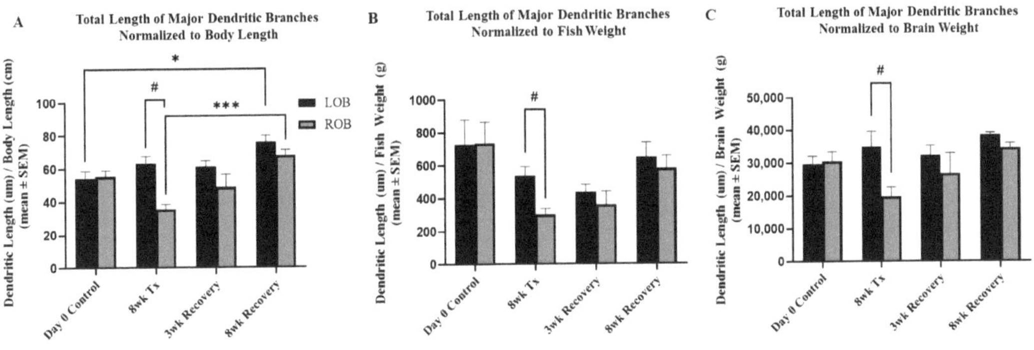

Figure 7. Total length of major dendritic branches normalized to body length, fish weight and brain weight following 8 weeks deafferentation and 3 weeks or 8 weeks of recovery. (**A**) After normalizing total length of major dendritic branches to fish body length, the treated right olfactory bulbs were still significantly different following 8 weeks of chronic, partial deafferentation compared to the internal control left olfactory bulb. Following 8 weeks of recovery, normalized data showed the left and right olfactory bulbs were increased significantly compared to day 0 controls and the 8-week injury timepoint, respectively. (**B**) When normalized to fish weight, total length of dendritic branches in the treated right olfactory bulbs were still significantly different following 8 weeks of chronic, partial deafferentation compared to the internal control left olfactory bulb. (**C**) Normalizing total length of major dendritic branches to brain weight still showed a significant difference in the treated right olfactory bulbs following 8 weeks of chronic, partial deafferentation compared to the internal control left olfactory bulb. LOB = left, internal-control olfactory bulb, ROB = right, treated olfactory bulb. Paired t-test compared to internal controls, # = $p < 0.05$; Two-way ANOVA, with Tukey's multiple comparisons test, *** = $p < 0.001$; * = $p < 0.05$. F-value is 10.54 (**A**), 5.22 (**B**), and 3.15 (**C**). Degrees of freedom is 3, for all. 95% Confidence interval is 5.54 to 18.26 (**A**), −51.39 to 237.20 (**B**), and 1327 to 10,596 (**C**).

3. Discussion

Plasticity of synaptic connections is essential for the development and maintenance of dendritic shape, which is critical for proper neural functioning [26–28]. Research involving the effects of afferent activity on dendrites has been conducted primarily in developmental systems [29,30]. Fewer studies have been performed in adult animals, and most of those examine the plasticity and response to deafferentation of dendritic spines [31–33]. Some researchers have taken advantage of the fact that the olfactory bulb is an adult brain structure that is continually developing, so developmental processes can be examined in an adult animal. Mizrahi [34] investigated dendritic development in adult-born periglomerular neurons. Among his findings was that the dendrites of aspinous periglomerular neurons are more dynamic, while spinous periglomerular and granule neurons show stability in the dendritic branches and plasticity in the spines. His group also examined mitral/tufted cell dendrites and found that they too are stable in adult mice, even with increased activity through odor training or with pharmacology [35]. Another group examined adult-born granule cells in olfactory bulb slices and found that the dendrites of these spinous neurons are more dynamic in early stages of formation but showed less filopodial activity as they mature [23].

In general, there is a lack of information on the effects of deafferentation and recovery on aspinous dendrites in adult brain structures, and we set out to fill in this gap using a popular model animal. Additionally, while it is known that adult zebrafish maintained in a constant, optimal temperature environment exhibit continued growth with age [25], we found no studies examining the role of growth on dendritic arbor structure and complexity. The fact that our study included long treatment times in an indeterminate grower allowed us also to examine effects of growth in adults. The purpose of this study was to examine the potential recovery of MC dendritic arbor structures that are reduced by chronic partial deafferentation and observe any potential growth-related changes to adult MC dendritic arbors over time.

Our model of reversible deafferentation allows us to investigate both degeneration and regeneration in the olfactory bulb of adult zebrafish. Chronic, partial deafferentation of the olfactory bulb results in severe alterations to MC dendritic arbor morphology [24]. It is known that some mitral cells project their axons through the anterior commissure to a similar location in the contralateral olfactory bulb [36]. Thus, in the current study, comparisons were made both to internal control olfactory bulbs and to the olfactory bulbs of untreated, external control fish. Following 8 weeks of deafferentation, MCs retain some of their major dendritic branches; however, smaller secondary branches are noticeably diminished, and the fine processes that establish the dendritic tuft of control cells are notably lacking in deafferented MCs. The morphology of deafferented MCs is in harsh contrast to the control morphology of unidendritic cell arbors, which maintain normal dendritic arbor characteristics. We show here that adult zebrafish MCs can recover their dendritic structure when afferent input is restored. When given time to recover for 3 weeks following deafferentation, dendritic arbors continued to lack fine processes and smaller branches, but appeared to be in a transitory state of returning to control levels. Following 8 weeks of recovery, MC dendritic morphology regained robust fine processes as well as control levels of primary and major branches. When allowed 3 weeks of recovery, the number of major branches, the distribution of fine processes and overall dendritic complexity were no longer significantly decreased. This indicates that the process of recovering dendritic structures occurs much more quickly than the process of losing them. When allowed 8 weeks to recover from chronic, partial deafferentation there were no differences between previously deafferented MCs in general morphology or in dendritic arbor quantifications compared to internal control cells; however, both were significantly increased from day 0 unlesioned control cells. Increased dendritic complexity in developing hippocampal neurons is activity dependent and uses a Semaphorin 3A pathway [37]. Relatedly, adult-born granule cells of the olfactory bulb show increased dendrite dynamics in response to activity as they are developing, but they do not require activity for filopodia extension when mature [23]. Interestingly, Mizrahi [34] found that sensory deprivation from naris occlusion in mice had no effect on dendrite complexity, at least during development of adult-born periglomerular cells. Our report of recovery of dendritic arborization following return of afferent input is consistent with some of these findings, and the fact that activity is not necessary in some instances is likely due to developmental stage of the neurons, where our study focused on mature mitral cells of the adult olfactory bulb.

The discovery of differences between day 0 control cells and internal control cells at recovery time periods led to the examination of potential growth-related changes to MC dendritic arbors. Unlesioned control fish were examined over time, and results revealed that over the course of a 16-week long experiment there were significant increases in animal length, weight, and brain weight. These changes may correspond to the significant increases in the number of major branches, the total length of those major branches, and the size of the dendritic field of MCs. Most notably, there were significant increases in the overall complexity of dendritic arbors at 16 weeks when compared to both day 0 and 8-week control arbors indicating that there was continued growth and elaboration of MC dendrites.

Many molecules involved in dendrite morphogenesis have been identified across different model systems and locations [38,39]; however, most of these have been shown in culture or in developing animals [18,40,41]. The process by which dendritic arbors elaborate and grow in proportion to animal growth during development is referred to as dendritic scaling [36], and this may continue throughout life. Growth and elaboration of dendritic arbors in correlation to animal size has been rarely studied in adult animals, although examples do exist. Goldfish demonstrate continued growth of dendritic arbors in retinal ganglion cells [42,43]. Examination of pyramidal neurons in the prefrontal cortex and basolateral amygdala of rats growing from juvenile to pubescent stages show an increase in overall dendritic length, complexity and spine density; growth from puberty into adulthood results in decreased dendritic spines and increased number of dendritic branches [44]. Many questions remain as to how and why dendrites coordinate and regulate adult arbor structures. It is interesting to find that the normalized data for MC total dendritic branch length, when considering individual growth-related physical characteristics, uniquely measures regenerative plasticity in the zebrafish olfactory bulb. For other measures of dendritic morphology, further experimentation may be required to elucidate whether the changes in complexity are functions of regenerative or growth-related plasticity. We report here that mitral cells in the adult zebrafish olfactory bulb recover from deafferentation-induced loss of dendritic branches and that fish growth likely influences the dendrite recovery process. Regardless of whether the restoration of dendritic arbors results from normal, age-related growth or dendritic regeneration, the capacity for the restoration of adult dendritic arbors to baseline morphology is novel, as this has not been previously shown in other studies. This study provides evidence of the plasticity of adult dendritic arbor structures in a complex model organism and further proves that the zebrafish is a superb model organism for understanding neuroregeneration.

4. Materials and Methods

4.1. Animals

A total of 19 adult male and 19 adult female zebrafish, *Danio rerio*, over 6 months of age were used, and 276 mitral cells were examined. The fish were maintained in 15-gallon aquaria filled with conditioned fish water (reverse osmosis water treated with dechlorinating solutions and conditioning salts) at 28.5 °C and were fed twice daily (morning and afternoon) with commercial flake food (Tetra) under natural lighting conditions. Characteristics including sex, length, weight, and brain weight were collected at the conclusion of the experiment. Fish were obtained from local commercial sources and all animal care protocols and experimental procedures were approved by the Institutional Animal Care and Use Committee.

4.2. Chronic Deafferentation

Chronic, partial deafferentation of the olfactory bulb was achieved via repeated chemical lesioning of the olfactory epithelium with detergent as described previously [7,24]. Zebrafish were anesthetized with 0.03% MS222 (3-amino benzoic acid ethyl ester, Sigma-Aldrich, St. Louis, MO, USA) in fish water. After anesthesia was confirmed with caudal fin pinch, fish were placed on a clay dish and a pulled Wiretrol capillary pipette tip was used to lavage the right nasal cavity with approximately 1 µL of 0.7% Triton X-100 (Sigma-Aldrich, St. Louis, MO, USA) and 0.005% methylene blue in 0.1 M phosphate-buffered saline, leaving the left olfactory epithelium to serve as the untreated internal control. The Triton X-100 solution was kept in contact with the tissue for two minutes. This procedure was repeated once every three days for eight weeks after which fish were euthanized or allowed to recover for three or eight weeks. Cohort control fish were euthanized along with the deafferented fish, and day 0 control fish were euthanized on the first day of the experiment.

4.3. Olfactory Tract Tracing

MCs from deafferented and control fish were labeled using retrograde tract tracing from the olfactory tracts [12,24]. For this technique, fish were over-anesthetized with 0.03% MS222 and then perfused with pH 7.4 phosphate-buffered saline before immediate brain dissection under a stereomicroscope. Approximately 0.05–0.1 µL of Texas Red Dextran (10,000 MW, 5 mg/mL in PBS, ThermoFisher Scientific, Waltham, MA, USA, catalog # D1863) was injected into both the medial and lateral olfactory tracts within the telencephalon. Following injection, brains were incubated at 28.5 °C in artificial fish cerebrospinal fluid [100 mM NaCl, 2.46 mM KCl, 1 mM $MgCl_2H_2O$, 0.44 mM $NaH_2PO_4H_2O$, 1.13 mM $CaCl_2H_2O$, 5 mM $NaHCO_3$; [45]] with 3–5% CO_2 for approximately four hours [12, 24]. Whole-brain specimens were then fixed in 4% paraformaldehyde for 24 h at 4 °C, rinsed in buffer, mounted between two coverslips, and viewed on a Nikon C2 confocal microscope.

4.4. Dendritic Analysis

Unidendritic MCs are the most predominant and uniform cell type within the zebrafish olfactory bulb [12]. For dendritic analysis, a minimum of five unidendritic cells were selected randomly from each right and left olfactory bulb from at least four fish from each treatment or control group. Confocal microscopy with z-stack imaging was used to gather fine optical sections of tissue throughout the olfactory bulb at 0.25 μm intervals. Cellular profiles were used to create two-dimensional projection images analyzed with Fiji, an open platform for biological image analysis [46]. Utilizing the Simple Neurite Tracer Plugin, quantification of major dendritic branches, total length of dendritic branches, size of dendritic field, optical density measures of fine processes of the dendritic arbor, and modified Sholl analysis were performed by tracing the two-dimensional outline of the arbor of the major primary, secondary, tertiary, and quaternary branches that possessed clearly defined borders. Optical density was measured with Fiji by converting the mean gray level from the area of the arbor using the formula OD = −log (intensity of background/intensity of area of interest). Examination of overall complexity was conducted using a modified Sholl analysis, where the concentric circles originated at the base of the dendritic arbor [24]. Data from cells from right and left olfactory bulbs were compared within groups using paired, two-tailed t-tests and between groups using ANOVA with Tukey's test for multiple comparisons. Data from the Sholl analysis was analyzed using paired t-tests to compare the number of intersections at each distance from the base of the arbor, and a Kolmogorov-Smirnov test was performed to determine significant differences in the lines formed by the number of intersections at each distance from the base of deafferented and control dendritic arbors; p values less than 0.05 were considered significant.

Author Contributions: Conceptualization, C.A.B.-J.; Methodology, C.A.B.-J., J.M.P. and J.P.R.; Investigation, J.M.P. and J.P.R.; Formal Analysis, C.A.B.-J., J.M.P. and J.P.R.; Resources, C.A.B.-J.; Data Curation, J.M.P. and J.P.R.; Validation, C.A.B.-J., J.M.P. and J.P.R.; Visualization, J.M.P. and J.P.R.; Writing: Original Draft Preparation, J.M.P. and J.P.R.; Writing: Review and Editing, C.A.B.-J.; Supervision, C.A.B.-J.; Project Administration, C.A.B.-J.; Funding Acquisition, C.A.B.-J. All authors have read and agreed to the published version of the manuscript.

Funding: This work was supported by Western Michigan University internal research funds (C.A.B.-J.).

Institutional Review Board Statement: The study was conducted according to the guidelines of the Declaration of Helsinki and approved by the Institutional Animal Care and Use Committee of Western Michigan University (IACUC Protocols #19-04-01, 10 April 2019 and #22-03-03, 14 March 2022).

Informed Consent Statement: Not applicable.

Data Availability Statement: Dataset available on request from the authors.

Acknowledgments: We are grateful to Rachel Warga of the Kane laboratory for the generous donation of fish and to the WMU Biological Imaging Center for use of the confocal microscope.

Conflicts of Interest: The authors declare no conflicts of interest.

References

1. Marques, I.J.; Lupi, E.; Mercader, N. Model Systems for Regeneration: Zebrafish. *Development* **2019**, *146*, dev167692. [CrossRef] [PubMed]
2. Murray, R.C.; Calof, A.L. Neuronal Regeneration: Lessons from the Olfactory System. *Semin. Cell Dev. Biol.* **1999**, *10*, 421–431. [CrossRef] [PubMed]

3. Calvo-Ochoa, E.; Byrd-Jacobs, C.A.; Fuss, S.H. Diving into the Streams and Waves of Constitutive and Regenerative Olfactory Neurogenesis: Insights from Zebrafish. *Cell Tissue Res.* **2021**, *383*, 227–253. [CrossRef] [PubMed]
4. Byrd, C.A.; Brunjes, P.C. Neurogenesis in the Olfactory Bulb of Adult Zebrafish. *Neuroscience* **2001**, *105*, 793–801. [CrossRef] [PubMed]
5. Calvo-Ochoa, E.; Byrd-Jacobs, C.A. The Olfactory System of Zebrafish as a Model for the Study of Neurotoxicity and Injury: Implications for Neuroplasticity and Disease. *Int. J. Mol. Sci.* **2019**, *20*, 1639. [CrossRef] [PubMed]
6. Byrd, C.A. Deafferentation-Induced Changes in the Olfactory Bulb of Adult Zebrafish. *Brain Res.* **2000**, *866*, 92–100. [CrossRef]
7. Paskin, T.R.; Iqbal, T.R.; Byrd-Jacobs, C.A. Olfactory Bulb Recovery Following Reversible Deafferentation with Repeated Detergent Application in the Adult Zebrafish. *Neuroscience* **2011**, *196*, 276–284. [CrossRef]
8. Hentig, J.T.; Byrd-Jacobs, C.A. Exposure to Zinc Sulfate Results in Differential Effects on Olfactory Sensory Neuron Subtypes in Adult Zebrafish. *Int. J. Mol. Sci.* **2016**, *17*, 1445. [CrossRef]
9. Scheib, J.J.; Pozzuto, J.M.; Byrd-Jacobs, C.A. Reversible Deafferentation of the Zebrafish Olfactory Bulb with Wax Plug Insertion. *J. Neurosci. Methods* **2019**, *311*, 47–56. [CrossRef]
10. Poling, K.R.; Brunjes, P.C. Sensory Deafferentation and Olfactory Bulb Morphology in the Zebrafish and Related Species. *Brain Res.* **2000**, *856*, 135–141. [CrossRef]
11. Paskin, T.R.; Byrd-Jacobs, C.A. Reversible Deafferentation of the Adult Zebrafish Olfactory Bulb Affects Glomerular Distribution and Olfactory-Mediated Behavior. *Behav. Brain Res.* **2012**, *235*, 293–301. [CrossRef] [PubMed]
12. Fuller, C.L.; Yettaw, H.K.; Byrd, C.A. Mitral Cells in the Olfactory Bulb of Adult Zebrafish (Danio Rerio): Morphology and Distribution. *J. Comp. Neurol.* **2006**, *499*, 218–230. [CrossRef] [PubMed]
13. Fuller, C.L.; Byrd, C.A. Ruffed Cells Identified in the Adult Zebrafish Olfactory Bulb. *Neurosci. Lett.* **2005**, *379*, 190–194. [CrossRef] [PubMed]
14. Miyasaka, N.; Morimoto, K.; Tsubokawa, T.; Higashijima, S.; Okamoto, H.; Yoshihara, Y. From the Olfactory Bulb to Higher Brain Centers: Genetic Visualization of Secondary Olfactory Pathways in Zebrafish. *J. Neurosci.* **2009**, *29*, 4756–4767. [CrossRef] [PubMed]
15. Gayoso, J.; Castro, A.; Anadón, R.; Manso, M.J. Crypt Cells of the Zebrafish Danio Rerio Mainly Project to the Dorsomedial Glomerular Field of the Olfactory Bulb. *Chem. Senses* **2012**, *37*, 357–369. [CrossRef] [PubMed]
16. Warren, B.L.; Sial, O.K.; Alcantara, L.F.; Greenwood, M.A.; Brewer, J.S.; Rozofsky, J.P.; Parise, E.M.; Bolaños-Guzmán, C.A. Altered Gene Expression and Spine Density in Nucleus Accumbens of Adolescent and Adult Male Mice Exposed to Emotional and Physical Stress. *Dev. Neurosci.* **2014**, *36*, 250–260. [CrossRef] [PubMed]
17. Holtmaat, A.; Wilbrecht, L.; Knott, G.W.; Welker, E.; Svoboda, K. Experience-Dependent and Cell-Type-Specific Spine Growth in the Neocortex. *Nature* **2006**, *441*, 979–983. [CrossRef] [PubMed]
18. Thompson, C.K.; Cline, H.T. Thyroid Hormone Acts Locally to Increase Neurogenesis, Neuronal Differentiation, and Dendritic Arbor Elaboration in the Tadpole Visual System. *J. Neurosci.* **2016**, *36*, 10356–10375. [CrossRef] [PubMed]
19. McGinn, T.E.; Mitchell, D.M.; Meighan, P.C.; Partington, N.; Leoni, D.C.; Jenkins, C.E.; Varnum, M.D.; Stenkamp, D.L. Restoration of Dendritic Complexity, Functional Connectivity, and Diversity of Regenerated Retinal Bipolar Neurons in Adult Zebrafish. *J. Neurosci.* **2018**, *38*, 120–136. [CrossRef]
20. Zhang, L.; Huang, Y.; Hu, B. Olfactory Experiences Dynamically Regulate Plasticity of Dendritic Spines in Granule Cells of Xenopus Tadpoles in Vivo. *Sci. Rep.* **2016**, *6*, 35009. [CrossRef]
21. Hawkins, S.J.; Weiss, L.; Offner, T.; Dittrich, K.; Hassenklöver, T.; Manzini, I. Functional Reintegration of Sensory Neurons and Transitional Dendritic Reduction of Mitral/Tufted Cells during Injury-Induced Recovery of the Larval Xenopus Olfactory Circuit. *Front. Cell. Neurosci.* **2017**, *11*, 380. [CrossRef] [PubMed]
22. Livneh, Y.; Feinstein, N.; Klein, M.; Mizrahi, A. Sensory Input Enhances Synaptogenesis of Adult-Born Neurons. *J. Neurosci.* **2009**, *29*, 86–97. [CrossRef] [PubMed]
23. Breton-Provencher, V.; Coté, D.; Saghatelyan, A. Activity of the Principal Cells of the Olfactory Bulb Promotes a Structural Dynamic on the Distal Dendrites of Immature Adult-Born Granule Cells via Activation of NMDA Receptors. *J. Neurosci.* **2014**, *34*, 1748–1759. [CrossRef] [PubMed]
24. Pozzuto, J.M.; Fuller, C.L.; Byrd-Jacobs, C.A. Deafferentation-Induced Alterations in Mitral Cell Dendritic Morphology in the Adult Zebrafish Olfactory Bulb. *J. Bioenerg. Biomembr.* **2019**, *51*, 29–40. [CrossRef] [PubMed]
25. Schaefer, J.; Ryan, A. Developmental Plasticity in the Thermal Tolerance of Zebrafish Danio Rerio. *J. Fish Biol.* **2006**, *69*, 722–734. [CrossRef]
26. Tavosanis, G. Dendritic Structural Plasticity. *Dev. Neurobiol.* **2012**, *72*, 73–86. [CrossRef]
27. Forrest, M.P.; Parnell, E.; Penzes, P. Dendritic Structural Plasticity and Neuropsychiatric Disease. *Nat. Rev. Neurosci.* **2018**, *19*, 215–234. [CrossRef]
28. d'Aquin, S.; Szonyi, A.; Mahn, M.; Krabbe, S.; Gründemann, J.; Lüthi, A. Compartmentalized Dendritic Plasticity during Associative Learning. *Science* **2022**, *376*, eabf7052. [CrossRef]
29. Cline, H.T. Dendritic Arbor Development and Synaptogenesis. *Curr. Opin. Neurobiol.* **2001**, *11*, 118–126. [CrossRef]
30. McAllister, A.K. Cellular and Molecular Mechanisms of Dendrite Growth. *Cereb. Cortex* **2000**, *10*, 963–973. [CrossRef]
31. Kim, B.G.; Dai, H.-N.; McAtee, M.; Vicini, S.; Bregman, B.S. Remodeling of Synaptic Structures in the Motor Cortex Following Spinal Cord Injury. *Exp. Neurol.* **2006**, *198*, 401–415. [CrossRef] [PubMed]

32. Kim, B.G.; Dai, H.-N.; McAtee, M.; Bregman, B.S. Modulation of Dendritic Spine Remodeling in the Motor Cortex Following Spinal Cord Injury: Effects of Environmental Enrichment and Combinatorial Treatment with Transplants and Neurotrophin-3. *J. Comp. Neurol.* **2008**, *508*, 473–486. [CrossRef] [PubMed]
33. Charych, E.I.; Akum, B.F.; Goldberg, J.S.; Jornsten, R.J.; Rongo, C.; Zheng, J.Q.; Firestein, B.L. Activity-Independent Regulation of Dendrite Patterning by Postsynaptic Density Protein PSD-95. *J. Neurosci.* **2006**, *26*, 10164–10176. [CrossRef]
34. Mizrahi, A. Dendritic Development and Plasticity of Adult-Born Neurons in the Mouse Olfactory Bulb. *Nat. Neurosci.* **2007**, *10*, 444–452. [CrossRef] [PubMed]
35. Mizrahi, A.; Katz, L.C. Dendritic Stability in the Adult Olfactory Bulb. *Nat. Neurosci.* **2003**, *6*, 1201–1207. [CrossRef]
36. Miyasaka, N.; Arganda-Carreras, I.; Wakisaka, N.; Masuda, M.; Sümbrül, U.; Seung, H.S.; Yoshihara, Y. Olfactory Projectome in the Zebrafish Forebrain Revealed by Genetic Single-neuron Labelling. *Nat. Commun.* **2014**, *5*, 3639. [CrossRef]
37. Cheadle, L.; Biederer, T. Activity-Dependent Regulation of Dendritic Complexity by Semaphorin 3A through Farp1. *J. Neurosci.* **2014**, *34*, 7999–8009. [CrossRef]
38. Jan, Y.N.; Jan, L.Y. Branching out: Mechanisms of Dendritic Arborization. *Nat. Rev. Neurosci.* **2010**, *11*, 316–328. [CrossRef] [PubMed]
39. Koyama, Y.; Hattori, T.; Nishida, T.; Hori, O.; Tohyama, M. Alterations in Dendrite and Spine Morphology of Cortical Pyramidal Neurons in DISC1-Binding Zinc Finger Protein (DBZ) Knockout Mice. *Front. Neuroanat.* **2015**, *9*, 52. [CrossRef]
40. Jaworski, J.; Spangler, S.; Seeburg, D.P.; Hoogenraad, C.C.; Sheng, M. Control of Dendritic Arborization by the Phosphoinositide-3′-Kinase- Akt-Mammalian Target of Rapamycin Pathway. *J. Neurosci.* **2005**, *25*, 11300–11312. [CrossRef]
41. Rajan, I.; Cline, H.T. Glutamate Receptor Activity Is Required for Normal Development of Tectal Cell Dendrites In Vivo. *J. Neurosci.* **1998**, *18*, 7836–7846. [CrossRef] [PubMed]
42. Bloomfield, S.; Hitchcock, P. Dendritic Arbors of Large-Field Ganglion Cells Show Scaled Growth during Expansion of the Goldfish Retina: A Study of Morphometric and Electrotonic Properties. *J. Neurosci.* **1991**, *11*, 910–917. [CrossRef] [PubMed]
43. Lee, S.; Stevens, C.F. General Design Principle for Scalable Neural Circuits in a Vertebrate Retina. *Proc. Natl. Acad. Sci. USA* **2007**, *104*, 12931–12935. [CrossRef]
44. Koss, W.A.; Belden, C.E.; Hristov, A.D.; Juraska, J.M. Dendritic Remodeling in the Adolescent Medial Prefrontal Cortex and the Basolateral Amygdala of Male and Female Rats. *Synapse* **2014**, *68*, 61–72. [CrossRef] [PubMed]
45. Tomizawa, K.; Kunieda, J.-I.; Nakayasu, H. Ex Vivo Culture of Isolated Zebrafish Whole Brain. *J. Neurosci. Methods* **2001**, *107*, 31–38. [CrossRef]
46. Schindelin, J.; Arganda-Carrera, I.; Frise, E.; Verena, K.; Mark, L.; Tobias, P.; Stephan, P.; Curtis, R.; Stephan, S.; Benjamin, S.; et al. Fiji—An Open Platform for Biological Image Analysis. *Nat. Methods* **2009**, *9*, 10–38. [CrossRef]

International Journal of
Molecular Sciences

MDPI

Review

Neurogenic Effects of Inorganic Arsenic and Cdk5 Knockdown in Zebrafish Embryos: A Perspective on Modeling Autism

Qiang Gu and Jyotshna Kanungo *

Division of Neurotoxicology, National Center for Toxicological Research, U.S. Food and Drug Administration, 3900 NCTR Road, Jefferson, AR 72079, USA
* Correspondence: jyotshnabala.kanungo@fda.hhs.gov; Tel.: +1-870-543-7591; Fax: +1-870-543-7143

Abstract: The exact mechanisms of the development of autism, a multifactorial neurological disorder, are not clear. The pathophysiology of autism is complex, and investigations at the cellular and molecular levels are ongoing to provide clarity. Mutations in specific genes have been identified as risk factors for autism. The role of heavy metals in the pathogenesis of autism is subject to many studies and remains debatable. Although no exact neuronal phenotypes have been identified linked to autistic symptoms, overproduction and reduction of specific neurons have been implicated. A growing literature on generating genetic and non-genetic models of autism aims to help with understanding mechanistic studies that can explain the complexity of the disorder. Both genetic and non-genetic methods of zebrafish have been used to model autism. For several human autism risk genes, validated zebrafish mutant models have been generated. There is growing evidence indicating a potential link between autism and inorganic arsenic exposure. We have previously shown that inorganic arsenic induces supernumerary spinal motor neurons via Sonic hedgehog (Shh) signaling pathway, and Cdk5 knockdown causes an overproduction of cranial and spinal motor neurons in zebrafish. Here, in this review, we provide a perspective on what these findings of neurogenic phenotypes mean in terms of dysregulated pathways of motor neuron development and their applicability to understanding cellular and molecular underpinnings of autism.

Keywords: arsenic; zebrafish; Sonic hedgehog; autism; motor neuron

1. Introduction

Autism spectrum disorder (ASD), commonly known as autism, is a complex neurodevelopmental disorder [1,2]. Due to the broad spectrum of this neurodevelopmental disorder, a diagnosis of autism remains a challenge and is based on the individual's behavioral patterns and developmental history, while the severity and variability of the symptoms can vary among individuals [3,4]. Autism causes motor defects such as difficulty in walking, postural irregularities causing clumsiness, and balance issues, while non-motor defects include memory and cognitive deficits, irritation, anxiety, and aggressive behavior [5].

Over the past two decades, the prevalence of autism reported worldwide has been steadily increasing. In 2000, according to Autism and Developmental Disabilities Monitoring (ADDM), the incidence of autism was estimated to be 1 in 150 children. In 2006, the incidence was 1 in 110 children, and by 2008, the incidence had increased to 1 in 88 children [6]. A recent estimate shows that more than 70 million people, i.e., 1.5% to 2% worldwide, suffer from autism [7]. In 2023, the Center for Disease Control (CDC) reported that the incidence increased to 1 in 36 (2.8%) [8].

With hitherto unknown specific causes, autism, a multifactorial neurodevelopmental disorder, is found to be highly heritable, and many studies reveal that genetic factors (the involvement of many genes) as well as environmental factors are the major contributors/risk factors for the development of autism [2,9,10]. Environmental chemicals can contribute to human diseases, including autism [11]. Mounting evidence indicates that autism results from complex interactions between genes and the environment [12,13]. A systematic review and meta-analyses of 53 studies involving 5054 children described an association between autism and heavy metal exposure including arsenic, cadmium, mercury, and lead [14]. Arsenic is considered a potential contributor to the development of autism [15]. Multiple studies have shown an association of arsenic with autism (reviewed in [14]). An epidemiological study on 397 autism cases and 1034 controls under the Norwegian Mother, Father and Child Cohort Study showed a positive association between prenatal exposure to arsenic and autism risk [16] A report showed that populations living closer to industrial facilities that emit heavy metals such as arsenic, lead, and mercury to the air had a higher occurrence of autism [17]. Epidemiological studies show that arsenic exposure during critical periods of neurodevelopment could pose as an environmental risk factor for autism development [18,19]. Significantly higher levels of arsenic in the urine [20–23], blood [24], and hair [22] of children with autism have also been reported [20,21]. Arsenic exposure through drinking water at 10–50 ppb has been shown to cause peripheral neuropathy in humans [25]. In children, central nervous system (CNS) impairment may occur at $\geq$ 50 ppb [26]. Mice, after prenatal exposure to arsenic, showed an increased number of pyramidal neurons of the prelimbic cortex, which has been linked to behavioral inflexibility in adulthood due to cortical disarrangement [27]. Although arsenic's effects on specific neurons in the brain have been well studied, very few studies have focused on its effects on motor neurons. An epidemiological study in arsenic-contaminated regions showed a 16.7% higher risk of mortality associated with motor neuron disease [28]. Additional epidemiological studies show a potential association of heavy metals, including arsenic, with autism (reviewed in [14]) that warrants further studies in order to determine whether there is a direct link between these heavy metals and autism.

Mutations in cyclin-dependent kinase 5 (Cdk5) have been reported in patients with non-syndromic intellectual disability [29]. Selective loss of Cdk5 in the dorsolateral striatum of mice caused increased locomotor activity with attenuated motor learning [30]. Valproic acid, prenatal exposure to which causes autism-like behavioral abnormalities and brain malformation in animal models including zebrafish [31,32], downregulates Cdk5 activity in cultured mouse neurons [33]. The effects of the downregulation of Cdk5 activity on specific neuron development (a specific neuronal phenotype) can help unravel cellular and molecular mechanisms behind autism-like symptoms in animal models with follow-up studies.

One of the pathological mechanisms of autism underlies impaired functions of specific brain regions and dysfunctional neural circuits [13]. For example, functional studies of an autism-associated gene, *Shank3*, a synaptic scaffold protein that is enriched at the postsynaptic excitatory synapses [34], show that mice lacking *Shank3* not only exhibit hypertrophy of the striatum but also experience decreased cortico-striatal excitatory synaptic transmission and show repetitive behaviors [35]. Lately, various animal modeling studies have revealed several types of viable mutations, which can shed light on the underlying mechanisms of autism pathogenesis [13]. Due to the evolutionary conservation of the developmental processes of the nervous system between zebrafish and mammals, zebrafish are used to investigate autism using both genetic and non-genetic methods (reviewed in [2]). Zebrafish exhibit similar behavioral responses as in mammals, such as social interactions and preference, as well as repetitive behaviors, making it possible to model phenotypes with ASD-like symptoms [36]. A list of zebrafish mutant lines for twelve autism risk genes has been curated (https://www.sfari.org/resource/zebrafish-models/ (accessed on 10 December 2023)) by the Simons Foundation for Autism Research Initiative (SFARI). As an alternative animal model, data from zebrafish autism studies can add to the knowledge gap that exists in mammalian studies, reveal mechanistic pathways, and help with drug discovery.

Chronic arsenic exposure altered social behavior, a characteristic of autism, in juvenile zebrafish, which was ameliorated by the antioxidant N-acetylcysteine [37]. In zebrafish larvae, arsenic caused motor behavioral deficit as well as mild impairment in behavior towards color preference [38]. Transgenerational changes in motor activity and anxiety-like behavior upon arsenic exposure, accompanied by a reduction in brain-derived neurotrophic factor level and increased histone methylation, have been reported in zebrafish [39]. Arsenic caused hypoactivity of zebrafish larvae in a photomotor response assay [40]. Anxiety-like behavior and alteration in long-term memory have also been reported in adult zebrafish upon arsenic exposure [41]. Based on existing studies both in mammals and zebrafish, this review provides a perspective on what the neurogenic phenotypes indicate in terms of modeling autism.

2. Risk Genes of Autism and Zebrafish

Many reliable risk genes for autism development have been discovered (reviewed in [2]), and approximately 5% of autism cases result from single-nucleotide polymorphisms (SNPs) in genes such as *NLGN3, NLGN4, NRXN1, MECP2, SHANK3, FMR1, TSC1/2*, and *UBE3A* (reviewed in [2]). Genetic alterations that can increase the risk of autism include changes in UBE3A, a ubiquitin protein ligase E3A [42], MAPK3 (mitogen-activated protein kinase 3) [43], as well as an increase in the copy number variants, such as single nucleotide polymorphisms (SNPs), for example, in the chromosomal region 15q11-q13.3 [42]. In addition to this, epigenetic mechanisms that include histone modification, DNA methylation, chromatin remodeling, and micro-RNA activity are involved in the regulation of social behavior in autism [44].

A study on zebrafish using high-throughput functional analysis of 10 autism risk genes identified convergence of dopaminergic and neuroimmune pathways [45]. The functions of 12 autism genes (*ARID1B, CHD8, CNTNAP2, DYRK1A, GRIN2B, FMR1, MECP2, NRXN1, PTEN, SCN2A, SHANK3*, and *SYNGAP1*) have been studied in zebrafish [2]. In zebrafish embryos, morpholino (MO)-mediated knockdown of CHD8, a chromatin-binding protein that targets many other autism-related genes, results in macrocephaly consistent with human autism cases with CHD8 loss of function [46]. Knockdown of FMR1 in zebrafish larvae resulted in autism-like behavior [47] similar to the valproic-acid treated zebrafish larvae [48]. *MECP2* knockout in zebrafish caused behavioral and motor deficits [49], and MECP2 knockdown suppressed neural precursor cell differentiation [50]. Double mutation in *CNTNAP2a/b* zebrafish caused reduced GABAergic neurons [51]. Zebrafish *DYRK1A* mutants have microcephaly [52]. Homozygous recessive loss-of-function mutation in *scn1alab*, a voltage-gated sodium ion channel, caused abnormal neuronal firing, hyperactivity, and convulsive behaviors in zebrafish that are consistent with effects shown in mice and humans [53]. *Shank3a/b* knockout zebrafish embryos/larvae as well as adults had reduced revels of synaptic proteins and displayed robust autism-like behaviors with reduced locomotor activity [54]. *Syngap1a/b* knockdown embryos had significantly decreased GABAergic neurons [55]. These studies emphasize the utilization of the zebrafish model for autism studies that can reveal useful information on this complex neurodevelopmental disorder.

3. Autism and Overproduction and Reduction of Specific Neurons

A preliminary study reported brain overgrowth and an excess number of neurons in the pre-frontal cortex of autistic male children [56]. An overproduction of upper-layer neurons in the neocortex in mice has been shown to lead to autism-like features, suggesting a causal link between the overproduction of certain neurons and autism, which offers some insight into the etiology of the disorder [57]. It has been reported that although the number of mature neurons of the human amygdala increases from childhood into adulthood under normal development, in autistic individuals, an initial excess of neurons in the amygdala during childhood is followed by a reduction of neurons in adulthood [58]. Such developmental anomalies might offer critical information on the etiology of autism. Decreased cortical interneurons in autism have also been reported, indicating that interneuron hypofunction could be a primary driver of erroneous circuit engagement and dysfunction in autism [59]. Furthermore, a study on organoids derived from induced pluripotent stem cells from patients with Fragile X Syndrome (a known cause of autism) showed a lower density of GABA-expressing neurons [60].

Autopsies of patients with autism have shown significant structural changes of their brains, e.g., altered grey and white matter ratios, increased neuronal numbers accompanied by reduced neuronal body volume, increased numbers of glia, and changes in dendritic spines and cerebral vasculature [61]. Longitudinal imaging studies on toddlers (18 and 60 months old) with autism revealed an enlarged amygdala [62], and children with autism had 67% more neurons in the prefrontal cortex [56]. On the other hand, Purkinje cells were decreased in the cerebellar hemispheres of autistic individuals, who also had reduced numerical density of neurons in the putamen and nucleus accumbens [63]. Brain tissues of individuals with autism have supernumerary neurons in the cerebral cortical subplate [64]. The increased brain size in subjects with autism [65] has been attributed to an increased number of neurons or increased neuropil when there was no change in neuron numbers [66]. Malformations of the CNS resulting from such abnormal neurodevelopment (lack of or over-abundance of specific neurons) can lead to autism, cognitive delay, and intractable epilepsy [67,68].

Moreover, early assessments of autism show striatal hypertrophy with reduced amygdala volume albeit increased neuronal density in the region covering the medial, central, and lateral nuclei that plays critical roles in anxiety, fear conditioning, and social behavior [13,69,70]. Additionally, prenatal exposure to valproic acid, which, clinical evidence indicate, has a strong association with autism [71,72], enhanced untimely embryonic neurogenesis in mice, leading to a depletion of the neural precursors and resulting in decreased levels of adult hippocampal neurogenesis [73]. In zebrafish embryos, valproic acid adversely affected neurogenesis in the optic tectum [74], reduced midbrain size, and reduced the number of neuronal progenitors, along with perturbations in the secondary motor neuron neurite development [75]. Modeling the genetic as well as environmental aspects in zebrafish embryos can offer an ideal system for an in-depth investigation of the potential mechanisms of autism development, since manipulation of individual risk genes in these embryos may lead to the identification of phenotype-based mechanistic pathways.

4. Arsenic and Zebrafish Motor Neurons: Relevance to Autism

Recently, we reported that arsenic induced supernumerary spinal motor neurons in transgenic (*hb9-GFP*) zebrafish that express green fluorescent protein (GFP) in the motor neurons via Sonic hedgehog pathway (Figure 1A–C) and also increased the density of tyrosine hydroxylase-positive dopaminergic neurons [76]. However, arsenic did not alter the density of serotonergic neurons [76]. In vertebrates, the formation of motor neurons depends on Hedgehog (Hh) signaling, which is mediated by Gli zinc finger proteins [77].

There are three Hh family members, Sonic Hedgehog (Shh), Indian Hedgehog (Ihh) and Desert Hedgehog (Dhh). These three proteins activate a common signaling pathway, called Hh signaling, and arsenic activates Hh signaling [78]. Shh, a secretory protein acts as a developmental morphogen, and Shh signaling plays an integral role in embryogenesis including neurodevelopment and neurodegeneration [79]. The Shh signaling pathway plays an important role in development [80]. Vertebrate Patched (a receptor) binds to the Shh ligand [81]. Such binding relieves the inhibitory effect of Patched on a seven-transmembrane protein, Smoothened, resulting in transcription of the Gli transcription factors, including Gli2 [82,83]. Gli2, a positive regulator of Shh signaling, is then activated [84–86]. However, supernumerary motor neuron development is inhibited by the Shh signaling inhibitor Gant61 [76]. While Gli1 can induce Nkx2.1-positive ventral forebrain neuron development, both Gli1 and Gli2 can induce Hb9-positive spinal motor neuron development [87]. In arsenic-treated zebrafish, Patched gene expression was not altered [88]. Whether protein levels of Patched and Shh changed in arsenic-treated embryos remains under investigation.

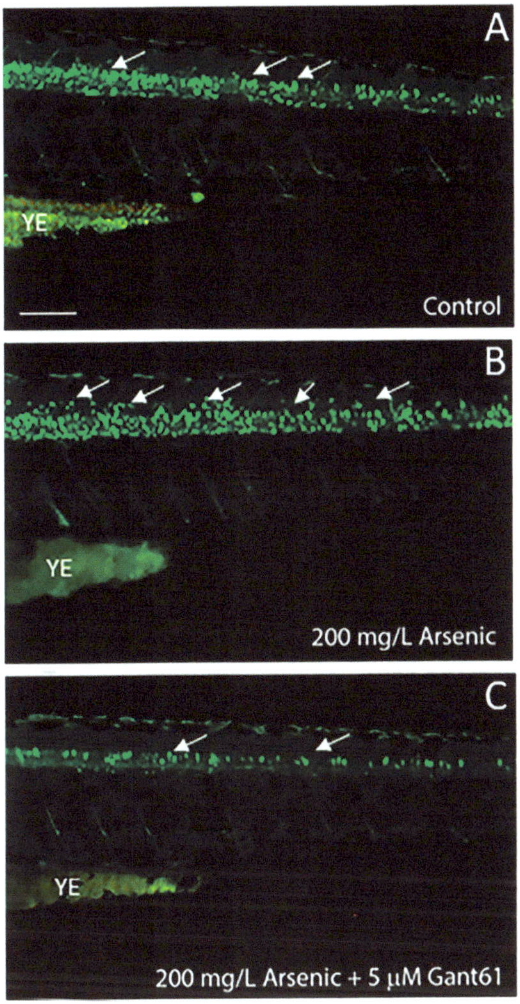

Figure 1. Inorganic arsenic-induced supernumerary motor neuron development is inhibited by the Shh inhibitor, Gant61. Five hours post fertilization (hpf), transgenic embryos (*hb9-GFP*) that express green fluorescent protein (GFP) in the motor neurons were exposed to 200 mg/L of sodium arsenite (internal concentration of 387.8 ± 26.9 pg/embryo). Fluorescent images of spinal cord regions of the 72 hpf embryos are shown for control (**A**), 200 mg/L sodium arsenite-treated (**B**), 200 mg/L sodium arsenite with 5 μM Gant61-treated (**C**). Arrows indicate GFP-expressing motor neurons. YE indicates yolk extension (Adapted from Kanungo et al. [76]).

Shh is secreted from the notochord and is critical for the development of the motor neurons in vertebrates [87] (Figure 2). Shh signaling activates Gli genes, which are known to affect motor neuron development and positioning in the spinal cord during early development of vertebrates [89,90]. In the CNS, Shh plays a critical role in ventral specification along the neural axis. Overexpression of Shh in the spinal cord has been shown to alter the positioning of the motor neurons and results in the aberrant structure of the motor column [90]. Misexpression of Shh can induce the differentiation of floor plate cells including motor neuron differentiation at ectopic locations in the spinal cord in vertebrate embryos [91–93]. A schematic presentation of motor neuron development in zebrafish pertaining to the Shh signaling pathway is shown in Figure 2. The cross-sectional view of the neural tube flanked dorsally by the ectoderm and ventrally by the endoderm shows the location of the roof plate, sensory neurons, interneurons, motor neurons, floor plate, and notochord, the latter producing the Shh that induces motor neuron development (Figure 2). This pathway of motor neuron development is conserved in vertebrates [94].

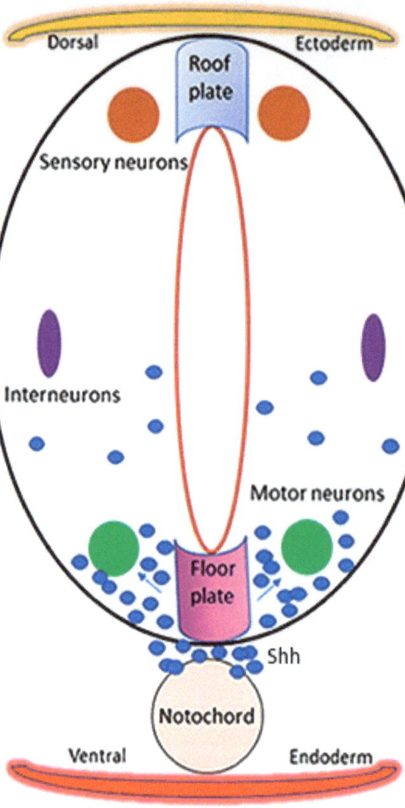

Figure 2. Schematic presentation of motor neuron development in zebrafish. The neural tube develops from the neural plate with the dorsal location of the sensory neurons, intermediate location of the interneurons, and ventral location of the motor neurons. The dorsal ectoderm above the roof plate and the ventral endoderm below the notochord are shown. Sonic hedgehog (Shh) is expressed in the floor plate and notochord, which triggers the development of motor neurons from the neuronal progenitors of the floor plate.

Based on rodent studies, it has been postulated that the motor dysfunction caused by arsenic ingestion may be a consequence of arsenic's direct influence on motor neurons rather than other processes, such as demyelination [95]. In children with autism, significantly higher levels of serum Shh protein have been reported [96]. A low dose of arsenic can induce Hh signaling in vitro and in vivo [78]. Dysregulation of Shh signaling leads to many physiological changes that precede neurological disorders such as autism and cognitive decline (reviewed in [97]). These data suggest a mechanistic link between arsenic, motor dysfunction, and some of the symptoms commonly observed in ASD. In mouse models of autism, mechanisms involving cellular and synaptic functions of the neurons of the peripheral somatosensory system, as well as spinal cord neurons, have been shown to contribute to tactile over-reactivity [98–102]. Further studies on arsenic-induced supernumerary neurons in zebrafish are needed to explore the molecular and behavioral changes related to autism. The questions that remain to be answered are whether arsenic alters Shh expression at the gene and/or protein level and whether downstream Gli genes are induced by arsenic.

In the *hb9:GFP* zebrafish embryos, it was difficult to discern brain motor neurons as opposed to those in the spinal cord [76]. Future research using high-resolution microscopy might reveal a clearer picture.

5. Cdk5 Knockdown and Zebrafish Motor Neurons: Relevance to Autism

Cdk5 is a member of the family of serine/threonine cyclin-dependent kinase, which is highly expressed in neurons [103]. A multi-functional protein kinase, Cdk5 regulates a wide range of neuronal functions, including neuronal survival and migration and plays a critical role in neuronal differentiation [103]. Additionally, Cdk5 is essential in regulating a number of cellular processes of the nervous system including protein trafficking, neurite and synapse development, dopaminergic function, learning, and memory [104,105]. Dysregulation of Cdk5 activity can cause a wide range of pathological processes affecting the nervous system development, leading to neurodegeneration [103]. Suppressing Cdk5 activity in cultured cortical neurons leads to compromised neurite outgrowth, whereas ectopic expression of exogenous Cdk5 and its regulator p35 produce longer neurites [106]. Cdk5-null mice that are embryonically lethal show an aberrant development of the cortex and cerebellum [107]. We have previously shown that the suppression of Cdk5 activity through MO-mediated Cdk5 knockdown or overexpression of the dominant negative human Cdk5 (hCdk5 DN) mRNA generated supernumerary motor neurons in vivo in zebrafish [108] (Figure 3A–F). In the *islet-1-GFP* transgenic zebrafish embryos that express GFP in the motor neurons, morpholino-mediated Cdk5 knockdown (translational inhibition) and hCdk5 DN-mediated suppression of Cdk5 activity caused supernumerary motor neuron generation in both cranial and spinal regions (Figure 3) [108]. In these embryos, Cdk5 activity was significantly reduced [108]. Although using a single morpholino may not be sufficient to provide a conclusive statement about a gene function in studies conducted lately, as the specificity of the morpholinos needs more controls, overexpression of the kinase-dead hCdk5 DN mRNA increasing motor neuron density further strengthens the finding (Figure 3). In agreement with this, the study also showed that overexpression of Cdk5 mRNA reduced the motor neuron density in the zebrafish embryos compared to the control [108].

Cdk5 has been implicated in the pathogenesis of various neurological disorders including autism [109]. Downregulation of Cdk5 has been associated with attention deficit and hyperactivity disorder [110], epilepsy [111], and schizophrenia [112]. Cdk5 rescued hippocampal synaptic plasticity in a mouse model of Fragile X Syndrome, a genetic form of intellectual disability associated with epilepsy, autism, and mood disorders, suggesting that activation of Cdk5 activity might be a pharmacological tool to treat Fragile X Syndrome [113]. However, no association between polymorphisms in Cdk5 with autism was found in a Chinese Han population [114]. On the other hand, studies explored a severe neurodevelopmental disorder that was characterized by intellectual disability, early-onset seizures, and autistic features resulting from mutations in the X-linked cyclin-dependent kinase-like 5 (CDKL5) gene [115–117]. The mechanism behind this link between the CDKL5 mutation and autistic behavior is not known.

We have shown that Cdk5 activity is significantly reduced in zebrafish *mindbomb 1* (*Mindbomb E3 ubiquitin protein ligase 1*) mutants [118]. The *mindbomb 1* (*Mib1*) gene was first identified as an E3 ubiquitin ligase in zebrafish through genetic mutagenesis screens [119]. In zebrafish, *Mib1* positively regulates the Notch pathway [119] necessary for cell fate specification [120]. While *Mib1*-null mice are embryonically lethal [121], the loss-of-function zebrafish mutant (*mindbomb 1*) exhibits developmental defects due to a loss of Notch signaling-induced lateral inhibition, thus resulting in a neurogenic phenotype characterized by increased supernumerary primary neurons [119]. In addition to defects in neurogenesis, severe defects in angiogenesis and somitogenesis occur in zebrafish *Mib1* mutants [122–124]. In humans, *Mib1* mutations contribute to congenital heart disease through disruptions in heart development [125]. *Mib1* homozygous mutant zebrafish do not live beyond four days, whereas *Mib1* heterozygotes (a recessive mutation) are able to survive and breed [126].

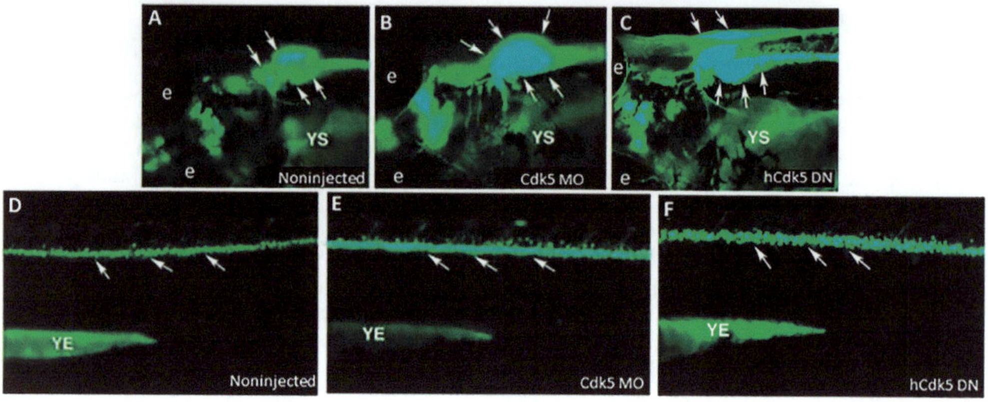

Figure 3. Cdk5 knockdown through microinjection of Cdk5 morpholino and dominant negative human Cdk5 (hCdk5 DN) mRNA caused supernumerary motor neuron generation in zebrafish embryos. Images of live 72 hpf *islet-1*–GFP transgenic zebrafish embryos show motor neurons in the brain; (**A**) noninjected (vehicle only) control, (**B**) Cdk5 morpholino (MO)-injected, and hCdk5 DN mRNA-injected (**C**); spinal regions of (**D**) noninjected (vehicle only) control; (**E**) Cdk5 morpholino (MO)-injected embryos; and (**F**) hCdk5 DN mRNA-injected embryos. Arrows indicate the GFP-expressing motor neuron populations in the brain (upper panel) and spinal cord (lower panel). YS indicates the yolk sac; e indicates the eye; YE indicates yolk extension. (Adapted from Kanungo et al. [108]).

Mindbomb mutant zebrafish exhibit spontaneous seizures accompanied by altered gene expression in the GABA signaling pathways [127]. Loss of function due to point mutations in human ubiquitin E3A ligase has been reported in patients with autism [128]. Additionally, small deletions or mutations in the human ubiquitin E3A ligase gene have been linked to autism [129]. *Mib1* has been shown to regulate neurite morphogenesis by interacting with Cdk5 and its regulator p35 [130]; however, it is not clear why *mindbomb* mutant zebrafish have reduced Cdk5 activity [118]. A potential pathway of the reduction in Cdk5 activity in the *mindbomb* mutant zebrafish has been proposed, which suggests that overexpression of Cdk5 beyond a threshold limit can reduce its own activity [118]. This study indicated that reduction of Cdk5 activity but not Cdk5 mRNA level itself is critical for the overproduction of primary neurons, and Notch inhibition (*mindbomb/Mib1* mutant) is upstream of the downregulation of Cdk5 activity [118].

Mib1 ubiquitinates and induces the degradation of survival of motor neuron proteins (SMNs), and *Mib1* knockdown increases SMN protein levels in HEK-293T cells, suggesting a beneficial effect on the survival of motor neurons [131]. Similar to *Mib1* mutation, arsenic has been shown to block Notch signaling in a human small-cell lung cancer cell line [132,133]. Whether arsenic inhibits Notch signaling in zebrafish, which could be responsible for the neurogenic phenotype we have reported [76], warrants further studies that would reveal divergent or convergent pathways linking the phenotypes to the upstream events.

6. Conclusions

An overproduction and reduction of specific neurons have been reported in autism, which can potentially explain the excitation/inhibition imbalance displayed in individuals with this serious and complex disorder [59]. In the phenotypes discussed above, data are lacking on whether the overproduction of tyrosine hydroxylase positive neurons in the brain by arsenic or motor neurons in the spinal cord by Cdk5 knockdown and arsenic in zebrafish occurred while there was reduction in other types of neurons (e.g., interneurons). Arsenic and Cdk5 knockdown, inducing supernumerary motor neurons and producing a similar outcome in motor neuron development, indicates that such phenotypes can occur through many different mechanisms and may cause an imbalance in specific neuron functions, e.g., excitatory to inhibitory imbalance—a hallmark of autism (Figure 4). Arsenic also increased tyrosine hydroxylase positive neurons in the brains of zebrafish embryos [76]. These phenotypes may be utilized as preclinical models for in-depth studies to demonstrate whether the alteration in the neuronal development patterns predisposes the organism to exhibit autism-like symptoms (Figure 4). Zebrafish have been utilized to model phenotypes related to autism either through genetic manipulation or chemical exposure (e.g., valproic acid) (reviewed in [2]). Although these phenotypes of zebrafish cannot completely simulate the pathological processes of autism reported in human beings, they will help to understand the triggers and molecular precursors of the development of autism. Therefore, alterations in early development of specific neurons in autism risk-gene mutants or those that are induced by chemicals need to be investigated. Furthermore, exposing specific autism risk-gene mutants of zebrafish to arsenic and examining the effects on specific neuron types would reveal deeper understanding of the multifactorial nature of the disease. While stem cell models have been able to reveal that disruptions in specific molecular processes, such as calcium and Wnt signaling, and chromatin remodeling can contribute to the pathogenesis of autism [46,134], being a vertebrate with conserved genetic and physiologic pathways [135], zebrafish carry an advantage in phenocopying cellular and behavioral aspects of autism that can reveal hitherto unknown mechanisms. The early detection of defective neuronal development would help delineate the mechanism of the role of environmental factors in autism development, which would shed light on gene/environment interactions and provide opportunities for therapeutic drug discovery. The current perspective presents a scenario that warrants further investigation of the zebrafish phenotypes with an overproduction of specific neurons in order to determine whether autism can be modeled to a certain extent, if not completely, using these embryos.

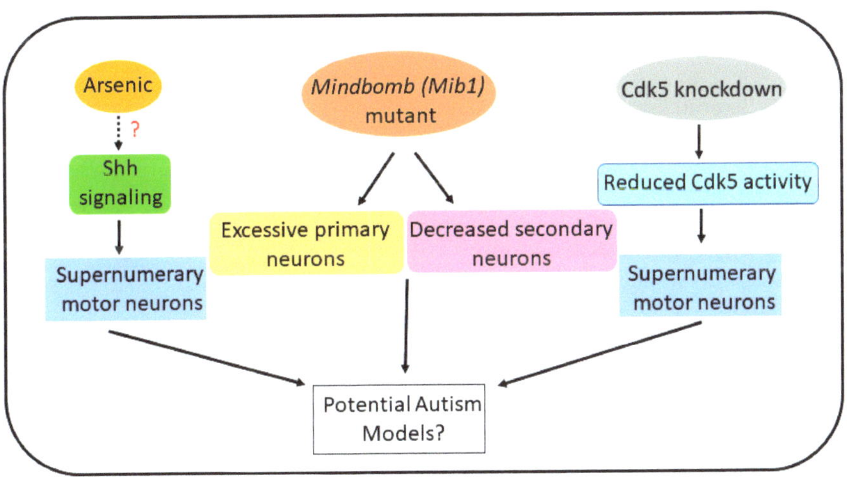

Figure 4. Schematic presentation of scenarios occurring in zebrafish treated with arsenic, the *mindbomb (Mib1)* mutant, and morpholino (MO)-mediated Cdk5 knockdown or human dominant negative Cdk5 mRNA expression that caused decreased Cdk5 activity. How arsenic could modulate Shh signaling is not known. Nonetheless, supernumerary (excessive) neurons (primary and motor neurons) resulting from these cases may be used to model and study cellular and molecular mechanisms of autism in a lower vertebrate like zebrafish.

Author Contributions: Conceptualization: J.K.; investigation: J.K. and Q.G.; visualization: J.K.; writing—original draft: J.K.; writing—review and editing: J.K. and Q.G. All authors have read and agreed to the published version of the manuscript.

Funding: This work was supported in part by the National Center for Toxicological Research within the Food and Drug Administration.

Institutional Review Board Statement: Not applicable.

Informed Consent Statement: Not applicable.

Data Availability Statement: No new data are reported in this review article.

Acknowledgments: The authors would like to thank Sangeeta Khare, Sumit Sarkar, and John Talpos for their critical review of the manuscript.

Conflicts of Interest: The authors declare that they have no known competing financial interests or personal relationships that could have appeared to influence the work reported in this paper.

Abbreviations

ASD	Autism spectrum disorder
Cdk5	Cyclin-dependent kinase 5
CHD8	Chromodomain helicase DNA binding protein 8
FMR1	Fragile X Messenger Ribonucleoprotein 1
GFP	Green fluorescent protein
Gli	Glioma-associated oncogene, a zinc finger protein
MECP2	Methyl CpG binding protein 2
Mib1	Mindbomb E3 ubiquitin protein ligase 1
NLGN3	Neuroligin-3
NLGN4	Neuroligin-4
NRXN1	Neurexin 1
SHANK3	SH3 and multiple ankyrin repeat domains 3 (also known as proline-rich synapse-associated protein 2)
TSC1/2	Tuberous sclerosis complex 1/2
UBE3A	Ubiquitin-protein ligase E3A (also known as E6AP ubiquitin-protein ligase)

References

1. Pierce, K.; Gazestani, V.H.; Bacon, E.; Barnes, C.C.; Cha, D.; Nalabolu, S.; Lopez, L.; Moore, A.; Pence-Stophaeros, S.; Courchesne, E. Evaluation of the Diagnostic Stability of the Early Autism Spectrum Disorder Phenotype in the General Population Starting at 12 Months. *JAMA Pediatr.* **2019**, *173*, 578–587. [CrossRef] [PubMed]
2. Rea, V.; Van Raay, T.J. Using Zebrafish to Model Autism Spectrum Disorder: A Comparison of ASD Risk Genes between Zebrafish and Their Mammalian Counterparts. *Front. Mol. Neurosci.* **2020**, *13*, 575575. [CrossRef] [PubMed]
3. Dyches, T.T.; Wilder, L.K.; Sudweeks, R.R.; Obiakor, F.E.; Algozzine, B. Multicultural issues in autism. *J. Autism Dev. Disord.* **2004**, *34*, 211–222. [CrossRef]
4. Szatmari, P.; Georgiades, S.; Bryson, S.; Zwaigenbaum, L.; Roberts, W.; Mahoney, W.; Goldberg, J.; Tuff, L. Investigating the structure of the restricted, repetitive behaviours and interests domain of autism. *J. Child Psychol. Psychiatry* **2006**, *47*, 582–590. [CrossRef]
5. Sharma, A.; Bhalla, S.; Mehan, S. PI3K/AKT/mTOR signalling inhibitor chrysophanol ameliorates neurobehavioural and neurochemical defects in propionic acid-induced experimental model of autism in adult rats. *Metab. Brain Dis.* **2022**, *37*, 1909–1929. [CrossRef] [PubMed]
6. Sharma, S.R.; Gonda, X.; Tarazi, F.I. Autism Spectrum Disorder: Classification, diagnosis and therapy. *Pharmacol. Ther.* **2018**, *190*, 91–104. [CrossRef] [PubMed]
7. Morales Hidalgo, P.; Voltas Moreso, N.; Canals Sans, J. Autism spectrum disorder prevalence and associated sociodemographic factors in the school population: EPINED study. *Autism* **2021**, *25*, 1999–2011. [CrossRef]
8. CDC. *Autism Prevalence Higher, According to Data from 11 ADDM Communities;* Center for Disease Control, CDC News: Atlanta, GA, USA, 2023.
9. Fakhoury, M. Autistic spectrum disorders: A review of clinical features, theories and diagnosis. *Int. J. Dev. Neurosci.* **2015**, *43*, 70–77. [CrossRef]
10. Mandy, W.; Lai, M.C. Annual Research Review: The role of the environment in the developmental psychopathology of autism spectrum condition. *J. Child Psychol. Psychiatry* **2016**, *57*, 271–292. [CrossRef]

11. Yasuda, H.; Tsutsui, T. Assessment of infantile mineral imbalances in autism spectrum disorders (ASDs). *Int. J. Environ. Res. Public Health* **2013**, *10*, 6027–6043. [CrossRef]
12. Huang, J.Y.; Tian, Y.; Wang, H.J.; Shen, H.; Wang, H.; Long, S.; Liao, M.H.; Liu, Z.R.; Wang, Z.M.; Li, D.; et al. Functional Genomic Analyses Identify Pathways Dysregulated in Animal Model of Autism. *CNS Neurosci. Ther.* **2016**, *22*, 845–853. [CrossRef]
13. Jiang, C.C.; Lin, L.S.; Long, S.; Ke, X.Y.; Fukunaga, K.; Lu, Y.M.; Han, F. Signalling pathways in autism spectrum disorder: Mechanisms and therapeutic implications. *Signal Transduct. Target. Ther.* **2022**, *7*, 229. [CrossRef]
14. Ding, M.; Shi, S.; Qie, S.; Li, J.; Xi, X. Association between heavy metals exposure (cadmium, lead, arsenic, mercury) and child autistic disorder: A systematic review and meta-analysis. *Front. Pediatr.* **2023**, *11*, 1169733. [CrossRef] [PubMed]
15. Baj, J.; Flieger, W.; Flieger, M.; Forma, A.; Sitarz, E.; Skórzyńska-Dziduszko, K.; Grochowski, C.; Maciejewski, R.; Karakuła-Juchnowicz, H. Autism spectrum disorder: Trace elements imbalances and the pathogenesis and severity of autistic symptoms. *Neurosci. Biobehav. Rev.* **2021**, *129*, 117–132. [CrossRef]
16. Skogheim, T.S.; Weyde, K.V.F.; Engel, S.M.; Aase, H.; Surén, P.; Øie, M.G.; Biele, G.; Reichborn-Kjennerud, T.; Caspersen, I.H.; Hornig, M.; et al. Metal and essential element concentrations during pregnancy and associations with autism spectrum disorder and attention-deficit/hyperactivity disorder in children. *Environ. Int.* **2021**, *152*, 106468. [CrossRef]
17. Dickerson, A.S.; Rahbar, M.H.; Han, I.; Bakian, A.V.; Bilder, D.A.; Harrington, R.A.; Pettygrove, S.; Durkin, M.; Kirby, R.S.; Wingate, M.S.; et al. Autism spectrum disorder prevalence and proximity to industrial facilities releasing arsenic, lead or mercury. *Sci. Total Environ.* **2015**, *536*, 245–251. [CrossRef]
18. Rodríguez-Barranco, M.; Gil, F.; Hernández, A.F.; Alguacil, J.; Lorca, A.; Mendoza, R.; Gómez, I.; Molina-Villalba, I.; González-Alzaga, B.; Aguilar-Garduño, C.; et al. Postnatal arsenic exposure and attention impairment in school children. *Cortex* **2016**, *74*, 370–382. [CrossRef] [PubMed]
19. Zuo, Z.; Liu, Z.; Gao, T.; Yin, Y.; Wang, Z.; Hou, Y.; Fu, J.; Liu, S.; Wang, H.; Xu, Y.; et al. Prolonged inorganic arsenic exposure via drinking water impairs brown adipose tissue function in mice. *Sci. Total Environ.* **2019**, *668*, 310–317. [CrossRef]
20. Rezaei, M.; Rezaei, A.; Esmaeili, A.; Nakhaee, S.; Azadi, N.A.; Mansouri, B. A case-control study on the relationship between urine trace element levels and autism spectrum disorder among Iranian children. *Environ. Sci. Pollut. Res. Int.* **2022**, *29*, 57287–57295. [CrossRef] [PubMed]
21. Shiani, A.; Sharafi, K.; Omer, A.K.; Kiani, A.; Karamimatin, B.; Massahi, T.; Ebrahimzadeh, G. A systematic literature review on the association between exposures to toxic elements and an autism spectrum disorder. *Sci. Total Environ.* **2023**, *857 Pt 2*, 159246. [CrossRef]
22. Nabgha, E.A.; Eqani, S.; Khuram, F.; Alamdar, A.; Tahir, A.; Shah, S.T.A.; Nasir, A.; Javed, S.; Bibi, N.; Hussain, A.; et al. Environmental exposure pathway analysis of trace elements and autism risk in Pakistani children population. *Sci. Total Environ.* **2020**, *712*, 136471. [CrossRef] [PubMed]
23. Lonsdale, D.; Shamberger, R.J.; Audhya, T. Treatment of autism spectrum children with thiamine tetrahydrofurfuryl disulfide: A pilot study. *Neuro Endocrinol. Lett.* **2002**, *23*, 303–308.
24. Amadi, C.N.; Orish, C.N.; Frazzoli, C.; Orisakwe, O.E. Association of autism with toxic metals: A systematic review of case-control studies. *Pharmacol. Biochem. Behav.* **2022**, *212*, 173313. [CrossRef]
25. Mochizuki, H.; Phyu, K.P.; Aung, M.N.; Zin, P.W.; Yano, Y.; Myint, M.Z.; Thit, W.M.; Yamamoto, Y.; Hishikawa, Y.; Thant, K.Z.; et al. Peripheral neuropathy induced by drinking water contaminated with low-dose arsenic in Myanmar. *Environ. Health Prev. Med.* **2019**, *24*, 23. [CrossRef]
26. Vibol, S.; Hashim, J.H.; Sarmani, S. Neurobehavioral effects of arsenic exposure among secondary school children in the Kandal Province, Cambodia. *Environ. Res.* **2015**, *137*, 329–337. [CrossRef]
27. Aung, K.H.; Kyi-Tha-Thu, C.; Sano, K.; Nakamura, K.; Tanoue, A.; Nohara, K.; Kakeyama, M.; Tohyama, C.; Tsukahara, S.; Maekawa, F. Prenatal Exposure to Arsenic Impairs Behavioral Flexibility and Cortical Structure in Mice. *Front. Neurosci.* **2016**, *10*, 137. [CrossRef]
28. Sanchez-Diaz, G.; Escobar, F.; Badland, H.; Arias-Merino, G.; Posada de la Paz, M.; Alonso-Ferreira, V. Geographic Analysis of Motor Neuron Disease Mortality and Heavy Metals Released to Rivers in Spain. *Int. J. Environ. Res. Public Health* **2018**, *15*, 2522. [CrossRef]
29. Moncini, S.; Castronovo, P.; Murgia, A.; Russo, S.; Bedeschi, M.F.; Lunghi, M.; Selicorni, A.; Bonati, M.T.; Riva, P.; Venturin, M. Functional characterization of CDK5 and CDK5R1 mutations identified in patients with non-syndromic intellectual disability. *J. Hum. Genet.* **2016**, *61*, 283–293. [CrossRef] [PubMed]
30. Hernandez, A.; Tan, C.; Mettlach, G.; Pozo, K.; Plattner, F.; Bibb, J.A. Cdk5 Modulates Long-Term Synaptic Plasticity and Motor Learning in Dorsolateral Striatum. *Sci. Rep.* **2016**, *6*, 29812. [CrossRef]
31. Chomiak, T.; Hu, B. Alterations of neocortical development and maturation in autism: Insight from valproic acid exposure and animal models of autism. *Neurotoxicology Teratol.* **2013**, *36*, 57–66. [CrossRef] [PubMed]
32. Chen, J.; Lei, L.; Tian, L.; Hou, F.; Roper, C.; Ge, X.; Zhao, Y.; Chen, Y.; Dong, Q.; Tanguay, R.L.; et al. Developmental and behavioral alterations in zebrafish embryonically exposed to valproic acid (VPA): An aquatic model for autism. *Neurotoxicology Teratol.* **2018**, *66*, 8–16. [CrossRef]
33. Takahashi, M.; Ishida, M.; Saito, T.; Ohshima, T.; Hisanaga, S. Valproic acid downregulates Cdk5 activity via the transcription of the p35 mRNA. *Biochem. Biophys. Res. Commun.* **2014**, *447*, 678–682. [CrossRef]
34. Uchino, S.; Waga, C. SHANK3 as an autism spectrum disorder-associated gene. *Brain Dev.* **2013**, *35*, 106–110. [CrossRef] [PubMed]

35. Peça, J.; Feliciano, C.; Ting, J.T.; Wang, W.; Wells, M.F.; Venkatraman, T.N.; Lascola, C.D.; Fu, Z.; Feng, G. Shank3 mutant mice display autistic-like behaviours and striatal dysfunction. *Nature* **2011**, *472*, 437–442. [CrossRef] [PubMed]
36. Stewart, A.M.; Nguyen, M.; Wong, K.; Poudel, M.K.; Kalueff, A.V. Developing zebrafish models of autism spectrum disorder (ASD). *Prog. Psychopharmacol. Biol. Psychiatry* **2014**, *50*, 27–36. [CrossRef]
37. Wang, Y.-H.; Wang, Y.-Q.; Yu, X.-G.; Lin, Y.; Liu, J.-X.; Wang, W.-Y.; Yan, C.-H. Chronic environmental inorganic arsenic exposure causes social behavioral changes in juvenile zebrafish (*Danio rerio*). *Sci. Total. Environ.* **2023**, *867*, 161296. [CrossRef] [PubMed]
38. Abu Bakar, N.; Ibrahim, W.N.W.; Abdullah, C.A.C.; Ramlan, N.F.; Shaari, K.; Shohaimi, S.; Mediani, A.; Nasruddin, N.S.; Kim, C.-H.; Faudzi, S.M.M. Embryonic Arsenic Exposure Triggers Long-Term Behavioral Impairment with Metabolite Alterations in Zebrafish. *Toxics* **2022**, *10*, 493. [CrossRef] [PubMed]
39. Valles, S.; Hernández-Sánchez, J.; Dipp, V.R.; Huerta-González, D.; Olivares-Bañuelos, T.N.; González-Fraga, J.; Bardullas, U. Exposure to low doses of inorganic arsenic induces transgenerational changes on behavioral and epigenetic markers in zebrafish (*Danio rerio*). *Toxicol. Appl. Pharmacol.* **2020**, *396*, 115002. [CrossRef] [PubMed]
40. Olivares, C.I.; Field, J.A.; Simonich, M.; Tanguay, R.L.; Sierra-Alvarez, R. Arsenic (III, V), indium (III), and gallium (III) toxicity to zebrafish embryos using a high-throughput multi-endpoint in vivo developmental and behavioral assay. *Chemosphere* **2016**, *148*, 361–368. [CrossRef]
41. Baldissarelli, L.A.; Capiotti, K.M.; Bogo, M.R.; Ghisleni, G.; Bonan, C.D. Arsenic alters behavioral parameters and brain ectonucleotidases activities in zebrafish (*Danio rerio*). *Comp. Biochem. Physiol. Part C Toxicol. Pharmacol.* **2012**, *155*, 566–572. [CrossRef]
42. Elamin, M.; Dumarchey, A.; Stoddard, C.; Robinson, T.M.; Cowie, C.; Gorka, D.; Chamberlain, S.J.; Levine, E.S. The role of UBE3A in the autism and epilepsy-related Dup15q syndrome using patient-derived, CRISPR-corrected neurons. *Stem Cell Rep.* **2023**, *18*, 884–898. [CrossRef]
43. Kalkman, H.O. Potential opposite roles of the extracellular signal-regulated kinase (ERK) pathway in autism spectrum and bipolar disorders. *Neurosci. Biobehav. Rev.* **2012**, *36*, 2206–2213. [CrossRef]
44. Bludau, A.; Royer, M.; Meister, G.; Neumann, I.D.; Menon, R. Epigenetic Regulation of the Social Brain. *Trends Neurosci.* **2019**, *42*, 471–484. [CrossRef] [PubMed]
45. Weinschutz Mendes, H.; Neelakantan, U.; Liu, Y.; Fitzpatrick, S.E.; Chen, T.; Wu, W.; Pruitt, A.; Jin, D.S.; Jamadagni, P.; Carlson, M.; et al. High-throughput functional analysis of autism genes in zebrafish identifies convergence in dopaminergic and neuroimmune pathways. *Cell Rep.* **2023**, *42*, 112243. [CrossRef] [PubMed]
46. Sugathan, A.; Biagioli, M.; Golzio, C.; Erdin, S.; Blumenthal, I.; Manavalan, P.; Ragavendran, A.; Brand, H.; Lucente, D.; Miles, J.; et al. CHD8 regulates neurodevelopmental pathways associated with autism spectrum disorder in neural progenitors. *Proc. Natl. Acad. Sci. USA* **2014**, *111*, E4468–E4477. [CrossRef]
47. Medishetti, R.; Rani, R.; Kavati, S.; Mahilkar, A.; Akella, V.; Saxena, U.; Kulkarni, P.; Sevilimedu, A. A DNAzyme based knockdown model for Fragile-X syndrome in zebrafish reveals a critical window for therapeutic intervention. *J. Pharmacol. Toxicol. Methods* **2020**, *101*, 106656. [CrossRef]
48. Dwivedi, S.; Medishetti, R.; Rani, R.; Sevilimedu, A.; Kulkarni, P.; Yogeeswari, P. Larval zebrafish model for studying the effects of valproic acid on neurodevelopment: An approach towards modeling autism. *J. Pharmacol. Toxicol. Methods* **2019**, *95*, 56–65. [CrossRef] [PubMed]
49. Pietri, T.; Roman, A.C.; Guyon, N.; Romano, S.A.; Washbourne, P.; Moens, C.B.; de Polavieja, G.G.; Sumbre, G. The first mecp2-null zebrafish model shows altered motor behaviors. *Front. Neural Circuits* **2013**, *7*, 118. [CrossRef] [PubMed]
50. Gao, H.; Bu, Y.; Wu, Q.; Wang, X.; Chang, N.; Lei, L.; Chen, S.; Liu, D.; Zhu, X.; Hu, K.; et al. Mecp2 regulates neural cell differentiation by suppressing the Id1 to Her2 axis in zebrafish. *J. Cell Sci.* **2015**, *128*, 2340–2350. [CrossRef] [PubMed]
51. Hoffman, E.J.; Turner, K.J.; Fernandez, J.M.; Cifuentes, D.; Ghosh, M.; Ijaz, S.; Jain, R.A.; Kubo, F.; Bill, B.R.; Baier, H.; et al. Estrogens Suppress a Behavioral Phenotype in Zebrafish Mutants of the Autism Risk Gene, CNTNAP2. *Neuron* **2016**, *89*, 725–733. [CrossRef]
52. Kim, O.H.; Cho, H.J.; Han, E.; Hong, T.I.; Ariyasiri, K.; Choi, J.H.; Hwang, K.S.; Jeong, Y.M.; Yang, S.Y.; Yu, K.; et al. Zebrafish knockout of Down syndrome gene, DYRK1A, shows social impairments relevant to autism. *Mol. Autism* **2017**, *8*, 50. [CrossRef] [PubMed]
53. Baraban, S.C.; Dinday, M.T.; Hortopan, G.A. Drug screening in Scn1a zebrafish mutant identifies clemizole as a potential Dravet syndrome treatment. *Nat. Commun.* **2013**, *4*, 2410. [CrossRef] [PubMed]
54. Liu, C.X.; Li, C.Y.; Hu, C.C.; Wang, Y.; Lin, J.; Jiang, Y.H.; Li, Q.; Xu, X. CRISPR/Cas9-induced shank3b mutant zebrafish display autism-like behaviors. *Mol. Autism* **2018**, *9*, 23. [CrossRef] [PubMed]
55. Kozol, R.A.; Cukier, H.N.; Zou, B.; Mayo, V.; De Rubeis, S.; Cai, G.; Griswold, A.J.; Whitehead, P.L.; Haines, J.L.; Gilbert, J.R.; et al. Two knockdown models of the autism genes SYNGAP1 and SHANK3 in zebrafish produce similar behavioral phenotypes associated with embryonic disruptions of brain morphogenesis. *Hum. Mol. Genet.* **2015**, *24*, 4006–4023. [CrossRef]
56. Courchesne, E.; Mouton, P.R.; Calhoun, M.E.; Semendeferi, K.; Ahrens-Barbeau, C.; Hallet, M.J.; Barnes, C.C.; Pierce, K. Neuron number and size in prefrontal cortex of children with autism. *JAMA* **2011**, *306*, 2001–2010. [CrossRef] [PubMed]
57. Fang, W.Q.; Chen, W.W.; Jiang, L.; Liu, K.; Yung, W.H.; Fu, A.K.Y.; Ip, N.Y. Overproduction of upper-layer neurons in the neocortex leads to autism-like features in mice. *Cell Rep.* **2014**, *9*, 1635–1643. [CrossRef] [PubMed]

58. Avino, T.A.; Barger, N.; Vargas, M.V.; Carlson, E.L.; Amaral, D.G.; Bauman, M.D.; Schumann, C.M. Neuron numbers increase in the human amygdala from birth to adulthood, but not in autism. *Proc. Natl. Acad. Sci. USA* **2018**, *115*, 3710–3715. [CrossRef]
59. Contractor, A.; Ethell, I.M.; Portera-Cailliau, C. Cortical interneurons in autism. *Nat. Neurosci.* **2021**, *24*, 1648–1659. [CrossRef]
60. Kang, Y.; Zhou, Y.; Li, Y.; Han, Y.; Xu, J.; Niu, W.; Li, Z.; Liu, S.; Feng, H.; Huang, W.; et al. A human forebrain organoid model of fragile X syndrome exhibits altered neurogenesis and highlights new treatment strategies. *Nat. Neurosci.* **2021**, *24*, 1377–1391. [CrossRef]
61. Casanova, M.F. Neuropathological and genetic findings in autism: The significance of a putative minicolumnopathy. *Neuroscientist* **2006**, *12*, 435–441. [CrossRef] [PubMed]
62. Schumann, C.M.; Barnes, C.C.; Lord, C.; Courchesne, E. Amygdala enlargement in toddlers with autism related to severity of social and communication impairments. *Biol. Psychiatry* **2009**, *66*, 942–949. [CrossRef] [PubMed]
63. Wegiel, J.; Flory, M.; Kuchna, I.; Nowicki, K.; Ma, S.Y.; Imaki, H.; Wegiel, J.; Cohen, I.L.; London, E.; Wisniewski, T.; et al. Stereological study of the neuronal number and volume of 38 brain subdivisions of subjects diagnosed with autism reveals significant alterations restricted to the striatum, amygdala and cerebellum. *Acta Neuropathol. Commun.* **2014**, *2*, 141. [CrossRef] [PubMed]
64. Avino, T.; Hutsler, J.J. Supernumerary neurons within the cerebral cortical subplate in autism spectrum disorders. *Brain Res.* **2021**, *1760*, 147350. [CrossRef] [PubMed]
65. Sacco, R.; Gabriele, S.; Persico, A.M. Head circumference and brain size in autism spectrum disorder: A systematic review and meta-analysis. *Psychiatry Res.* **2015**, *234*, 239–251. [CrossRef] [PubMed]
66. Varghese, M.; Keshav, N.; Jacot-Descombes, S.; Warda, T.; Wicinski, B.; Dickstein, D.L.; Harony-Nicolas, H.; De Rubeis, S.; Drapeau, E.; Buxbaum, J.D.; et al. Autism spectrum disorder: Neuropathology and animal models. *Acta Neuropathol.* **2017**, *134*, 537–566. [CrossRef] [PubMed]
67. Guerrini, R.; Marini, C. Genetic malformations of cortical development. *Exp. Brain Res.* **2006**, *173*, 322–333. [CrossRef] [PubMed]
68. Schwartzkroin, P.A.; Walsh, C.A. Cortical malformations and epilepsy. *Ment. Retard. Dev. Disabil. Res. Rev.* **2000**, *6*, 268–280. [CrossRef] [PubMed]
69. Wegiel, J.; Flory, M.; Kuchna, I.; Nowicki, K.; Ma, S.Y.; Imaki, H.; Wegiel, J.; Cohen, I.L.; London, E.; Brown, W.T.; et al. Brain-region-specific alterations of the trajectories of neuronal volume growth throughout the lifespan in autism. *Acta Neuropathol. Commun.* **2014**, *2*, 28. [CrossRef]
70. Sah, P. Fear, Anxiety, and the Amygdala. *Neuron* **2017**, *96*, 1–2. [CrossRef]
71. Hertz-Picciotto, I.; Croen, L.A.; Hansen, R.; Jones, C.R.; van de Water, J.; Pessah, I.N. The CHARGE study: An epidemiologic investigation of genetic and environmental factors contributing to autism. *Environ. Health Perspect.* **2006**, *114*, 1119–1125. [CrossRef]
72. Dietert, R.R.; Dietert, J.M.; Dewitt, J.C. Environmental risk factors for autism. *Emerg. Health Threat. J.* **2011**, *4*, 7111. [CrossRef]
73. Juliandi, B.; Tanemura, K.; Igarashi, K.; Tominaga, T.; Furukawa, Y.; Otsuka, M.; Moriyama, N.; Ikegami, D.; Abematsu, M.; Sanosaka, T.; et al. Reduced Adult Hippocampal Neurogenesis and Cognitive Impairments following Prenatal Treatment of the Antiepileptic Drug Valproic Acid. *Stem Cell Rep.* **2015**, *5*, 996–1009. [CrossRef] [PubMed]
74. Dixon, S.C.; Calder, B.J.; Lilya, S.M.; Davies, B.M.; Martin, A.; Peterson, M.; Hansen, J.M.; Suli, A. Valproic acid affects neurogenesis during early optic tectum development in zebrafish. *Biol. Open* **2023**, *12*, bio059567. [CrossRef]
75. Muhsen, M.; Youngs, J.; Riu, A.; Gustafsson, J.; Kondamadugu, V.S.; Garyfalidis, E.; Bondesson, M. Folic acid supplementation rescues valproic acid-induced developmental neurotoxicity and behavioral alterations in zebrafish embryos. *Epilepsia* **2021**, *62*, 1689–1700. [CrossRef]
76. Kanungo, J.; Twaddle, N.C.; Silva, C.; Robinson, B.; Wolle, M.; Conklin, S.; MacMahon, S.; Gu, Q.; Edhlund, I.; Benjamin, L.; et al. Inorganic arsenic alters the development of dopaminergic neurons but not serotonergic neurons and induces motor neuron development via Sonic hedgehog pathway in zebrafish. *Neurosci. Lett.* **2023**, *795*, 137042. [CrossRef]
77. Ruiz i Altaba, A. Gli proteins and Hedgehog signaling: Development and cancer. *Trends Genet.* **1999**, *15*, 418–425. [CrossRef] [PubMed]
78. Fei, D.L.; Li, H.; Kozul, C.D.; Black, K.E.; Singh, S.; Gosse, J.A.; DiRenzo, J.; Martin, K.A.; Wang, B.; Hamilton, J.W.; et al. Activation of Hedgehog signaling by the environmental toxicant arsenic may contribute to the etiology of arsenic-induced tumors. *Cancer Res.* **2010**, *70*, 1981–1988. [CrossRef]
79. Peterson, R.; Turnbull, J. Sonic hedgehog is cytoprotective against oxidative challenge in a cellular model of amyotrophic lateral sclerosis. *J. Mol. Neurosci.* **2012**, *47*, 31–41. [CrossRef] [PubMed]
80. Briscoe, J. Agonizing hedgehog. *Nat. Chem. Biol.* **2006**, *2*, 10–11. [CrossRef] [PubMed]
81. Tukachinsky, H.; Petrov, K.; Watanabe, M.; Salic, A. Mechanism of inhibition of the tumor suppressor Patched by Sonic Hedgehog. *Proc. Natl. Acad. Sci. USA* **2016**, *113*, E5866–E5875. [CrossRef]
82. Robbins, D.J.; Hebrok, M. Hedgehogs: La dolce vita. Workshop on Hedgehog-Gli Signaling in Cancer and Stem Cells. *EMBO Rep.* **2007**, *8*, 451–455. [CrossRef] [PubMed]
83. Ruiz i Altaba, A.; Mas, C.; Stecca, B. The Gli code: An information nexus regulating cell fate, stemness and cancer. *Trends Cell Biol.* **2007**, *17*, 438–447. [CrossRef] [PubMed]
84. Bai, C.B.; Joyner, A.L. Gli1 can rescue the in vivo function of Gli2. *Development* **2001**, *128*, 5161–5172. [CrossRef] [PubMed]

85. Bai, C.B.; Stephen, D.; Joyner, A.L. All mouse ventral spinal cord patterning by hedgehog is Gli dependent and involves an activator function of Gli3. *Dev. Cell* **2004**, *6*, 103–115. [CrossRef] [PubMed]
86. Briscoe, J.; Thérond, P.P. The mechanisms of Hedgehog signalling and its roles in development and disease. *Nat. Rev. Mol. Cell Biol.* **2013**, *14*, 416–429. [CrossRef]
87. Ruiz i Altaba, A. Combinatorial Gli gene function in floor plate and neuronal inductions by Sonic hedgehog. *Development* **1998**, *125*, 2203–2212. [CrossRef]
88. Silva, C.S.; Kudlyk, T.; Tryndyak, V.P.; Twaddle, N.C.; Robinson, B.; Gu, Q.; Beland, F.A.; Fitzpatrick, S.C.; Kanungo, J. Gene expression analyses reveal potential mechanism of inorganic arsenic-induced apoptosis in zebrafish. *J. Appl. Toxicol.* **2023**, *43*, 1872–1882. [CrossRef]
89. Andrews, M.G.; Kong, J.; Novitch, B.G.; Butler, S.J. New perspectives on the mechanisms establishing the dorsal-ventral axis of the spinal cord. *Curr. Top. Dev. Biol.* **2019**, *132*, 417–450. [CrossRef]
90. Yang, C.; Li, S.; Li, X.; Li, H.; Li, Y.; Zhang, C.; Lin, J. Effect of sonic hedgehog on motor neuron positioning in the spinal cord during chicken embryonic development. *J. Cell. Mol. Med.* **2019**, *23*, 3549–3562. [CrossRef]
91. Echelard, Y.; Epstein, D.J.; St-Jacques, B.; Shen, L.; Mohler, J.; McMahon, J.A.; McMahon, A.P. Sonic hedgehog, a member of a family of putative signaling molecules, is implicated in the regulation of CNS polarity. *Cell* **1993**, *75*, 1417–1430. [CrossRef]
92. Krauss, S.; Concordet, J.P.; Ingham, P.W. A functionally conserved homolog of the Drosophila segment polarity gene hh is expressed in tissues with polarizing activity in zebrafish embryos. *Cell* **1993**, *75*, 1431–1444. [CrossRef]
93. Roelink, H.; Augsburger, A.; Heemskerk, J.; Korzh, V.; Norlin, S.; Ruiz i Altaba, A.; Tanabe, Y.; Placzek, M.; Edlund, T.; Jessell, T.M.; et al. Floor plate and motor neuron induction by vhh-1, a vertebrate homolog of hedgehog expressed by the notochord. *Cell* **1994**, *76*, 761–775. [CrossRef]
94. Catela, C.; Kratsios, P. Transcriptional mechanisms of motor neuron development in vertebrates and invertebrates. *Dev. Biol.* **2021**, *475*, 193–204. [CrossRef]
95. Dhar, P.; Jaitley, M.; Kalaivani, M.; Mehra, R.D. Preliminary morphological and histochemical changes in rat spinal cord neurons following arsenic ingestion. *Neurotoxicology* **2005**, *26*, 309–320. [CrossRef] [PubMed]
96. Al-Ayadhi, L.Y. Relationship between Sonic hedgehog protein, brain-derived neurotrophic factor and oxidative stress in autism spectrum disorders. *Neurochem. Res.* **2012**, *37*, 394–400. [CrossRef]
97. Rahi, S.; Mehan, S. Understanding Abnormal SMO-SHH Signaling in Autism Spectrum Disorder: Potential Drug Target and Therapeutic Goals. *Cell. Mol. Neurobiol.* **2022**, *42*, 931–953. [CrossRef] [PubMed]
98. Chirila, A.M.; Rankin, G.; Tseng, S.Y.; Emanuel, A.J.; Chavez-Martinez, C.L.; Zhang, D.; Harvey, C.D.; Ginty, D.D. Mechanoreceptor signal convergence and transformation in the dorsal horn flexibly shape a diversity of outputs to the brain. *Cell* **2022**, *185*, 4541–4559.e23. [CrossRef] [PubMed]
99. Orefice, L.L.; Mosko, J.R.; Morency, D.T.; Wells, M.F.; Tasnim, A.; Mozeika, S.M.; Ye, M.; Chirila, A.M.; Emanuel, A.J.; Rankin, G.; et al. Targeting Peripheral Somatosensory Neurons to Improve Tactile-Related Phenotypes in ASD Models. *Cell* **2019**, *178*, 867–886.e24. [CrossRef] [PubMed]
100. Orefice, L.L.; Zimmerman, A.L.; Chirila, A.M.; Sleboda, S.J.; Head, J.P.; Ginty, D.D. Peripheral Mechanosensory Neuron Dysfunction Underlies Tactile and Behavioral Deficits in Mouse Models of ASDs. *Cell* **2016**, *166*, 299–313. [CrossRef] [PubMed]
101. Tasnim, A.; Alkislar, I.; Hakim, R.; Turecek, J.; Abdelaziz, A.; Orefice, L.L.; Ginty, D.D. The developmental timing of spinal touch processing alterations predicts behavioral changes in genetic mouse models of autism spectrum disorders. *Nat. Neurosci.* **2024**, *27*, 484–496. [CrossRef]
102. Zimmerman, A.L.; Kovatsis, E.M.; Pozsgai, R.Y.; Tasnim, A.; Zhang, Q.; Ginty, D.D. Distinct Modes of Presynaptic Inhibition of Cutaneous Afferents and Their Functions in Behavior. *Neuron* **2019**, *102*, 420–434.e8. [CrossRef]
103. Pao, P.C.; Tsai, L.H. Three decades of Cdk5. *J. Biomed. Sci.* **2021**, *28*, 79. [CrossRef]
104. Cheung, Z.H.; Fu, A.K.; Ip, N.Y. Synaptic roles of Cdk5: Implications in higher cognitive functions and neurodegenerative diseases. *Neuron* **2006**, *50*, 13–18. [CrossRef]
105. Dhavan, R.; Tsai, L.H. A decade of CDK5. *Nat. Rev. Mol. Cell Biol.* **2001**, *2*, 749–759. [CrossRef] [PubMed]
106. Nikolic, M.; Dudek, H.; Kwon, Y.T.; Ramos, Y.F.; Tsai, L.H. The cdk5/p35 kinase is essential for neurite outgrowth during neuronal differentiation. *Genes Dev.* **1996**, *10*, 816–825. [CrossRef] [PubMed]
107. Ohshima, T.; Ward, J.M.; Huh, C.G.; Longenecker, G.; Veeranna; Pant, H.C.; Brady, R.O.; Martin, L.J.; Kulkarni, A.B. Targeted disruption of the cyclin-dependent kinase 5 gene results in abnormal corticogenesis, neuronal pathology and perinatal death. *Proc. Natl. Acad. Sci. USA* **1996**, *93*, 11173–11178. [CrossRef] [PubMed]
108. Kanungo, J.; Zheng, Y.L.; Amin, N.D.; Kaur, S.; Ramchandran, R.; Pant, H.C. Specific inhibition of cyclin-dependent kinase 5 activity induces motor neuron development in vivo. *Biochem. Biophys. Res. Commun.* **2009**, *386*, 263–267. [CrossRef] [PubMed]
109. Ao, C.; Li, C.; Chen, J.; Tan, J.; Zeng, L. The role of Cdk5 in neurological disorders. *Front. Cell. Neurosci.* **2022**, *16*, 951202. [CrossRef] [PubMed]
110. Drerup, J.M.; Hayashi, K.; Cui, H.; Mettlach, G.L.; Long, M.A.; Marvin, M.; Sun, X.; Goldberg, M.S.; Lutter, M.; Bibb, J.A. Attention-deficit/hyperactivity phenotype in mice lacking the cyclin-dependent kinase 5 cofactor p35. *Biol. Psychiatry* **2010**, *68*, 1163–1171. [CrossRef]

111. Liu, X.X.; Yang, L.; Shao, L.X.; He, Y.; Wu, G.; Bao, Y.H.; Lu, N.N.; Gong, D.M.; Lu, Y.P.; Cui, T.T.; et al. Endothelial Cdk5 deficit leads to the development of spontaneous epilepsy through CXCL1/CXCR2-mediated reactive astrogliosis. *J. Exp. Med.* **2020**, *217*, e20180992. [CrossRef]
112. Engmann, O.; Hortobágyi, T.; Pidsley, R.; Troakes, C.; Bernstein, H.G.; Kreutz, M.R.; Mill, J.; Nikolic, M.; Giese, K.P. Schizophrenia is associated with dysregulation of a Cdk5 activator that regulates synaptic protein expression and cognition. *Brain* **2011**, *134*, 2408–2421. [CrossRef]
113. Costa, L.; Tempio, A.; Lacivita, E.; Leopoldo, M.; Ciranna, L. Serotonin 5-HT7 receptors require cyclin-dependent kinase 5 to rescue hippocampal synaptic plasticity in a mouse model of Fragile X Syndrome. *Eur. J. Neurosci.* **2021**, *54*, 4124–4132. [CrossRef] [PubMed]
114. Wang, L.; Li, J.; Jia, M.; Yue, W.; Ruan, Y.; Lu, T.; Zhang, J.; Liu, J.; Zhang, D. No association of polymorphisms in the CDK5, NDEL1, and LIS1 with autism in Chinese Han population. *Psychiatry Res.* **2011**, *190*, 369–371. [CrossRef]
115. Tao, J.; Van Esch, H.; Hagedorn-Greiwe, M.; Hoffmann, K.; Moser, B.; Raynaud, M.; Sperner, J.; Fryns, J.P.; Schwinger, E.; Gécz, J.; et al. Mutations in the X-linked cyclin-dependent kinase-like 5 (CDKL5/STK9) gene are associated with severe neurodevelopmental retardation. *Am. J. Hum. Genet.* **2004**, *75*, 1149–1154. [CrossRef] [PubMed]
116. Wang, I.T.; Allen, M.; Goffin, D.; Zhu, X.; Fairless, A.H.; Brodkin, E.S.; Siegel, S.J.; Marsh, E.D.; Blendy, J.A.; Zhou, Z. Loss of CDKL5 disrupts kinome profile and event-related potentials leading to autistic-like phenotypes in mice. *Proc. Natl. Acad. Sci. USA* **2012**, *109*, 21516–21521. [CrossRef] [PubMed]
117. Weaving, L.S.; Christodoulou, J.; Williamson, S.L.; Friend, K.L.; McKenzie, O.L.; Archer, H.; Evans, J.; Clarke, A.; Pelka, G.J.; Tam, P.P.; et al. Mutations of CDKL5 cause a severe neurodevelopmental disorder with infantile spasms and mental retardation. *Am. J. Hum. Genet.* **2004**, *75*, 1079–1093. [CrossRef]
118. Kanungo, J.; Goswami, M.T.; Pant, H.C. Notch and Cdk5 in Zebrafish Mindbomb Mutant: Co-regulation or Coincidence? *Folia Biol.* **2018**, *64*, 35–40. [CrossRef]
119. Itoh, M.; Kim, C.H.; Palardy, G.; Oda, T.; Jiang, Y.J.; Maust, D.; Yeo, S.Y.; Lorick, K.; Wright, G.J.; Ariza-McNaughton, L.; et al. Mind bomb is a ubiquitin ligase that is essential for efficient activation of Notch signaling by Delta. *Dev. Cell* **2003**, *4*, 67–82. [CrossRef]
120. Weinmaster, G.; Kintner, C. Modulation of notch signaling during somitogenesis. *Annu. Rev. Cell Dev. Biol.* **2003**, *19*, 367–395. [CrossRef]
121. Koo, B.K.; Lim, H.S.; Song, R.; Yoon, M.J.; Yoon, K.J.; Moon, J.S.; Kim, Y.W.; Kwon, M.C.; Yoo, K.W.; Kong, M.P.; et al. Mind bomb 1 is essential for generating functional Notch ligands to activate Notch. *Development* **2005**, *132*, 3459–3470. [CrossRef]
122. Jiang, Y.J.; Aerne, B.L.; Smithers, L.; Haddon, C.; Ish-Horowicz, D.; Lewis, J. Notch signalling and the synchronization of the somite segmentation clock. *Nature* **2000**, *408*, 475–479. [CrossRef] [PubMed]
123. Kang, K.; Lee, D.; Hong, S.; Park, S.G.; Song, M.R. The E3 ligase Mind bomb-1 (Mib1) modulates Delta-Notch signaling to control neurogenesis and gliogenesis in the developing spinal cord. *J. Biol. Chem.* **2013**, *288*, 2580–2592. [CrossRef] [PubMed]
124. Lawson, N.D.; Scheer, N.; Pham, V.N.; Kim, C.H.; Chitnis, A.B.; Campos-Ortega, J.A.; Weinstein, B.M. Notch signaling is required for arterial-venous differentiation during embryonic vascular development. *Development* **2001**, *128*, 3675–3683. [CrossRef] [PubMed]
125. Li, B.; Yu, L.; Liu, D.; Yang, X.; Zheng, Y.; Gui, Y.; Wang, H. MIB1 mutations reduce Notch signaling activation and contribute to congenital heart disease. *Clin. Sci.* **2018**, *132*, 2483–2491. [CrossRef] [PubMed]
126. Hortopan, G.A.; Baraban, S.C. Aberrant expression of genes necessary for neuronal development and Notch signaling in an epileptic mind bomb zebrafish. *Dev. Dyn.* **2011**, *240*, 1964–1976. [CrossRef] [PubMed]
127. Hortopan, G.A.; Dinday, M.T.; Baraban, S.C. Spontaneous seizures and altered gene expression in GABA signaling pathways in a mind bomb mutant zebrafish. *J. Neurosci.* **2010**, *30*, 13718–13728. [CrossRef]
128. Muhle, R.; Trentacoste, S.V.; Rapin, I. The genetics of autism. *Pediatrics* **2004**, *113*, e472–e486. [CrossRef]
129. Glessner, J.T.; Wang, K.; Cai, G.; Korvatska, O.; Kim, C.E.; Wood, S.; Zhang, H.; Estes, A.; Brune, C.W.; Bradfield, J.P.; et al. Autism genome-wide copy number variation reveals ubiquitin and neuronal genes. *Nature* **2009**, *459*, 569–573. [CrossRef]
130. Choe, E.A.; Liao, L.; Zhou, J.Y.; Cheng, D.; Duong, D.M.; Jin, P.; Tsai, L.H.; Peng, J. Neuronal morphogenesis is regulated by the interplay between cyclin-dependent kinase 5 and the ubiquitin ligase mind bomb 1. *J. Neurosci.* **2007**, *27*, 9503–9512. [CrossRef]
131. Kwon, D.Y.; Dimitriadi, M.; Terzic, B.; Cable, C.; Hart, A.C.; Chitnis, A.; Fischbeck, K.H.; Burnett, B.G. The E3 ubiquitin ligase mind bomb 1 ubiquitinates and promotes the degradation of survival of motor neuron protein. *Mol. Biol. Cell* **2013**, *24*, 1863–1871. [CrossRef]
132. Yang, M.H.; Chang, K.J.; Li, B.; Chen, W.S. Arsenic Trioxide Suppresses Tumor Growth through Antiangiogenesis via Notch Signaling Blockade in Small-Cell Lung Cancer. *BioMed Res. Int.* **2019**, *2019*, 4647252. [CrossRef] [PubMed]
133. Yang, M.H.; Li, B.; Chang, K.J. Notch pathway inhibition mediated by arsenic trioxide depletes tumor initiating cells in small cell lung cancer. *Mol. Biol. Rep.* **2022**, *49*, 2245–2253. [CrossRef] [PubMed]
134. Tian, Y.; Voineagu, I.; Paşca, S.P.; Won, H.; Chandran, V.; Horvath, S.; Dolmetsch, R.E.; Geschwind, D.H. Alteration in basal and depolarization induced transcriptional network in iPSC derived neurons from Timothy syndrome. *Genome Med.* **2014**, *6*, 75. [CrossRef] [PubMed]
135. Kanungo, J.; Cuevas, E.; Ali, S.F.; Paule, M.G. Zebrafish model in drug safety assessment. *Curr. Pharm. Des.* **2014**, *20*, 5416–5429. [CrossRef]

MDPI AG
Grosspeteranlage 5
4052 Basel
Switzerland
Tel.: +41 61 683 77 34

International Journal of Molecular Sciences Editorial Office
E-mail: ijms@mdpi.com
www.mdpi.com/journal/ijms